ADVANCED LEVEL PURE MATHEMATICS

C. J. TRANTER
C.B.E., M.A., D.Sc., F.I.M.A.

Bashforth Professor of Mathematical Physics,
Royal Military College of Science, Shrivenham

THE ENGLISH UNIVERSITIES PRESS LTD
ST. PAUL'S HOUSE WARWICK LANE
LONDON EC4

ISBN 0 340 04508 6

First Edition 1953. Reprinted 1955, 1956, 1957, 1958, 1960, 1961, 1963, 1965.
Second Edition 1967. Reprinted 1968.
Third Edition 1970.

Phototypeset by Keyspools Ltd, Golborne, Lancs.
Printed and bound in Great Britain for The English Universities Press Limited
by C. Tinling & Co. Ltd, London and Prescot.

GENERAL EDITOR'S FOREWORD

The present volume is one of a series on physics and mathematics, for the upper forms at school and the first year at the university. The books have been written by a team of experienced teachers at the Royal Military College of Science, and the series therefore forms an integrated course of study.

In preparing their manuscripts the writers have been mainly guided by the examination syllabuses of London University, the Joint Board of Oxford and Cambridge and the Joint Matriculation Board, but they have also taken a broad view of their tasks and have endeavoured to produce works which aim to give a student that solid foundation without which it is impossible to proceed to higher studies. The books are suitable either for class teaching or self study; there are many illustrative examples and large collections of problems for solution taken, in the main, from recent examination papers.

It is a truism too often forgotten in teaching that knowledge is acquired by a student only when his interest is aroused and maintained. The student must not only be shown how a class of problems in mathematics is solved but, within limits, why a particular method works and in physics, why a technique is especially well adapted for some particular measurement. Throughout the series special emphasis has been laid on illustrations which may be expected to appeal to the experience of the student in matters of daily life, so that his studies are related to what he sees, feels and knows of the world around him. Treated in this way, science ceases to be an arid abstraction and becomes vivid and real to the inquiring mind.

The books have therefore been written, not only to ensure the passing of examinations, but as a preparation for the exciting world which lies ahead of the reader. They incorporate many of the suggestions which have been made in recent years by other teachers and, it is hoped, will bring some new points of view into the classroom and the study. Last, but by no means least, they have been written by a team working together, so that the exchange of ideas has been constant and vigorous. It is to be hoped that the result is a series which is adequate for all examinations at this level and yet broad enough to satisfy the intellectual needs of teachers and students alike.

O. G. SUTTON

PREFACE TO THE THIRD EDITION

In preparing this edition, the opportunity has been taken to amend the text so that the units used are those of the Système International d'Unités (SI). This system, in which the basic units of length, mass and time are the metre, kilogramme and second, has many advantages; its use by schools, universities and industry is being actively encouraged and it is, in the words of the Royal Society Conference of Editors on Metrication in Scientific Journals, "destined to become the universal currency of science and commerce".

C.J.T.

PREFACE TO THE SECOND EDITION

Some extra material has been included in this edition to take account of recent (and forthcoming) changes in the requirements of the major examination boards. In particular, I have inserted sections on the determination of linear laws from experimental data and on elementary three-dimensional coordinate geometry and I have added a chapter on complex numbers. I have also taken the opportunity of making a number of minor alterations where these have been required.

C.J.T.

PREFACE

The needs of those taking Pure Mathematics at Advanced Level in the recently introduced General Certificates of Education have set the standard for this book. The contents should also prove suitable for candidates preparing themselves for the Intermediate Examination of London University, the Qualifying Examination for the Mechanical Science Tripos at the University of Cambridge and for several of the examinations set by the Civil Service Commission.

The starting point is the Ordinary Level for the General Certificates and I have included in a single volume the appropriate parts of Algebra, Trigonometry, Calculus and Geometry. With so wide a field of study, the order in which the subjects appear is not necessarily the order in which they should be read. I believe that an early start should be made with the Calculus and the chapters on this subject may well be studied concurrently with those on Algebra and Trigonometry. I have included a large number of worked examples and graded the exercises in a way which will, I hope, make the book equally suitable for class or private study.

In preparing this book, I have made great use of the reports of the Teaching Committees of the Mathematical Association. In particular, I have found their recent reports on the Teaching of Trigonometry and Calculus quite invaluable and I wish to acknowledge my debt to them. Only very occasionally have I differed from their recommendations and here, of course, I bear full responsibility.

Among nearly 1600 examples and exercises, I have included a large number taken from recent papers set by the various examining bodies. My thanks are due to the Senate of the University of London, the Oxford and Cambridge Schools Examination Board, the Joint Matriculation Board of the Universities of Manchester, Liverpool, Leeds, Sheffield and Birmingham and the Syndics of the Cambridge University Press for permission to use their questions. My thanks are also due to many friends and colleagues who read the manuscript and offered constructive criticism. I am particularly grateful to Dr. E. T. Davies, Professor of Mathematics, University of Southampton, Dr. D. R. Dickinson, Senior Mathematics Master, Bristol Grammar School, Mr. F. L. Heywood, Senior Mathematics Master, Manchester Grammar School and Mr. H. K. Prout, Head of the Department of Mathematics, Royal Naval College, Dartmouth, all of whom made most useful suggestions when the book was in its first draft.

C. J. TRANTER

Royal Military College of Science,
Shrivenham.

CONTENTS

ix

CONTENTS

CONTENTS

CHAPTER 1

THE THEORY OF QUADRATIC EQUATIONS. MISCELLANEOUS EQUATIONS

1.1. The roots of a quadratic equation

The general quadratic equation can be written

$$ax^2 + bx + c = 0, \qquad (1.1)$$

where a, b and c are numerical coefficients and x is the quantity to be found. Dividing by a and transposing the term not containing x to the right hand side

$$x^2 + \frac{b}{a}x = -\frac{c}{a}.$$

The left hand side can be made into the perfect square $\left(x + \frac{b}{2a}\right)^2$ by adding a term $b^2/(4a^2)$. If therefore such a term is added to each side

$$\left(x + \frac{b}{2a}\right)^2 = \frac{b^2}{4a^2} - \frac{c}{a}$$
$$= \frac{b^2 - 4ac}{4a^2}.$$

Taking the square root of each side

$$x + \frac{b}{2a} = \pm \frac{\sqrt{(b^2 - 4ac)}}{2a}.$$

giving the two roots

$$x = \frac{-b \pm \sqrt{(b^2 - 4ac)}}{2a}. \qquad (1.2)$$

If $b^2 > 4ac$ the two roots are real and different, if $b^2 = 4ac$ the roots are real and both equal to $-b/(2a)$. If $b^2 < 4ac$ the expression under the square root sign is negative and, since there is no real quantity whose square is negative, the roots are in this case said to be imaginary.

The formula (1.2) is quite general and can always be used to obtain the roots of a quadratic equation. If, however, factors of the left hand side of the equation $ax^2 + bx + c = 0$ can be found, the roots are more easily obtained by setting each of the factors in turn equal to zero and solving the resulting simple equations. This process is illustrated in the first example below.

Example 1. *Solve the equations (a)* $2x^2 + 5x - 12 = 0$, *(b)* $x^2 + 11 = 7x$.

The left hand side of equation (a) has factors $(2x - 3)(x + 4)$ so that the equation can be written

$$(2x - 3)(x + 4) = 0.$$

Hence either $2x - 3 = 0$ giving $x = 3/2$, or $x + 4 = 0$ giving $x = -4$. For equation (b), $a = 1$, $b = -7$, $c = 11$ and formula (1.2) gives

$$x = \frac{7 \pm \sqrt{\{7^2 - 4(1)(11)\}}}{2} = \frac{7 \pm \sqrt{5}}{2} = \frac{7 \pm 2 \cdot 236}{2},$$

giving $x = 4 \cdot 618$ or $2 \cdot 382$.

Example 2. *Find the value of k so that the equation* $4x^2 - 8x + k = 0$ *shall have equal roots.*

Here $a = 4$, $b = -8$, $c = k$. The condition for equal roots $(b^2 = 4ac)$ gives

$$(-8)^2 = 4(4)(k) \quad \text{or} \quad 16k = 64, \quad \text{giving } k = 4.$$

Example 3. *Prove that the roots of the equation*

$$(p - q - r)x^2 + px + q + r = 0$$

are real if p, q and r are real.

The condition for real roots $(b^2 \geqslant 4ac)$ is here that

$$p^2 \geqslant 4(p - q - r)(q + r),$$

i.e., that $\qquad p^2 - 4p(q + r) + 4(q + r)^2 \geqslant 0,$

or $\qquad \{p - 2(q + r)\}^2 \geqslant 0.$

This is always true for the left hand side is the square of a real quantity and therefore cannot be negative.

Example 4. *If x is real, show that the expression* $y = (x^2 + x + 1)/(x + 1)$ *can have no real value between* -3 *and* 1.
                                                                                        (L.U.)

Rearranging as a quadratic in x,

$$(x + 1)y = x^2 + x + 1$$

giving $\qquad x^2 + (1 - y)x + 1 - y = 0.$

For x to be real $\qquad (1 - y)^2 \geqslant 4(1 - y),$

or, $\qquad (1 - y)(-3 - y) \geqslant 0.$

Changing the signs, for x to be real

$$(y - 1)(y + 3) \geqslant 0.$$

If y lies between -3 and 1, $y + 3 > 0$, and $y - 1 < 0$ giving $(y - 1)(y + 3) < 0$ and the above inequality is not satisfied. Hence there is no real value between -3 and 1.

1.2. The sum and product of the roots of a quadratic equation

The general quadratic equation

$$ax^2 + bx + c = 0 \tag{1.1}$$

can be written as

$$x^2 + \frac{b}{a}x + \frac{c}{a} = 0. \tag{1.3}$$

If its roots are α and β, the left hand side of the equation can be written as the product of two factors $(x - \alpha)(x - \beta)$ and thus the equation can be written

$$(x - \alpha)(x - \beta) = 0,$$

or, $$x^2 - (\alpha + \beta)x + \alpha\beta = 0. \qquad (1.4)$$

Since equations (1.3) and (1.4) are identical

$$\left.\begin{aligned}
\alpha + \beta &= -\frac{b}{a} = -\frac{\text{coefficient of } x \text{ in equation (1.1)}}{\text{coefficient of } x^2 \text{ in equation (1.1)}}, \\
\alpha\beta &= \frac{c}{a} = \frac{\text{coefficient independent of } x \text{ in equation (1.1)}}{\text{coefficient of } x^2 \text{ in equation (1.1)}}.
\end{aligned}\right\} \qquad (1.5)$$

The formulae (1.5) enable the values of the sum and product of the roots to be written down in terms of the coefficients in the given equation.

Example 5. *Find the relation between a, b and c if one root of the equation* $ax^2 + bx + c = 0$ *is three times the other.*

Let the roots be α and 3α. Then formulae (1.5) give

$$4\alpha = -b/a \quad \text{and} \quad 3\alpha^2 = c/a.$$

Substituting $\alpha = -b/(4a)$ from the first of these relations in the second

$$3\left(-\frac{b}{4a}\right)^2 = \frac{c}{a},$$

giving $3b^2 = 16ac$ as the required relation.

Example 6. *If α, β are the roots of the equation* $x^2 - px + q = 0$, *form the equation whose roots are* α/β^2 *and* β/α^2.

From (1.5), $$\alpha + \beta = p, \quad \alpha\beta = q. \qquad (1.6)$$

The sum of the roots of the required equation $= \dfrac{\alpha}{\beta^2} + \dfrac{\beta}{\alpha^2} = \dfrac{\alpha^3 + \beta^3}{\alpha^2\beta^2}$.

Now,

$$\alpha^3 + \beta^3 = (\alpha + \beta)(\alpha^2 - \alpha\beta + \beta^2) = (\alpha + \beta)\{(\alpha + \beta)^2 - 3\alpha\beta\}$$
$$= p(p^2 - 3q), \text{ using (1.6)}.$$

Hence the sum of the roots of the required equation $= \dfrac{p(p^2 - 3q)}{q^2}$, since $\alpha^2\beta^2 = q^2$ from the second of (1.6).

The product of the roots of the required equation $= \dfrac{\alpha}{\beta^2} \cdot \dfrac{\beta}{\alpha^2} = \dfrac{1}{\alpha\beta} = \dfrac{1}{q}$.

From (1.5), when the coefficient of x^2 is unity,

the coefficient of $x = -$ the sum of the roots,

the coefficient independent of $x = $ the product of the roots.

Hence the required equation is

$$x^2 - \frac{p(p^2 - 3q)}{q^2}x + \frac{1}{q} = 0,$$

or,

$$q^2x^2 - p(p^2 - 3q)x + q = 0.$$

EXERCISES 1 (a)

1. Solve the equations, (i) $8x^2 - 2x - 3 = 0$, (ii) $5x^2 + 10 = 17x$.
2. Show that the equation $kx(1 - x) = 1$ has no real roots if $0 < k < 4$. (L.U.)
3. Find the range of values of x for which

$$0 \leqslant \frac{(x - 1)^2}{x - 2} \leqslant \frac{9}{2}.$$ (L.U.)

4. Find the relation between p, q and r if one root of the equation $px^2 + qx + r = 0$ is double the other.
5. If α, β are the roots of the quadratic equation $ax^2 + bx + c = 0$, obtain the equation whose roots are $1/\alpha^3$ and $1/\beta^3$.
   If, in the above equation $\alpha\beta^2 = 1$, prove that $a^3 + c^3 + abc = 0$. (L.U.)
6. In the equation $ax^2 + bx + c = 0$, one root is the square of the other. Without solving the equation, prove that $c(a - b)^3 = a(c - b)^3$. (L.U.)

1.3. Miscellaneous equations involving one unknown

The solution of certain types of equation can sometimes be made to depend on that of the ordinary quadratic equation. Some of the artifices employed in such solutions are illustrated in the examples which follow.

Example 7. *Solve the equation* $\sqrt{(3 - x)} - \sqrt{(7 + x)} = \sqrt{(16 + 2x)}$. (L.U.)

When, as here, one side of an equation contains a single term involving a square root, this can be removed by squaring both sides. If the terms in the resulting equation be transposed so that any radical term remaining is again by itself, this can be removed by squaring again.

Applying this process to the equation given here, squaring both sides gives

$$3 - x - 2\sqrt{\{(3 - x)(7 + x)\}} + 7 + x = 16 + 2x,$$

which, on rearrangement and division by 2, gives

$$3 + x = -\sqrt{\{(3 - x)(7 + x)\}}.$$

Squaring again we have

$$(3 + x)^2 = (3 - x)(7 + x),$$

leading to the ordinary quadratic equation

$$2x^2 + 10x - 12 = 0.$$

After division by 2, this can be written

$$(x + 6)(x - 1) = 0,$$

with roots $x = 6$ and $x = 1$.

Only one of these roots ($x = -6$) satisfies the given equation. The other value ($x = 1$) does not satisfy the equation for solution but satisfies

$$\sqrt{(3 - x)} + \sqrt{(7 + x)} = \sqrt{(16 + 2x)},$$

in which the radical terms on the left are separated by a plus instead of a minus sign. If the same process is applied to this equation it will be found to lead to the same quadratic $2x^2 + 10x - 12 = 0$ as before. Hence, in solving equations of this type it is essential to check the values found in the actual equation given. In the case of the equation given here the required solution is $x = -6$.

Example 8. *Solve the equation* $x^2 + 3x - 2 = 8/(x^2 + 3x)$.

Write $y = x^2 + 3x$ and we obtain

$$y - 2 = 8/y,$$

or, $$y^2 - 2y - 8 = 0.$$

This gives $(y - 4)(y + 2) = 0$, so that $y = 4$ or -2.

Since $y = x^2 + 3x$, $y = 4$ gives the quadratic $x^2 + 3x - 4 = 0$, or $(x + 4)(x - 1) = 0$ with roots $x = 1$, -4. The other value $(y = -2)$ gives $x^2 + 3x + 2 = 0$ or $(x + 2)(x + 1) = 0$ with roots $x = -1$, -2. Hence the roots of the original equation are -1, -2, -4 and 1.

1.4. Simultaneous equations

It is assumed that the student is familiar with the solution of pairs of equations such as $3x + 4y = 7, 2x - y = 1$, in which both equations are of the first degree. Here we shall consider pairs of equations in which at least one is of a higher degree than the first and where the solution can be made to depend on that of a quadratic equation. Few fixed rules can be laid down but some of the methods available are illustrated in the examples below.

Example 9. *Solve the pair of equations* $xy = 10, 3x + 2y = 16$.

When one of the equations is of the first degree, either unknown is easily expressed in terms of the other. Substitution in the second equation then results in a single equation in one unknown.

For the pair given here, the second equation gives $y = 8 - \frac{3}{2}x$.

Substituting in the first equation

$$x\left(8 - \frac{3}{2}x\right) = 10,$$

which, after multiplication by 2 and slight rearrangement, can be written

$$3x^2 - 16x + 20 = 0,$$

or, $$(3x - 10)(x - 2) = 0.$$

Thus $x = \frac{10}{3}$, and since $y = 8 - \frac{3}{2}x$, the corresponding value of y is 3; or $x = 2$ and $y = 5$.

Example 10. *Solve the pair of equations* $x^2 + 4xy + y^2 = 13$, $2x^2 + 3xy = 8$.

When the two equations are of the same degree in x and y and when the separate terms involving the unknowns are all of this degree, the solution can be obtained by writing $y = mx$ and proceeding as follows.

With $y = mx$, the two equations become

$$x^2(1 + 4m + m^2) = 13 \quad \text{and} \quad x^2(2 + 3m) = 8. \qquad (1.7)$$

By division,

$$\frac{1 + 4m + m^2}{2 + 3m} = \frac{13}{8}.$$

Cross multiplying we have $8(1 + 4m + m^2) = 13(2 + 3m)$, giving the quadratic in m,

$$8m^2 - 7m - 18 = 0.$$

This can be written $\qquad (m - 2)(8m + 9) = 0$,

so that $\qquad\qquad\qquad m = 2 \quad \text{or} \quad -9/8$.

The value of x can now be obtained by substitution in one of equations (1.7). Choosing the second of (1.7) here as it is rather simpler than the first, $m = 2$ gives

$$x^2\{2 + (3)(2)\} = 8,$$

or, $x^2 = 1$ so that $x = \pm 1$. Since $y = mx$ and $m = 2$, the corresponding values of y are ± 2.

The second value $-9/8$ for m gives similarly

$$x^2\left\{2 + (3)\left(-\frac{9}{8}\right)\right\} = 8,$$

leading to a negative value for x^2. There are thus no real solutions corresponding to this value of m.

EXERCISES 1 (*b*)

1. Find t from the equation $t - 1.324\sqrt{t} - 2.896 = 0$.

2. Solve $\sqrt{(x + 6)} - \sqrt{(x + 3)} = \sqrt{(2x + 5)}$. $\qquad\qquad$ (L.U.)

3. Solve the equation $x^2 + 2x + \dfrac{12}{x^2 + 2x} = 7$.

4. Solve the simultaneous equations, $2x - y = 5$, $x^2 + xy = 2$. $\qquad$ (L.U.)

5. Solve the equations $x^2 + y^2 = 5$, $xy = 2$.

6. Solve the simultaneous equations,

$$\left.\begin{array}{r} \dfrac{x + 2}{y - 4} + \dfrac{2(y - 4)}{x + 2} + 3 = 0, \\[2mm] x - y = 3. \end{array}\right\}$$

1.5. The square root of $(a + \sqrt{b})$

Simultaneous equations of a type considered in the last section appear in the calculation of the square root of the quantity $a + \sqrt{b}$ in which a is a rational and $\sqrt{b}$ an irrational quantity. Before proceeding to this calculation we consider two important results in connection with such quantities.

Firstly, *the square root of a rational quantity cannot be partly rational and partly irrational.*

To prove this, let c be a rational quantity and suppose it is possible that

$$\sqrt{c} = p + \sqrt{q},$$

p being rational and $\sqrt{q}$ irrational. Squaring

$$c = p^2 + q + 2p\sqrt{q},$$

leading to $\qquad \sqrt{q} = \dfrac{c - p^2 - q}{2p};$

this requires that an irrational quantity should be equal to a rational one, and is impossible.

Secondly, if $p + \sqrt{q} = a + \sqrt{b}$, where p and a are both rational and $\sqrt{q}$, $\sqrt{b}$ are both irrational, then $p = a$ and $q = b$.

If p is not equal to a, let $p = a + \alpha$. Then

$$a + \alpha + \sqrt{q} = a + \sqrt{b},$$

giving $\qquad\qquad \sqrt{b} = \alpha + \sqrt{q}.$

This, by our first result, is impossible, so that $p = a$. It then follows immediately that $q = b$.

To calculate the square root of $a + \sqrt{b}$ we suppose that

$$\sqrt{(a + \sqrt{b})} = \pm(\sqrt{x} + \sqrt{y}).$$

Squaring both sides,

$$a + \sqrt{b} = x + 2\sqrt{(xy)} + y.$$

Using the second of the above results, we have

$$x + y = a,$$
$$2\sqrt{(xy)} = \sqrt{b}.$$

The second of these can be written $4xy = b$ and we have therefore only to solve the simultaneous equations

$$x + y = a,$$
$$4xy = b$$

in order to find x, y and hence $\sqrt{x}$ and $\sqrt{y}$.

Example 11. *Find the square root of* $14 + 6\sqrt{5}$. $\qquad$ (L.U.)

Let $\qquad\qquad \sqrt{(14 + 6\sqrt{5})} = \pm(\sqrt{x} + \sqrt{y}).$

Squaring, $\qquad\qquad 14 + 6\sqrt{5} = x + 2\sqrt{(xy)} + y.$

Hence $\qquad\qquad x + y = 14, \quad 2\sqrt{(xy)} = 6\sqrt{5}.$

The second equation gives $\sqrt{(xy)} = 3\sqrt{5}$ or $xy = 45$. From the first equation we have $y = 14 - x$ and substitution in $xy = 45$ gives the quadratic equation

$$x(14 - x) = 45.$$

This can be written

$$x^2 - 14x + 45 = 0,$$

or,
$$(x - 9)(x - 5) = 0.$$

Hence $x = 9$ or 5, and, from $x + y = 14$, the corresponding values of y are 5 and 9. The required square root is therefore $\pm(\sqrt{9} + \sqrt{5})$ or $\pm(3 + \sqrt{5})$. It makes no difference to the final result whether we take the solution $x = 9$, $y = 5$, or the alternative one $x = 5$, $y = 9$.

Note. In finding the square root of $a - \sqrt{b}$ the original assumption is modified to $\sqrt{(a - \sqrt{b})} = \pm (\sqrt{x} - \sqrt{y})$.

EXERCISES 1 (c)

1. Find the square root of $5 + 2\sqrt{6}$.

2. Express the square root of $18 - 12\sqrt{2}$ in the form $\sqrt{x} - \sqrt{y}$ where x and y are rational. (L.U.)

3. Find the square root of $a + b + \sqrt{(2ab + b^2)}$.

4. Find rational numbers a and b such that

$$3 + \sqrt{2} = (a + b\sqrt{2})(6 - \sqrt{2})^2.$$ (O.C.)

EXERCISES 1 (d)

1. Show that for all real values of y, the expression

$$\frac{3y^2 - 2y - 1}{y^2 + y + 2}$$

always lies between $-4/7$ and 4.

2. Find the range of values of x for which

$$\frac{x(x - 2)}{x + 6} > 2.$$ (L.U.)

3. Find the values of λ for which the equation

$$10x^2 + 4x + 1 = 2\lambda x(2 - x)$$

has equal roots. (L.U.)

4. Find, in its simplest rational form, the equation whose roots are $\sqrt{7}/(\sqrt{7} \pm \sqrt{5})$. (L.U.)

5. Show that the roots of the equation $2bx^2 + 2(a + b)x + 3a = 2b$ are real when a and b are real.

   If one root of this equation is double the other, prove that either $a = 2b$ or $4a = 11b$. (L.U.)

6. If the roots of the equation $x^2 + bx + c = 0$ are α, β and the roots of the equation $x^2 + \lambda bx + \lambda^2 c = 0$ are γ, δ show that the equation whose roots are $\alpha\gamma + \beta\delta$ and $\alpha\delta + \beta\gamma$ is

$$x^2 - \lambda b^2 x + 2\lambda^2 c(b^2 - 2c) = 0.$$

   Show that the roots of this equation are always real.   (L.U.)

7. The roots of the quadratic equation $x^2 - px + q = 0$ are α and β. Determine the equation having the roots $\alpha^2 + \beta^{-2}$ and $\beta^2 + \alpha^{-2}$, expressing the coefficients in terms of p and q. Prove further that if p and q are both real, then this equation can have equal roots only if $p = 0$ or $p^2 = 4q$.   (L.U.)

8. Prove that the roots of the equation

$$(k + 3)x^2 + (6 - 2k)x + k - 1 = 0$$

   are real if, and only if, k is not greater than 3/2. Find the values of k if one root is six times the other.   (L.U.)

9. If the equation $a^2x^2 + 6abx + ac + 8b^2 = 0$ has equal roots, prove that the roots of the equation $ac(x + 1)^2 = 4b^2 x$ are also equal.   (L.U.)

10. For what values of λ has the equation $x^2 - 3x + 2 = \lambda(2x - 5)$ two equal roots?

11. The roots of the equation $x^2 + ax + b = 0$ are α, β. Find the equation whose roots are $p\alpha + q\beta, p\beta + q\alpha$. If the original equation is $x^2 - 4x - 5 = 0$ find the values of p/q in order that the new equation shall have one zero root.

12. Form the equation whose roots are the cubes of the roots of the equation $x^2 - 3x + 4 = 0$, without solving the equation, giving the numerical values of the coefficients of the new equation.   (L.U.)

13. Show that if the equations $x^2 + bx + c = 0$, $x^2 + px + q = 0$ have a common root, then $(c - q)^2 = (b - p)(cp - bq)$.   (L.U.)

14. Solve the equation $(3x^2 + 2x)^2 + 8 = 9(3x^2 + 2x)$.

15. Solve the equation $\sqrt{(x - 5)} + 2 = \sqrt{(x + 7)}$.

16. Solve the equation $\sqrt{(3x + 4)} - \sqrt{(x - 3)} = 3$.   (L.U.)

17. Solve the simultaneous equations,

$$\frac{x^2}{4} - \frac{1}{y + 1} = 1, \quad \frac{1}{3(y + 1)} + \frac{x}{2} = 3.$$ (L.U.)

18. Solve the simultaneous equations, $x + y = 6, x^2y^2 + 2xy - 35 = 0$.   (L.U.)

19. Solve the simultaneous equations,

$$\frac{x}{y} + \frac{y}{x} = \frac{5}{2}, \quad x^2 + 2y^2 = 9.$$

20. Solve the simultaneous equations,

$$\frac{x}{y + a} + \frac{y}{x + a} = 1, \quad x + y = 2a.$$ (L.U.)

21. Find u and v from the equations,

$$\frac{1}{u+v} + \frac{2}{u-v} = 8, \quad u^2 - v^2 = \frac{1}{6}.$$

22. If

$$\frac{x^2}{y} + \frac{y^2}{x} = 9, \quad \frac{1}{x} + \frac{1}{y} = \frac{3}{4},$$

find the values of the product xy and hence solve the given equations completely.   (L.U.)

23. If a, b, c, d are rational numbers and if neither b nor d is a perfect square, prove that the product $(a + \sqrt{b})(c + \sqrt{d})$ can be a rational number only if

$$\frac{c}{a} = -\frac{\sqrt{d}}{\sqrt{b}}.$$ (L.U.)

24. Express

$$\frac{1 + \sqrt{3}}{(\sqrt{3} - 1)^3},$$

in the form $a + b\sqrt{c}$, where a, b, c are rational.   (L.U.)

25. By putting $z = x + x^{-1}$, solve the equation

$$2x^4 - 9x^3 + 14x^2 - 9x + 2 = 0.$$ (L.U.)

CHAPTER 2

INDICES AND LOGARITHMS. THE REMAINDER THEOREM. THE PRINCIPLE OF UNDETERMINED COEFFICIENTS. PARTIAL FRACTIONS

2.1. The fundamental laws for positive integral indices

When a quantity a is multiplied by itself any number of times the product is called a power of a. Thus $a \times a$ is the second power of a and is written a^2. The number expressing the power is called the *index*. Thus the index of a^2 is the number 2. Generalising, we have the definition that if m is a positive integer, a^m *denotes the product of m factors each equal to a*. We give below three fundamental laws for the combination of indices. In all cases the indices m and n are assumed to be positive integers, and in (ii) we assume that $m > n$.

(i) $a^m \times a^n = a^{m+n}$.   (2.1)

By definition, $a^m = a.a.a \ldots$ to m factors and $a^n = a.a.a \ldots$ to n factors. Hence $a^m \times a^n = a.a.a \ldots$ to $(m + n)$ factors, $= a^{m+n}$, by definition.

(ii) $a^m \div a^n = a^{m-n}$.   (2.2)

From the definitions of a^m and a^n,

$$a^m \div a^n = \frac{a^m}{a^n} = \frac{a.a.a \ldots \text{to } m \text{ factors}}{a.a.a \ldots \text{to } n \text{ factors}}$$

$$= a.a.a \ldots \text{to } (m - n) \text{ factors} = a^{m-n}.$$

(iii) $(a^m)^n = a^{mn}$.   (2.3)

$(a^m)^n = a^m.a^m.a^m \ldots$ to n factors

$$= (a.a.a \ldots \text{to } m \text{ factors})(a.a.a \ldots \text{to } m \text{ factors}) \ldots$$

the bracketed terms being repeated n times, so that

$$(a^m)^n = a.a.a \ldots \text{to } mn \text{ factors} = a^{mn}.$$

2.2. Fractional, zero and negative indices

It is convenient to have available fractional, zero and negative indices and for one set of laws to apply in all cases. However, the definition of a^m as the product of m factors each equal to a is clearly meaningless except when m is a positive integer. We introduce fractional and negative indices by determining their meaning when the first fundamental law $a^m \times a^n = a^{m+n}$ is true. It is then possible to show that with the interpretations arrived at on this basis the other two laws of § 2.1 remain valid.

23

The interpretation of $a^{p/q}$, p and q being positive integers

Since the first rule of § 2.1 is to be true, $a^{p/q} \times a^{p/q} = a^{2p/q}$. Similarly $a^{p/q} \times a^{p/q} \times a^{p/q} = a^{2p/q} \times a^{p/q} = a^{3p/q}$ and so on. Hence

$$a^{p/q} \times a^{p/q} \times a^{p/q} \ldots \text{to } q \text{ factors} = a^{qp/q} = a^p.$$

This implies that $(a^{p/q})^q = a^p$, and taking the qth root

$$a^{p/q} = \sqrt[q]{a^p}, \tag{2.4}$$

i.e., that $a^{p/q}$ is the qth root of a^p.

The interpretation of a^0

Since $a^m \times a^n = a^{m+n}$ is to be true for all values of m and n we can take $m = 0$ and hence

$$a^0 \times a^n = a^n,$$

giving $\qquad\qquad a^0 = a^n/a^n = 1, \tag{2.5}$

i.e., any quantity with zero index is equivalent to unity.

The interpretation of a^{-n}

The rule $a^m \times a^n = a^{m+n}$ is to hold for all m, n and we can therefore take $m = -n$. The rule then gives

$$a^{-n} \times a^n = a^{-n+n} = a^0 = 1,$$

giving, $\qquad\qquad a^{-n} = 1/a^n, \tag{2.6}$

showing that a^{-n} is the reciprocal of a^n.

With these interpretations it remains to show that the two laws $a^m \div a^n = a^{m-n}$ and $(a^m)^n = a^{mn}$ remain true for all values of m and n. To prove the first we have

$$a^m \div a^n = a^m \times \frac{1}{a^n}$$

$$= a^m \times a^{-n}, \text{ since } a^{-n} \text{ means } 1/a^n,$$

$$= a^{m-n}, \text{ by the fundamental law.}$$

To show that $(a^m)^n = a^{mn}$ for all m and n we take the value of m to be unrestricted and consider in turn the cases in which n is a positive integer, a positive fraction and any negative quantity. For any m and positive integral n,

$$(a^m)^n = a^m . a^m . a^m \ldots \text{to } n \text{ factors}$$

$$= a^{m+m+m+ \ldots \text{to } n \text{ terms}} = a^{mn}.$$

For any m and $n = p/q$ where p, q are positive integers, $(a^m)^n = (a^m)^{p/q}$. Now the qth power of $(a^m)^{p/q}$ is $\{(a^m)^{p/q}\}^q$ or $(a^m)^p$. This is a^{mp}. Hence we have, on taking the qth root

$$(a^m)^{p/q} = \sqrt[q]{a^{mp}} = a^{mp/q}.$$

Finally, for unrestricted m and n any negative quantity, we replace n by $-\lambda$. Then

$$(a^m)^n = (a^m)^{-\lambda} = \frac{1}{(a^m)^{\lambda}} = \frac{1}{a^{m\lambda}} = a^{-m\lambda} = a^{mn}.$$

Hence, with the interpretation of fractional, zero and negative indices given in equations (2.4), (2.5) and (2.6) the three fundamental laws for combination of indices given in (2.1), (2.2) and (2.3) remain valid.

Example 1. *Express with positive indices*, (i) $\dfrac{2b^{-3}x^2}{7c^{-4}y^2}$, (ii) $\dfrac{\sqrt{(y^{-c})}}{\sqrt[3]{y^2}}$ *and evaluate*

(iii) $\left(\dfrac{81}{16}\right)^{3/4}$.

(i) $\dfrac{2b^{-3}x^2}{7c^{-4}y^2} = \dfrac{2x^2c^4}{7b^3y^2}$.   (ii) $\dfrac{\sqrt[3]{(y^{-c})}}{\sqrt[3]{(y^2)}} = \dfrac{y^{-c/3}}{y^{2/3}} = \dfrac{1}{y^{c/3+2/3}}$.

(iii) $\left(\dfrac{81}{16}\right)^{3/4} = \sqrt[4]{\left(\dfrac{81}{16}\right)^3} = \left(\dfrac{3}{2}\right)^3 = \dfrac{27}{8}$.

Example 2. *Show that $(xy)^n = x^n y^n$ for all values of n.*

If n is a positive integer, $(xy)^n = xy . xy . xy \ldots$ to n factors. This can be written as the product

$$(x . x . x \ldots \text{to } n \text{ factors}) \times (y . y . y \ldots \text{to } n \text{ factors})$$

which is $x^n y^n$.

If n is a positive fraction, say p/q where p and q are positive integers, the qth power of $(xy)^{p/q}$

$$= \{(xy)^{p/q}\}^q = (xy)^p = x^p y^p = (x^{p/q} y^{p/q})^q.$$

Taking the qth root, $(xy)^{p/q} = x^{p/q} y^{p/q}$.

If n is any negative quantity, say $-\lambda$.

$$(xy)^n = (xy)^{-\lambda} = \frac{1}{(xy)^{\lambda}} = \frac{1}{x^{\lambda} y^{\lambda}} = x^{-\lambda} y^{-\lambda} = x^n y^n.$$

EXERCISES 2 (*a*)

1. Express with positive integers (i) $b^{-3}x^{-2} \div 4x$, (ii) $\sqrt[4]{y^3} \times \sqrt{y^{1/2}}$.

2. Evaluate (i) $(64)^{-3/2}$, (ii) $\left(\dfrac{8}{27}\right)^{-1/3}$.

3. Simplify $(x^4 y z^{-3})^2 \times \sqrt{(x^{-5}y^2z)} \div (xz)^{7/2}$.   (L.U.)

4. Prove that $(a - a^{-1})(a^{4/3} + a^{-2/3}) = \dfrac{a^2 - a^{-2}}{a^{-1/3}}$.   (L.U.)

5. Evaluate $\dfrac{x^{3/2} + xy}{xy - y^3} - \dfrac{\sqrt{x}}{\sqrt{x} - y}$.

6. Simplify $\dfrac{3(2^{n+1}) - 4(2^{n-1})}{2^{n+1} - 2^n}$.

2.3. The theory of logarithms

The logarithm of a positive quantity N to a given *base* a is defined as the index of the power to which the base a must be raised to make it equal the given quantity N. Thus

$$a^x = N, \tag{2.7}$$

when x is the logarithm of N to the base a.

x is written
$$x = \log_a N \tag{2.8}$$

and the two formulae (2.7), (2.8) are equivalent statements expressing the relationship between x, a and N.

If we substitute for x from (2.8) in (2.7) we have

$$a^{\log_a N} = N, \tag{2.9}$$

a result which is often useful.

If we set $x = 0$ in (2.7) we have $N = a^0 = 1$. The equivalent formula (2.8) gives in this case

$$\log_a 1 = 0, \tag{2.10}$$

so that *the logarithm of unity is zero*. If we put $x = 1$ in (2.7) $N = a$ and formula (2.8) then gives

$$\log_a a = 1, \tag{2.11}$$

or, *the logarithm of the base itself is unity*.

To find the logarithm of the product of two positive numbers M and N, we have, using (2.9),

$$MN = a^{\log_a M} \cdot a^{\log_a N} = a^{\log_a M + \log_a N}$$

Hence, by the definition of a logarithm,

$$\log_a MN = \log_a M + \log_a N, \tag{2.12}$$

showing that *the logarithm of a product of two positive numbers is the sum of the logarithms of the separate numbers*. Similarly

$$\log_a MNP = \log_a M + \log_a N + \log_a P,$$

and so on for products of more factors.

For the logarithm of a quotient of two positive quantities M and N, equation (2.9) gives

$$\frac{M}{N} = \frac{a^{\log_a M}}{a^{\log_a N}} = a^{\log_a M - \log_a N},$$

showing that

$$\log_a \frac{M}{N} = \log_a M - \log_a N, \tag{2.13}$$

i.e., *the logarithm of a quotient is the difference between the logarithm of the numerator and that of the denominator*.

The logarithm of a positive quantity raised to a power can be found similarly. Thus, again from (2.9),

$$M^p = (a^{\log_a M})^p = a^{p\log_a M},$$

giving $$\log_a M^p = p \log_a M. \qquad (2.14)$$

If, in (2.14) we write $p = 1/r$,

$$\log_a M^{1/r} = \frac{1}{r} \log_a M. \qquad (2.15)$$

Thus *the logarithm of the pth power of a positive quantity is p times the logarithm of the quantity*, (2.14), and *the logarithm of the rth root of a positive quantity is 1/r times the logarithm of the quantity*, (2.15).

Example 3. *Prove that* $log_b N = \dfrac{1}{log_a b} \times log_a N.$

Let $x = \log_b N$ so that $b^x = N$. Taking logarithms to base a,

$$\log_a(b^x) = \log_a N,$$

giving, $$x \log_a b = \log_a N$$

or, $$x = \frac{1}{\log_a b} \times \log_a N.$$

The result proved here is of importance in that it relates logarithms to different bases. It shows that to transform logarithms from base a to base b we have to multiply by the quantity $1/(\log_a b)$.

Example 4. *Prove that* $2 \, log_c (a + b) = 2 \, log_c a + log_c \left(1 + \dfrac{2b}{a} + \dfrac{b^2}{a^2}\right)$ (L.U.)

$$2 \log_c (a + b) = \log_c (a + b)^2 = \log_c (a^2 + 2ab + b^2)$$

$$= \log_c \left\{ a^2 \left(1 + \frac{2b}{a} + \frac{b^2}{a^2}\right) \right\}$$

$$= \log_c a^2 + \log_c \left(1 + \frac{2b}{a} + \frac{b^2}{a^2}\right)$$

$$= 2 \log_c a + \log_c \left(1 + \frac{2b}{a} + \frac{b^2}{a^2}\right)$$

2.4. Common logarithms

The logarithms used in everyday calculations are those with base 10. Such logarithms are referred to as *common* logarithms and the base is often omitted in written work. Thus log 24 is generally taken to mean $\log_{10} 24$. The student is assumed to be familiar with their use in arithmetical work and only very few examples will be given in this section. Examples involving the use of common logarithms occur throughout the book, particularly in the chapters on Trigonometry.

Example 5. *Calculate log 5 and log 0·125 given that log 2 = 0·3010.*

$$\log 5 = \log\left(\frac{10}{2}\right) = \log 10 - \log 2 = 1 - \log 2, \text{ since } \log 10 = 1 \text{ (by 2.11)},$$
$$= 1 - 0·3010 = 0·6990.$$

$$\log 0·125 = \log\left(\frac{1}{8}\right) = \log 1 - \log 8 = -3\log 2, \text{ since } \log 1 = 0 \text{ (by 2.10)},$$
$$= -0·9030 = \bar{1}·0970, \text{ the last form meaning } -1 + 0·0970.$$

Example 6. *Given that log 3 = 0·4771, find the number of digits in the integral part of* $(\sqrt{3})^{89}$.

$$\log(\sqrt{3})^{89} = \frac{89}{2}\log 3 = \frac{89}{2} \times 0·4771 = 21·23\ldots$$

Thus 10 has to be raised to rather more than the twenty-first power to give $(\sqrt{3})^{89}$ and this quantity will therefore contain 22 digits in its integral part.

2.5. Equations in which the unknown occurs as an index

When the unknown quantity in an equation occurs as an index, the laws of combination of indices and the use of logarithms usually enable the solution to be found. Some of the artifices used are illustrated in the following examples.

Example 7. *Solve the equation* $2^{x^2} = 16^{x-1}$.   (L.U.)

Since $16 = 2^4$, the equation can be written

$$2^{x^2} = (2^4)^{x-1} = 2^{4x-4}.$$

Hence $x^2 = 4x - 4$, or $x^2 - 4x + 4 = 0$,

giving, $(x - 2)^2 = 0$, so that $x = 2$.

Example 8. *Find x from the equation* $3^{2x} = 5^{x+1}$.

Taking logarithms,

$$2x\log 3 = (x + 1)\log 5,$$

so that $(2\log 3 - \log 5)x = \log 5.$

Hence

No.	log
0·6990	$\bar{1}$·8445
0·2552	$\bar{1}$·4068
x	0·4377

$$x = \frac{\log 5}{2\log 3 - \log 5} = \frac{0·6990}{2 \times 0·4771 - 0·6990}$$
$$= \frac{0·6990}{0·2552} = 2·74.$$

Example 9. *Solve the equation* $5^{2x} - 5^{x+1} + 4 = 0$.

This can be written

$$(5^x)^2 - 5(5^x) + 4 = 0,$$

which in factor form is

$$(5^x - 1)(5^x - 4) = 0.$$

No.	log
0·6021	$\bar{1}$·7797
0·6990	$\bar{1}$·8445
x	$\bar{1}$·9352

Hence either $5^x = 1$ leading to $x = 0$ [by (2.5)], or $5^x = 4$. Taking logarithms this gives

$$x\log 5 = \log 4,$$

so that

$$x = \frac{\log 4}{\log 5} = \frac{0·6021}{0·6990} = 0·861.$$

EXERCISES 2 (b)

1. If $a = \log_b c$, $b = \log_c a$, $c = \log_a b$, prove that $abc = 1$. (L.U.)
2. Without using tables, show that
$$\frac{\log \sqrt{27} + \log \sqrt{8} - \log \sqrt{125}}{\log 6 - \log 5} = \frac{3}{2}.$$ (L.U.)
3. If $\log_x 10\cdot24 = 2$, find x. (L.U.)
4. Using logarithms, evaluate the following:
$$\text{(i) } \log_{25} 3\cdot142; \quad \text{(ii) } \left(\frac{1}{12}\right)^{1\cdot405}.$$ (Q.E.)

5. Solve the equations
$$\text{(a) } \log(x^2 + 2x) = 0\cdot9031, \quad \text{(b) } (2\cdot4)^x = 0\cdot59.$$ (L.U.)
6. Find x from the equation $3^x - 3^{-x} = 6\cdot832$.
7. Solve the equation $2^{2x+8} - 32(2^x) + 1 = 0$.
8. Solve the simultaneous equations $2^{x+y} = 6^y$, $3^x = 6(2^y)$.

2.6. The remainder theorem

The polynomial expression
$$c_0 x^n + c_1 x^{n-1} + c_2 x^{n-2} + \ldots + c_{n-1}x + c_n, \quad (2.16)$$
can be written
$$c_0(x^n - a^n) + c_1(x^{n-1} - a^{n-1}) + c_2(x^{n-2} - a^{n-2}) + \ldots + c_{n-1}(x - a)$$
$$+ c_0 a^n + c_1 a^{n-1} + c_2 a^{n-2} + \ldots + c_{n-1}a + c_n. \quad (2.17)$$
Since, as can be verified by actual multiplication,
$$x^m - a^m = (x - a)(x^{m-1} + ax^{m-2} + a^2 x^{m-3} + \ldots + a^{m-2}x + a^{m-1}),$$
each of the terms in the first line of the expression (2.17) is divisible by $x - a$. Hence we can write
$$c_0 x^n + c_1 x^{n-1} + c_2 x^{n-2} + \ldots + c_{n-1}x + c_n$$
$$= \text{a multiple of } (x - a) + c_0 a^n + c_1 a^{n-1} + c_2 a^{n-2} + \ldots + c_{n-1}a + c_n.$$
Hence, *the remainder when a polynomial expression is divided by* $(x - a)$ *is obtained by writing a for x in the given expression.* This result is known as the *remainder theorem* and it enables the remainder to be found without having to perform the division.

An alternative proof of the remainder theorem can be given as follows. Let $P(x)$ denote a polynomial expression in x, let $Q(x)$ be the quotient when $P(x)$ is divided by $(x - a)$ and let R be the remainder. Then, for all values of x,
$$P(x) = (x - a)Q(x) + R,$$
and R is independent of x. Putting $x = a$ we have
$$P(a) = R,$$

since the first term on the right hand side vanishes because of the factor $(x - a)$. Hence the remainder is obtained by writing a for x in the given expression.

An immediate and important consequence of the remainder theorem is that if a polynomial expression in x vanishes for a certain value a of x, then $(x - a)$ is a factor of the expression.

Example 10. *Find the value of k if the remainder when the polynomial $2x^4 + kx^3 - 11x^2 + 4x + 12$ is divided by $(x - 3)$ is 60.*   (L.U.)

By the remainder theorem, the remainder after division by $(x - 3)$ is obtained by writing $x = 3$ in the expression $2x^4 + kx^3 - 11x^2 + 4x + 12$. The remainder is therefore

$$2(3)^4 + k(3)^3 - 11(3)^2 + 4(3) + 12,$$

or, $162 + 27k - 99 + 12 + 12.$

This reduces to $27k + 87$, and setting it equal to 60 we have

$$27k + 87 = 60.$$

giving $k = -1$.

Example 11. *Factorise $a^2(b - c) + b^2(c - a) + c^2(a - b)$.*   (L.U.)

If we set $a = b$ the given expression vanishes; in other words, there is no remainder when the expression is divided by $(a - b)$. Hence $(a - b)$ is a factor. Similarly $(b - c)$ and $(c - a)$ are factors.

The given expression is of the third degree so that, beside $(a - b)$, $(b - c)$, $(c - a)$, there can be no further factor involving a, b or c. There may, however, be a numerical factor so we write

$$a^2(b - c) + b^2(c - a) + c^2(a - b) = N(a - b)(b - c)(c - a),$$

where N is the numerical factor. To determine N we can give a, b and c any values we find convenient. Choosing $a = 0$, $b = 1$, $c = 2$ the left hand side becomes -2 while the right hand side is $2N$. Thus $2N = -2$ giving $N = -1$ and the required factors are

$$-(a - b)(b - c)(c - a).$$

2.7. The principle of undetermined coefficients

We start by showing that if a polynomial expression of degree n in x vanishes for more than n different values of x, the coefficients of each power of x must be zero. We write

$$P(x) = c_0 x^n + c_1 x^{n-1} + c_2 x^{n-2} + \ldots + c_{n-1} x + c_n,$$

and suppose that $P(x) = 0$ when x equals each of the unequal values $\alpha_1, \alpha_2, \ldots, \alpha_n$. Then $(x - \alpha_1)$, $(x - \alpha_2)$, $\ldots$, $(x - \alpha_n)$ are all factors of $P(x)$ and we can write

$$P(x) = c_0(x - \alpha_1)(x - \alpha_2) \ldots (x - \alpha_n).$$

Let β be another value of x which makes $P(x)$ vanish, then

$$c_0(\beta - \alpha_1)(\beta - \alpha_2) \ldots (\beta - \alpha_n) = 0,$$

and since none of the factors $(\beta - \alpha_1)$, $(\beta - \alpha_2)$, ..., $(\beta - \alpha_n)$ vanish, c_0 must vanish. The expression $P(x)$ now reduces to

$$P(x) = c_1 x^{n-1} + c_2 x^{n-2} + \ldots + c_{n-1} x + c_n,$$

and since this vanishes for more than n values of x we can show similarly that $c_1 = 0$. In a similar way we can show that each of the coefficients $c_2, c_3, \ldots, c_n$ must also vanish.

We can now show that if two polynomials of degree n in x are equal for more than n values of x, they are equal for all values of x. If we suppose that the two expressions

$$c_0 x^n + c_1 x^{n-1} + c_2 x^{n-2} + \ldots + c_{n-1} x + c_n,$$
$$d_0 x^n + d_1 x^{n-1} + d_2 x^{n-2} + \ldots + d_{n-1} x + d_n,$$

are equal for more than n values of x, then the polynomial

$$(c_0 - d_0) x^n + (c_1 - d_1) x^{n-1} + \ldots + (c_{n-1} - d_{n-1}) x + c_n - d_n,$$

vanishes for more than n values of x and therefore all the coefficients must be zero. Hence

$$c_0 - d_0 = 0, c_1 - d_1 = 0, \ldots, c_{n-1} - d_{n-1} = 0, c_n - d_n = 0,$$

leading to

$$c_0 = d_0, c_1 = d_1, \ldots, c_{n-1} = d_{n-1}, c_n = d_n.$$

The two expressions are thus identical and therefore equal for all values of x. Hence we have established the important result that *if two polynomial expressions in x are equal for all values of x we may equate the coefficients of the like powers of x.*

The result remains valid if the two expressions are not of the same degree. For example if one is of degree n and the other of degree $n - 1$ we should have

$$c_0 x^n + c_1 x^{n-1} + c_2 x^{n-2} + \ldots + c_{n-1} x + c_n$$
$$= d_1 x^{n-1} + d_2 x^{n-2} + \ldots + d_{n-1} x + d_n$$

and hence

$$c_0 = 0, c_1 = d_1, c_2 = d_2, \ldots, c_{n-1} = d_{n-1}, c_n = d_n.$$

The result given above is often called the *principle of undetermined coefficients* and it has important applications. Some examples of its use are given below.

Example 12. *Find constants a, b, c such that*
$$2x^2 - 9x + 14 \equiv a(x - 1)(x - 2) + b(x - 1) + c. \qquad \text{(L.U.)}$$

The sign $\equiv$ is used to denote equality between two expressions for all values of the variable involved. When two expressions are separated by such a sign we can equate the coefficients of like powers of the variable. Here we have

$$2x^2 - 9x + 14 \equiv a(x^2 - 3x + 2) + b(x - 1) + c$$
$$\equiv ax^2 - (3a - b)x + 2a - b + c.$$

Equating the coefficients of x^2, x and the term independent of x in turn gives

$$a = 2, \quad 3a - b = 9, \quad 2a - b + c = 14.$$

Substituting $a = 2$ in the second equation we have $6 - b = 9$ giving $b = -3$ and substituting $a = 2, b = -3$ in the third equation, $4 + 3 + c = 14$ leading to $c = 7$.

Example 13. *Find the relation between q and r so that $x^3 + 3px^2 + qx + r$ shall be a perfect cube for all values of x.*

Let
$$x^3 + 3px^2 + qx + r \equiv (x + a)^3$$
$$\equiv x^3 + 3ax^2 + 3a^2x + a^3.$$

Equating the coefficients of x and the term independent of x, $3a^2 = q$, $a^3 = r$. Cubing the first, squaring the second and dividing we have

$$\frac{27a^6}{a^6} = \frac{q^3}{r^2},$$

giving $q^3 = 27r^2$ as the required relation between q and r.

EXERCISES 2 (c)

1. Find the values of p and q so that $(x + 1)$ and $(x - 2)$ shall be factors of $x^3 + px^2 + 2x + q$. What is then the third factor?

2. Use the remainder theorem to find the factors of
$$(a - b)^3 + (b - c)^3 + (c - a)^3.$$

3. Find the values of λ and μ if the expression
$$3x^4 + \lambda x^3 + 12x^2 + \mu x + 4$$
is (i) exactly divisible by $(x - 1)$ and (ii) leaves remainder 18 when divided by $(x + 2)$.

4. Find the values of a and b in terms of n in
$$(x - n + 1)^3 - (x - n)^3 = 3x^2 + ax + b$$
for all values of x. (L.U.)

5. Find the values of A, B and C if the expression
$$Ax(x - 2)(x + 3) + Bx(x - 2) + Cx(x + 3) + (x - 2)(x + 3)$$
has a constant value for all values of x. (L.U.)

6. If $4x^3 + kx^2 + px + 2$ is divisible by $x^2 + \lambda^2$, prove that $kp = 8$.

2.8. Partial fractions

The student will already be familiar with the process of simplifying a group of fractions separated by addition or subtraction signs into a single fraction. For instance the expression

$$\frac{1}{x - 2} + \frac{1}{x + 2} - \frac{2x}{x^2 + 4}$$

can be simplified to give the single fraction $16x/(x^4 - 16)$ in which the denominator is the lowest common denominator of the separate fractions. It is often desirable, for instance in expansions and in the integral calculus (chapters 3 and 13), to be able to perform the reverse process. In other words, we require to be able to split up a single fraction whose denominator has factors into two or more *partial fractions*.

This reverse process, the resolution into partial fractions, depends on the following simple rules:—

(i) If the degree of the numerator of the given fraction is equal to or greater than that of the denominator, divide the numerator by the denominator until a remainder is obtained which is of lower degree than the denominator.

(ii) To every linear factor like $(x - a)$ in the denominator there corresponds a partial fraction of the form $A/(x - a)$.

(iii) To every repeated linear factor like $(x - a)^2$ in the denominator there corresponds two partial fractions of the form $A/(x - a)$ and $B/(x - a)^2$. Similarly for factors like $(x - a)^3$ we have three partial fractions $A/(x - a)$, $B/(x - a)^2$ and $C/(x - a)^3$ and so on.

(iv) To every quadratic factor like $x^2 + ax + b$ there corresponds a partial fraction $(Cx + D)/(x^2 + ax + b)$. Repeated quadratic factors require additional partial fractions as in (iii) above. Thus a factor $(x^2 + ax + b)^2$ would require partial fractions $(Cx + D)/(x^2 + ax + b)$ and $(Ex + F)/(x^2 + ax + b)^2$.

The application of these rules is illustrated in examples 14 to 17 below.

Example 14. *Resolve into partial fractions* $5/(x^2 + x - 6)$.

The factors of $x^2 + x - 6$ being $(x + 3)(x - 2)$ we assume that

$$\frac{5}{x^2 + x - 6} \equiv \frac{A}{x + 3} + \frac{B}{x - 2} \equiv \frac{A(x - 2) + B(x + 3)}{(x + 3)(x - 2)}.$$

The denominators of the expressions on the left and right being the same, the numerators must be the same. Hence we have

$$A(x - 2) + B(x + 3) \equiv 5. \qquad (2.18)$$

A and B can be found from this identity by applying the principle of undetermined coefficients. Thus, equating the coefficients of x and the term not containing x, we have

$$A + B = 0 \quad \text{and} \quad -2A + 3B = 5.$$

The solution of this pair of simultaneous equations is $A = -1, B = 1$. Another and, in the case of linear factors such as we have here, rather simpler method of determining A and B from the identity (2.18) is to give x suitable numerical values so that A and B can be found separately. Thus by putting $x = 2$ in (2.18) we have $(2 + 3)B = 5$, giving $B = 1$ and by putting $x = -3$, we have $(-3 - 2)A = 5$, giving $A = -1$.

Hence $$\frac{5}{x^2 + x - 6} \equiv \frac{1}{x - 2} - \frac{1}{x + 3}.$$

Example 15. *Separate* $\dfrac{9}{(x-1)(x+2)^2}$ *into partial fractions.*

Here, because of the repeated factor $(x + 2)$, the correct assumption is

$$\frac{9}{(x-1)(x+2)^2} \equiv \frac{A}{x-1} + \frac{B}{x+2} + \frac{C}{(x+2)^2}$$

$$\equiv \frac{A(x+2)^2 + B(x-1)(x+2) + C(x-1)}{(x-1)(x+2)^2}.$$

This identity requires that

$$A(x+2)^2 + B(x-1)(x+2) + C(x-1) \equiv 9.$$

A can be found immediately by taking $x = 1$. Thus $(1 + 2)^2 A = 9$ giving $A = 1$. To find C, take $x = -2$ and we have $(-2 - 1)C = 9$ leading to $C = -3$. To find B we can equate the coefficients of, say, x^2. This gives $A + B = 0$ so that, as $A = 1$, $B = -1$. Hence

$$\frac{9}{(x-1)(x+2)^2} \equiv \frac{1}{x-1} - \frac{1}{x+2} - \frac{3}{(x+2)^2}.$$

Example 16. *Resolve* $16x/(x^4 - 16)$ *into partial fractions.*

The factors of the denominator are $(x - 2)$, $(x + 2)$ and $(x^2 + 4)$. In view of rule (iv) and the quadratic factor $(x^2 + 4)$, we assume

$$\frac{16x}{x^4 - 16} \equiv \frac{A}{x-2} + \frac{B}{x+2} + \frac{Cx+D}{x^2+4}$$

$$\equiv \frac{A(x+2)(x^2+4) + B(x-2)(x^2+4) + (Cx+D)(x-2)(x+2)}{(x-2)(x+2)(x^2+4)}.$$

This requires that

$$A(x+2)(x^2+4) + B(x-2)(x^2+4) + (Cx+D)(x-2)(x+2) \equiv 16x.$$

Putting $x = 2$,

$$(2+2)(2^2+4)A = 16 \times 2,$$

giving $32A = 32$ or $A = 1$.
Putting $x = -2$,

$$(-2-2)\{(-2)^2+4\}B = 16 \times (-2),$$

giving $-32B = -32$ or $B = 1$.
Equating coefficients of x^3, $A + B + C = 0$, or, since $A = B = 1$, $C = -2$; and equating coefficients of the term independent of x, $8A - 8B - 4D = 0$, which with $A = B = 1$, gives $D = 0$.

Hence

$$\frac{16x}{x^4 - 16} \equiv \frac{1}{x-2} + \frac{1}{x+2} - \frac{2x}{x^2+4}.$$

Example 17. *Separate* $x^3/(x^2 - 3x + 2)$ *into partial fractions.*

In all our examples so far, the numerator of the given fraction has been of lower degree than the denominator. Here the numerator is of the third while the denominator is of the second degree. Dividing $x^2 - 3x + 2$ into x^3 we find that the quotient is $x + 3$ and that the remainder is $7x - 6$. Hence

$$\frac{x^3}{x^2 - 3x + 2} \equiv x + 3 + \frac{7x-6}{x^2 - 3x + 2}.$$

We now proceed to separate $(7x - 6)/(x^2 - 3x + 2)$ into two partial fractions as in Example 14. Thus we assume that

$$\frac{7x - 6}{x^2 - 3x + 2} \equiv \frac{A}{x - 2} + \frac{B}{x - 1} \equiv \frac{A(x - 1) + B(x - 2)}{(x - 2)(x - 1)}.$$

Hence $\qquad\qquad A(x - 1) + B(x - 2) \equiv 7x - 6.$

Setting $x = 1$ we find that $B = -1$ and taking $x = 2$ leads to $A = 8$. Finally we therefore have

$$\frac{x^3}{x^2 - 3x + 2} \equiv x + 3 + \frac{8}{x - 2} - \frac{1}{x - 1}.$$

EXERCISES 2 (d)

Resolve into partial fractions:—

1. $\dfrac{5(x + 1)}{25 - x^2}$

2. $\dfrac{x^3}{x^2 + x - 2}.$

3. $\dfrac{x^2}{(x + 1)^3}$

4. $\dfrac{x^2 + x + 1}{x^2 + 2x + 1}.$

5. $\dfrac{10 - 11x}{(x - 4)(x^2 + 1)}.$

6. $\dfrac{3x^3 + x}{x^4 - 81}.$

7. $\dfrac{2x^2 - 11x + 5}{(x^2 + 2x - 5)(x - 3)}.$

8. $\dfrac{2y + 1}{(y + 1)^2(y - 2)}.$

EXERCISES 2 (e)

1. Simplify the expression $5 \times 4^{3n+1} - 20 \times 8^{2n}$.

2. If $a = 2, b = 3$ show that $(a^3b^2c^3)^2\sqrt{(a^{-3}b^{-4}c^6)} = 144\sqrt{2c^9}$.

3. Simplify $9^{2n+2} \times 6^{2n-3} \div (3^{5n} \times 6 \times 4^{n-2})$.

4. Evaluate (i) $\dfrac{8^{2/3} + 4^{3/2}}{16^{3/4}}$, (ii) $\dfrac{\sqrt{a^3} \times \sqrt[3]{b^2}}{\sqrt[4]{a^6} \times \sqrt[6]{b^{-2}}}$ when $b = 3$.

5. If $\log_a n = x$ and $\log_c n = y$, where $n \neq 1$, prove that

$$\frac{x - y}{x + y} = \frac{\log_b c - \log_b a}{\log_b c + \log_b a}.$$

Verify this result, without using any tables, when $a = 4$, $b = 2$, $c = 8$, $n = 4096$.   (L.U.)

6. Using logarithm tables evaluate (i) $(0.0371)^{\frac{1}{1.2}}$, (ii) $\log_4 0.65$.   (Q.E.)

7. If $\log_a (1 + \tfrac{1}{8}) = l$, $\log_a (1 + \tfrac{1}{15}) = m$ and $\log_a (1 + \tfrac{1}{24}) = n$, show that $\log_a (1 + \tfrac{1}{80}) = l - m - n$.   (L.U.)

8. Find x from the equation $9^x - 12(3^x) + 27 = 0$.   (L.U.)

9. Solve the equation $4^x + 2 = 3 \times 2^x$.   (L.U.)

10. If $y = a + bx^n$ is satisfied by the values

$x =$	1	2	4
$y =$	7	10	15

show that $n = \log_2 (5/3)$ and deduce the values of a and b.   (L.U.)

11. If $2 \log_8 N = p$, $\log_2 2N = q$, $q - p = 4$, find N.   (L.U.)

12. If $x = \log_a (bc)$, $y = \log_b (ca)$ and $z = \log_c (ab)$ prove that
$$x + y + z = xyz - 2.$$

13. Find the values of a and b if the expression $2x^3 - 15x^2 + ax + b$ is divisible both by $x - 4$ and by $2x - 1$.   (O.C.)

14. A polynomial expression $P(x)$, when divided by $(x - 1)$ leaves a remainder 3 and, when divided by $(x - 2)$ leaves a remainder 1. Show that when divided by $(x - 1)(x - 2)$ it leaves a remainder $-2x + 5$.   (O.C.)

15. Express $a^4(b - c) + b^4(c - a) + c^4(a - b)$ as the product of four factors.   (L.U.)

16. If $x^3 \equiv a(x + 1)(x + 2)(x + 3) + b(x + 1)(x + 2) + c(x + 1) + d$, find the numerical values of a, b, c and d.   (L.U.)

17. Show that $x^2 + 6x - 10$ can be expressed in the form
$$(x - \alpha)(x - \beta) + 2(x - \alpha) + 3x$$
in two different ways and find the values of α and β in each case.   (L.U.)

18. Find a and b so that
$$x^4 - 7x^3 + 17x^2 - 17x + 6 \equiv (x - 1)^2(x^2 + ax + b).$$
Hence find all the factors of the quartic expression.

19. Find the value of c if the expression
$$2x^2 + xy - 6y^2 + 4x + y + c$$
can be expressed as the product of two linear factors.

20. Use the principle of undetermined coefficients to find the square root of the expression $x^4 + 4x^3 + 8x^2 + 8x + 4$.

21. Separate $\dfrac{5x + 3}{(x + 1)^2(2x + 1)}$ into partial fractions.

22. Express $\dfrac{x^2 - x - 1}{x^3 - 8}$ in partial fractions.

23. Express $\dfrac{5x^3 + 2x^2 + 5x}{x^4 - 1}$ in partial fractions.   (Q.E.)

24. Express in partial fractions $\dfrac{1 + x^2}{(1 + x)(1 + x^3)}$ (Q.E.)

25. Use the remainder theorem to find the three factors of $x^4 + 3x^2 - 4$ and hence resolve
$$\frac{2x^3 - x^2 - 7x - 14}{x^4 + 3x^2 - 4}$$
into partial fractions.

CHAPTER 3

ARITHMETICAL AND GEOMETRICAL PROGRESSIONS. PERMUTATIONS AND COMBINATIONS. THE BINOMIAL THEOREM

3.1. Series

A set of numbers each of which can be obtained from some definite law is called a *series* or *progression*. Each of the numbers forming the set is called a *term* of the series. Thus the sets

(i) $1, 3, 5, 7, \ldots,$

(ii) $1, 2, 4, 8, \ldots,$

(iii) $1^2, 2^2, 3^2, 4^2, \ldots,$

are all series. In the first set, each number is obtained by adding 2 to the preceding one, in the second each term is twice the preceding one and in the third each number is the square of successive integers. It is possible to give a formula for the general or nth term of each of the above series. Thus for (i) the nth term is $2n - 1$, for (ii) it is 2^{n-1} and for (iii) it is n^2. If an expression for the nth term of a series is known it is possible to write down successive terms by giving successive integral values to n Thus the series whose nth term is $\left(1 + \dfrac{1}{n}\right)^n$ is one whose terms are $2, \frac{9}{4}, \frac{64}{27}, \frac{625}{256}, \ldots$, these being the values obtained by putting $n = 1, 2, 3, 4, \ldots$ in the formula for the nth term.

Series play a very important part in mathematical analysis; in this chapter we shall be concerned with a few of the simpler ones.

3.2. The arithmetical progression

A series in which each term is obtained from the preceding one by adding (or subtracting) a constant quantity is called an *arithmetical progression* (A.P.). Thus the series

$$1, 3, 5, 7, \ldots,$$
$$a, a + d, a + 2d, a + 3d, \ldots$$

are arithmetical progressions. The difference between each term and the preceding one is called the *common difference*. When three quantities are in arithmetical progression the middle one is called the *arithmetic mean* of the other two. Thus a is the arithmetic mean between $a - d$ and $a + d$.

In the series

$$a, a + d, a + 2d, a + 3d, \ldots, \tag{3.1}$$

the coefficient of d in any term is one less than the number of the term in the series. Thus $a + 3d$ is the fourth term. If then the series consists of n terms and l denotes the last or nth term

$$l = a + (n - 1)d. \qquad (3.2)$$

To obtain the sum s_n of n terms of the series (3.1) we have

$$s_n = a + (a + d) + (a + 2d) + \ldots + (l - 2d) + (l - d) + l,$$

for if l is the last term, the next to last will be $l - d$ and the preceding one will be $l - 2d$ and so on. If now we write the series in the reverse order

$$s_n = l + (l - d) + (l - 2d) + \ldots + (a + 2d) + (a + d) + a.$$

Adding and noticing that the sums of terms in corresponding positions are all $a + l$ we have

$$2s_n = (a + l) + (a + l) + (a + l) + \ldots \text{ to } n \text{ terms,}$$
$$= n(a + l).$$

Hence

$$s_n = \frac{n}{2}(a + l), \qquad (3.3)$$

or, using (3.2),

$$s_n = \frac{n}{2}\{2a + (n - 1)d\}. \qquad (3.4)$$

Example 1. *Insert seven arithmetic means between 2 and 26.*

It is always possible to insert any number of terms between two given quantities such that the resulting series shall be an arithmetical progression. Terms inserted in this way are called *arithmetic means*, an extension of the meaning of an arithmetic mean between two given quantities.

Including the first and last terms, the number of terms will here be nine, so we have to find an arithmetical progression of nine terms of which the first is 2 and the last is 26. Let d be the common difference. Then

$$26 = 9\text{th term} = 2 + 8d,$$

so that $d = 3$. The second term is therefore $2 + 3$ or 5, the third $2 + 6$ or 8 and so on. Hence the required means are 5, 8, 11, 14, 17, 20 and 23.

Example 2. *Find three numbers in arithmetical progression such that their sum is 27 and their product is 504.*

Let the three numbers in arithmetical progression be $a - d$, a and $a + d$. Then the sum of the numbers is $3a$ and since this is 27, $a = 9$. Their product is $(a - d)a(a + d)$ or $a(a^2 - d^2)$. Hence

$$a(a^2 - d^2) = 504,$$

and since $a = 9$, $81 - d^2 = 504/9 = 56$, leading to $d^2 = 25$ and $d = \pm 5$. Hence the required numbers are 9, 9 ± 5, i.e., 4, 9 and 14.

Example 3. *The first term of an arithmetical progression is 25 and the third term is 19. Find the number of terms in the progression if its sum is 82.*

Here $a = 25$, $a + 2d = 19$, so that $2d = 19 - a = 19 - 25 = -6$ giving

$d = -3$. With $s_n = 82$, formula (3.4) for the sum of n terms of the series gives

$$82 = \frac{n}{2}\{50 + (n - 1)(-3)\}$$

as the equation for the number of terms (n). This reduces to the quadratic equation

$$3n^2 - 53n + 164 = 0,$$

or

$$(n - 4)(3n - 41) = 0.$$

Hence $n = 4$ or $n = \frac{41}{3} = 13\frac{2}{3}$. The fractional result means that the sum of 13 terms will be greater than 82 and that of 14 terms will be less than 82.

3.3. The geometrical progression

A series in which each term is obtained from the preceding one by multiplying (or dividing) by a constant quantity is called a *geometrical progression* (G.P.). Examples are

$$1, 2, 4, 8, \ldots,$$
$$a, ar, ar^2, ar^3, \ldots$$

The ratio between each term and the preceding one is called the *common ratio*. When three quantities are in geometrical progression the middle one is called the *geometric mean* between the other two. Thus a is the geometric mean between a/r and ar.

In the series,

$$a, ar, ar^2, ar^3, \ldots, \tag{3.5}$$

the index of r in any term is one less than the number of the term in the series. Thus ar^3 is the fourth term. The last or nth term of the series is given by

$$l = ar^{n-1}. \tag{3.6}$$

To obtain the sum s_n of n terms of the series (3.5) we have

$$s_n = a + ar + ar^2 + \ldots + ar^{n-2} + ar^{n-1}.$$

Multiplying throughout by r

$$rs_n = ar + ar^2 + ar^3 + \ldots + ar^{n-1} + ar^n.$$

If we subtract, all the terms on the right hand side except a and ar^n cancel in pairs. Hence $s_n - rs_n = a - ar^n$, leading to

$$s_n = \frac{a(1 - r^n)}{1 - r}. \tag{3.7}$$

Example 4. *Insert three geometric means between 162 and 1250.*

As with arithmetical progressions, it is possible to insert any number of terms between two given quantities such that the resulting series shall be a geometrical progression. Such terms are referred to as *geometric means*, an extension of a geometric mean between two given quantities. Here, including the first and last terms, the number of terms will be five, so we have to find a geometrical progression of five terms of which the first is 162 and the last is 1250. Let r be the common ratio. Then

$$1250 = \text{5th term} = 162r^4.$$

Hence $r^4 = 1250/162 = 625/81$, so that $r = \pm 5/3$. The second term is therefore $(\pm 5/3) \times 162$ or ± 270, the third is $(\pm 5/3)^2 \times 162$ or 450 and the fourth is $(\pm 5/3)^3 \times 162$ or ± 750.

Example 5. *Find three numbers in geometrical progression such that their sum is 39 and their product is 729.*

Let the required numbers be a/r, a and ar. Then their product is a^3 and hence $a^3 = 729$ giving $a = 9$. Since the sum is 39, we have

$$\frac{9}{r} + 9 + 9r = 39,$$

so that $9r^2 - 30r + 9 = 0$. This can be written $3r^2 - 10r + 3 = 0$ or $(3r - 1)(r - 3) = 0$, giving $r = 1/3$ or 3. The required numbers are therefore $9/3$, 9 and 9×3, or 3, 9 and 27.

Example 6. *Find the sum of ten terms of the geometrical series* 2, -4, 8, $\dots$

Here the first term is 2 and the common ratio is -2. Hence in the formula (3.7), $a = 2, r = -2, n = 10$. Hence the required sum

$$= \frac{2\{1 - (-2)^{10}\}}{1 - (-2)} = -\tfrac{2}{3}\{(2)^{10} - 1\} = -\tfrac{2}{3}(1024 - 1) = -682.$$

EXERCISES 3 (*a*)

1. Write down the first three and the 8th term of the series whose nth terms are:—

   (i) $4n - 5$, (ii) 3^{n-1}, (iii) $(-1)^n$.

2. Show that the arithmetic and geometric means between the two quantities a, b are respectively $\tfrac{1}{2}(a + b)$ and $\sqrt{(ab)}$.

3. Find the sum of ten terms of an arithmetical progression of which the first term is 60 and the last is -104.

4. If the first, third and sixth terms of an arithmetical progression are in geometrical progression, find the common ratio of the geometrical progression.

5. The sum of the last three terms of a geometrical progression having n terms is 1024 times the sum of the first three terms of the progression. If the third term is 5, find the last term.   (L.U.)

6. Find two numbers whose arithmetic mean is 39 and geometric mean 15.
                                                                      (L.U.)

7. Prove that the series $\log a$, $\log (ar)$, $\log (ar^2)$, $\dots$, is an arithmetical progression whose sum to n terms is $\tfrac{1}{2}n \log (a^2 r^{n-1})$.

8. The second and third terms of a geometrical progression are 24 and $12(b + 1)$ respectively. Find b if the sum of the first three terms of the progression is 76.

3.4. Simple and compound interest

If a sum of money P (the principal) is invested at *simple* interest of r per cent. per annum, the amount A (principal plus interest) after n years is given by

$$A = P\left(1 + \frac{nr}{100}\right), \tag{3.8}$$

for the interest for one year is $Pr/100$ and for n years $nPr/100$. The various amounts after one, two, three, ... years therefore form an arithmetical progression.

If, on the other hand, the same principal is invested at *compound* interest of r per cent. per annum, the interest being added annually, the amount after one year is $P\left(1 + \frac{r}{100}\right)$, and this is the principal for the second year. Hence after two years the amount is

$$P\left(1 + \frac{r}{100}\right)\left(1 + \frac{r}{100}\right) \text{ or } P\left(1 + \frac{r}{100}\right)^2$$

and so on. Thus after n years the amount will be given by

$$A = P\left(1 + \frac{r}{100}\right)^n. \tag{3.9}$$

In this case the amounts after one, two, three, ... years form a geometrical progression.

If, with compound interest the interest is added half yearly, the interest is half as much as when added yearly but it is added twice as often. Hence in this case

$$A = P\left(1 + \frac{r}{200}\right)^{2n}, \tag{3.10}$$

and similarly for cases where the interest is added at other intervals.

Suppose we wish to find the present value (V) of a given sum (S) due n years hence. Then V is the sum put out to interest at the present time which in n years will amount to S. Thus at simple interest

$$S = V\left(1 + \frac{nr}{100}\right),$$

giving,

$$V = \frac{S}{1 + \frac{nr}{100}}, \tag{3.11}$$

and at compound interest (added yearly),

$$S = V\left(1 + \frac{r}{100}\right)^n,$$

giving,

$$V = S\left(1 + \frac{r}{100}\right)^{-n}. \tag{3.12}$$

Example 7. *Find the amount at the end of* 10 *years when* £400 *is invested at 4 per cent. compound interest,* (i) *the interest being added annually and* (ii) *the interest being added twice a year.*

Here $P = 400, r = 4, n = 10$ and from (3.9), (3.10) we have

(i) interest added annually, amount $= P\left(1 + \dfrac{r}{100}\right)^n = 400(1{\cdot}04)^{10}$

$$= \text{£591{\cdot}7.}$$

(ii) interest added twice a year, amount $= P\left(1 + \dfrac{r}{200}\right)^{2n} = 400(1{\cdot}02)^{20}$

$$= \text{£594{\cdot}4.}$$

Example 8. *Find what sum a man has to invest on his fortieth birthday so that he may be able to draw out a lump sum of* £2000 *on his sixtieth birthday, the investment being made at 5 per cent. per annum compound interest.*

Let $V =$ sum required. Then V has to amount to £2000 in 20 years.

i.e., $V = 2000(1{\cdot}05)^{-20} = \dfrac{2000}{2{\cdot}655} = \text{£753{\cdot}3.}$

3.5. The convergence of the geometric series

Consider the geometrical progression $1 + \frac{1}{2} + \frac{1}{4} + \frac{1}{8} + \ldots$ If we stop at the third term the sum is $1\frac{3}{4}$ and this is less than 2 by the third term $\frac{1}{4}$. Similarly the sum of four terms is $1\frac{7}{8}$ which differs from 2 by the fourth term $\frac{1}{8}$. Similarly wherever we stop the sum is less than 2 by precisely the last term added. Thus the sum of this series never exceeds 2, never reaches 2 but may be made as near to 2 as we please by taking a sufficient number of terms. The value 2 is called the *limit of the sum* of this series. Series for which such a limit exists are said to be *convergent*.

Consider now the general geometrical progression

$$a + ar + ar^2 + \ldots$$

By (3.7) the sum to n terms, denoted by s_n, is given by

$$s_n = \frac{a(1 - r^n)}{1 - r} = \frac{a}{1 - r} - \frac{ar^n}{1 - r}.$$

Suppose r lies between 0 and 1. Then r^n decreases as n increases and, since it cannot be negative, it must tend to some positive limit l. Since $r^{n+1} = r \cdot r^n$, r^{n+1} and r^n are both ultimately equal to l and we have $l = rl$, showing, since r is not equal to unity, that l must be zero. In a similar way, if r lies between -1 and 0, we can show that the limiting value of r^n is also zero. Thus the value of the term $ar^n/(1 - r)$ becomes nearer and nearer to zero as n increases and the limit of the sum of the progression, denoted by s is, for $-1 < r < 1$, given by

$$s = \frac{a}{1 - r}, \tag{3.13}$$

for we may make s_n as near to $a/(1 - r)$ as we please by making n sufficiently large. We say that the *geometric series whose first term is a and common ratio r converges when* $-1 < r < 1$ *and the limit of the sum is* $a/(1 - r)$.

There is no limit of the sum of a geometrical progression whose common ratio lies outside the range -1 to 1. For instance the sum of the progression $1 + 2 + 4 \ldots$ gets more and more unmanageable as more terms are taken. Each term in fact exceeds the sum of all the preceding ones. The series is in this case said to be *divergent*.

Rather loose expressions are sometimes used in connection with convergent series. Thus the limit of the sum is sometimes called "the sum to infinity" and a convergent geometrical progression is sometimes referred to as an "infinite" geometrical progression. It is preferable to avoid the words infinity and infinite as far as possible.

Example 9. *The limit of the sum of a convergent geometrical progression is k and the limit of the sum of the squares of its terms is l. Find the first term and the common ratio of the progression.*

Let the first term be a and the common ratio r. Then by (3.13) $k = a/(1 - r)$. For the series comprised of the squares of these terms the first term is a^2 and the common ratio r^2. Hence $l = a^2/(1 - r^2)$. Squaring the expression for k and dividing by that for l we have

$$\frac{k^2}{l} = \frac{1 - r^2}{(1 - r)^2} = \frac{1 + r}{1 - r}.$$

Solving for r, we find $r = (k^2 - l)/(k^2 + l)$.
Substituting this value of r in the expression for k,

$$a = k(1 - r) = k\left[1 - \frac{k^2 - l}{k^2 + l}\right] = \frac{2kl}{k^2 + l}$$

Example 10. *Evaluate* $0\dot{6}$ *as a fraction.*

$0\dot{6}$ means $\frac{6}{10} + \frac{6}{100} + \frac{6}{1000} + \ldots$ and this can be written

$$\frac{6}{10}\left[1 + \frac{1}{10} + \frac{1}{10^2} + \ldots\right].$$

The series inside the square brackets is a convergent geometrical progression with first term unity and common ratio $\frac{1}{10}$. By (3.13) the limit of its sum is $1/(1 - \frac{1}{10})$ or $\frac{10}{9}$. Hence the value of

$$0\dot{6} = \frac{6}{10} \times \frac{10}{9} = \frac{2}{3}.$$

3.6. Series involving the natural numbers

The positive integers 1, 2, 3, ... are often referred to as the *natural numbers*. They form an arithmetical progression with first term and common difference unity and the sum of the first n natural numbers is therefore given by

$$S_1 = \frac{n}{2}\{2 + (n - 1)\} = \frac{n(n + 1)}{2}. \tag{3.14}$$

The sum of the squares of the first n natural numbers can be found by starting from the identity

$$n^3 - (n - 1)^3 \equiv 3n^2 - 3n + 1.$$

Changing n successively into $(n - 1), (n - 2), \ldots, 2, 1$, we have

$$(n - 1)^3 - (n - 2)^3 = 3(n - 1)^2 - 3(n - 1) + 1,$$
$$(n - 2)^3 - (n - 3)^3 = 3(n - 2)^2 - 3(n - 2) + 1,$$
$$\cdots\cdots\cdots\cdots\cdots\cdots\cdots\cdots\cdots\cdots\cdots$$
$$2^3 - 1^3 = 3.2^2 - 3.2 + 1,$$
$$1^3 - 0^3 = 3.1^2 - 3.1 + 1.$$

By addition and noticing that apart from n^3 and 0^3 all the terms on the left hand side cancel in pairs,

$$n^3 = 3\{1^2 + 2^2 + \ldots + (n - 1)^2 + n^2\}$$
$$- 3\{1 + 2 + \ldots + (n - 1) + n\} + n.$$

If we denote the sum of $1^2 + 2^2 + \ldots + n^2$ by S_2, this gives

$$n^3 = 3S_2 - 3S_1 + n.$$

Using the value of S_1 given in (3.14) we have

$$3S_2 = n^3 + \frac{3}{2}n(n + 1) - n = \frac{n}{2}(2n^2 + 3n + 1),$$

or, $$S_2 = \frac{n(n + 1)(2n + 1)}{6}. \tag{3.15}$$

The sums of the cubes and higher powers of the natural numbers can be found in a similar way but the process gets more and more tedious.

Example 11. *Find the sum of n terms of the series* $1.2 + 2.3 + 3.4 + \ldots$

The nth term is $n(n + 1)$ and by writing this as $(n^2 + n)$, the sum of the series is the sum of the squares of the first n natural numbers plus the sum of these numbers. Hence the required sum

$$= \tfrac{1}{6}n(n + 1)(2n + 1) + \tfrac{1}{2}n(n + 1)$$

$$= \tfrac{1}{2}n(n + 1)\left\{\frac{2n + 1}{3} + 1\right\} = \frac{n(n + 1)(n + 2)}{3}.$$

EXERCISES 3 (b)

1. In how many years will a sum of money double itself (a) at 5 per cent. simple interest, (b) at 5 per cent. compound interest?

2. A man arranges to purchase a house, valued now at £1000, by paying £500 in ten years' time and spreading the remaining payments in 10 equal annual instalments of £X, the first being paid now. If compound interest on all outstanding amounts is payable at 4 per cent. per annum, calculate the value of X.
(L.U.)

3. Prove that the geometrical progression,

$$1 + \frac{2x}{3 + x^2} + \left(\frac{2x}{3 + x^2}\right)^2 + \dots,$$

is convergent for all values of x and find the limit of its sum. (L.U.)

4. Show that there are two geometrical progressions in which the second term is $-4/3$ and the sum of the first three terms is $28/9$. Show also that one of these progressions is convergent and, in this case, find the limit of its sum. (L.U.)

5. Find the first term and the common ratio of a convergent geometrical progression in which (i) the limit of the sum is 4 and (ii) the limit of the sum of the series formed by the cubes of the terms of the geometrical progression is 192.

6. When does the series $a + \dfrac{a}{1 + a} + \dfrac{a}{(1 + a)^2} + \dots$ converge and what is then the limit of its sum?

7. Starting from the identity

$$(2n + 1)^4 - (2n - 1)^4 \equiv 64n^3 + 16n,$$

show that the sum of the cubes of the first n natural numbers is equal to the square of the sum of these numbers.

8. The first, second, third and nth terms of a series are 4, -3, -16 and $(an^2 + bn + c)$ respectively. Find a, b, c and the sum of n terms of the series.

3.7. Permutations and combinations

Suppose we have four objects denoted by A, B, C and D and we select groups of two. Possible selections are AB, AC, AD, BC, BD and CD. Each *selection* is called a *combination* and it is possible to make six different combinations from four objects taken two at a time. If, however, we are concerned with the arrangements of the four objects taken two at a time we can do this in twelve different ways, viz.,

$$AB, AC, AD, BC, BD, CD,$$

and
$$BA, CA, DA, CB, DB, DC.$$

Each *arrangement* is called a *permutation* and it is possible to make twelve different permutations from four objects taken two at a time. Thus in forming combinations we are concerned only with the number of things each selection contains whereas in forming permutations we are concerned with the order of the component objects as well.

A formula giving the number of permutations which can be made from n unlike things taken r at a time can be obtained as follows. We have to fill up r places from n things. The first place can be filled in n ways for we have n things at our disposal. When it has been filled the second place can be filled in $(n - 1)$ ways for now we have only $(n - 1)$ things available to fill it. Each way of filling the first place may be associated with each way of filling the second, so that the first two places may be filled in $n(n-1)$ different ways. Proceeding in this way the first three places can be filled in $n(n-1)(n-2)$ ways and all the r places can therefore be filled in

$$n(n-1)(n-2)\ldots(n-r+1)$$

ways. A convenient notation for the number of permutations of n things taken r at a time is ^nP_r and we therefore have

$$^nP_r = n(n-1)(n-2)\ldots(n-r+1), \qquad (3.16)$$

the number of factors being r (the number in the suffix of the symbol ^nP_r). Putting $r=n$ we have for the number of permutations of n things taken all at a time (or the number of ways of arranging n things among themselves)

$$^nP_n = n(n-1)(n-2)\ldots3.2.1,$$

there now being n factors. The product $n(n-1)(n-2)\ldots3.2.1$ is called "factorial n" and written $(n)!$, or sometimes, $\underline{|n}$.

To find the number of combinations which can be made from n unlike things taken r at a time, let ^nC_r (a notation similar to that for the number of permutations) be the required number. Then each of these ^nC_r combinations consists of a group of r things. These can be arranged among themselves in $(r)!$ ways. Hence the product of ^nC_r and $(r)!$ is the number of arrangements of n things taken r at a time, so that

$$^nC_r \times (r)! = {}^nP_r$$
$$= n(n-1)(n-2)\ldots(n-r+1).$$

Hence
$$^nC_r = \frac{n(n-1)(n-2)\ldots(n-r+1)}{(r)!}. \qquad (3.17)$$

An alternative form of (3.17) can be obtained by multiplying numerator and denominator by $(n-r)!$. Since $(n-r)! = (n-r)(n-r-1)\ldots3.2.1$ the numerator will now contain all the numbers n, $(n-1)$, $(n-2)$, down to unity and will therefore be $(n)!$.

Hence
$$^nC_r = \frac{(n)!}{(r)!(n-r)!}. \qquad (3.18)$$

Example 12. *Find how many different numbers can be made by using four out of the nine digits* 1, 2, 3, ..., 9.

The required number is the number of permutations of nine things taken four at a time and is therefore

$$^9P_4 = 9 \times 8 \times 7 \times 6 = 3024.$$

Example 13. *In how many ways can an escort of four soldiers be chosen from nine soldiers and in how many of these escorts will a particular soldier be included?*
(L.U.)

The required number is the number of selections which can be made from nine things taken four at a time. This is $^9C_4$ which by (3.17) is

$$\frac{9 \times 8 \times 7 \times 6}{4 \times 3 \times 2 \times 1} = 126.$$

When a particular soldier is always to be included, we have to find the number of ways in which selections for the other three places in the escort can be made from the remaining eight men. This is $^8C_3$ or

$$\frac{8 \times 7 \times 6}{3 \times 2 \times 1} = 56.$$

The method of the last part of example 13 can be employed to obtain a formula which we shall use later in this chapter (page 50). Suppose we have $(n + 1)$ objects: the number of combinations of these objects taken r at a time such that a particular object is always *excluded* is ^nC_r for we have to select from only n objects. The number of combinations of the objects taken r at a time such that a particular object is always *included* is $^nC_{r-1}$ for we have to select from only n objects for the remaining $(r - 1)$ places in a selection. Since the object must be either included or excluded, the sum is the total number of combinations of $(n + 1)$ objects taken r at a time or $^{n+1}C_r$. Hence

$$^{n+1}C_r = {}^nC_r + {}^nC_{r-1}. \tag{3.19}$$

So far we have based our work on the assumption that the objects of which arrangements have been made or from which selections have been taken are all dissimilar. Formulae for the number of permutations or combinations when the objects are not all unlike are rather complicated. Such cases are best treated on their merits and we consider the following as an example.

To find the number of arrangements of n things taken all at a time when p are exactly alike of one kind and q are exactly alike of another kind, let x be the required number of permutations. Then if the p like objects were replaced by p unlike objects different from any of the rest, from any one of the x arrangements we could form $(p)!$ new permutations without altering the position of any of the remaining objects. If then this change were made in each of the x arrangements, we should obtain $x \times (p)!$ permutations. Similarly if the q like objects were replaced by q unlike ones, the number of permutations would be $x \times (p)! \times (q)!$. But the objects are now all different and can be arranged among themselves in $(n)!$ ways. Hence

$$x \times (p)! \times (q)! = (n)!,$$

giving,
$$x = \frac{(n)!}{(p)!\,(q)!} \tag{3.20}$$

Similarly the number of arrangements of n things taken all at a time when p are alike of one kind, q alike of a second kind and r alike of a third kind and so on is

$$\frac{(n)!}{(p)!\,(q)!\,(r)!\ldots} \tag{3.21}$$

Example 14. *How many different arrangements of letters can be made by using all the letters of the word* contact? *In how many of these arrangements are the vowels separated?* (L.U.)

Here we have seven letters including two c's and two t's. The required number of arrangements by (3.20) is

$$\frac{(7)!}{(2)!\,(2)!} = \frac{7 \times 6 \times 5 \times 4 \times 3 \times 2 \times 1}{2 \times 1 \times 2 \times 1} = 1260.$$

If we treat the vowels o, a as one letter, the number of arrangements with the vowels together is

$$\frac{2(6)!}{(2)!\,(2)!} = \frac{2 \times 6 \times 5 \times 4 \times 3 \times 2 \times 1}{2 \times 1 \times 2 \times 1} = 360,$$

the multiplier 2 being introduced in the numerator to allow for the two possible arrangements ao, oa of the vowels among themselves. The number of arrangements with the vowels separated is the difference, $1260 - 360$ or 900.

Sometimes the number of permutations of n things taken r at a time is required when each thing may be repeated any number of times (up to r) in any arrangement. Here the first place may be filled in n ways and, when it has been filled, the second place may also be filled in n ways for we are able, if we wish, to use the same thing again. Thus the first two places can be filled in $n \times n$ or n^2 ways. Similarly the first three places can be filled in $n \times n \times n$ or n^3 ways and so on. The total number of arrangements is therefore n^r.

Example 15. *How many entries must be made in a football pool consisting of twelve matches to ensure a correct forecast?*

The result of each match may be win, lose or draw so that the forecast of the first match can be made in three ways. The result of the second game can similarly be entered in three ways so that a correct forecast of the first two matches will require 3^2 entries. For the first three matches 3^3 entries will be required and so on. Hence for all twelve matches the required number of entries will be 3^{12} or 531441.

3.8. Probability or chance

Suppose any one of $m + n$ events is just as likely to happen as any other and that one event is certain to happen. Then if m of the events are considered favourable and n unfavourable the *probability* or *chance* of a favourable event is said to be $m/(m + n)$. For example, in tossing a coin, heads or tails are equally likely and either a head or tail must occur. The chance of throwing a head is therefore $1/2$. Similarly the chance of throwing a two with a six-sided die is $1/6$ for only the two is favourable, any one of the numbers one to six is equally likely and one number must occur. The phrase "just as likely to happen" in the above definition is open to criticism but the general sense is clear. Thus the belief that a coin will be equally likely to fall heads or tails is generally accepted in the absence of any distinct proof to the contrary.

If the probability of an event happening is p and of its not happening is q, we have from the above.

$$p = \frac{m}{m+n}, \quad q = \frac{n}{m+n}.$$

Thus $p + q = 1$ and $q = 1 - p$. Probabilities can therefore range between 0 and 1, 0 indicating impossibility and 1 certainty. Sometimes percentages are used, a one per cent. chance means a chance of one in a hundred. If the chances for and against an event are p and q, the odds against an event happening are q to p. Thus odds of 5 to 2 against an event implies that the chance of the event happening is $2/7$ and of it not happening is $5/7$.

Example 16. *What is the chance that a hand of thirteen cards dealt from a pack shall contain only red cards?*

The total number of possible hands is $^{52}C_{13}$ and the total number of favourable hands is $^{26}C_{13}$ for in this case selections of 13 have to be made from the 26 red cards. The required chance is therefore

$$^{26}C_{13}/^{52}C_{13} = 1/61055$$

approximately.

If the chance of one event happening is p and that of another independent event happening is p' the chance of both happening is pp'. The chance of the first happening and not the second is $p(1 - p')$, that of the second happening and not the first is $p'(1 - p)$ and the chance of neither happening is $(1 - p)(1 - p')$. These results follow directly from the definition of probability. For example the chance of throwing a six with one die is $1/6$, the chance of two sixes when two dice are thrown is $1/36$. The chance of one six only from two dice is

$$\tfrac{1}{6} \times \tfrac{5}{6} + \tfrac{5}{6} \times \tfrac{1}{6} = \tfrac{10}{36},$$

and the chance of no sixes is $25/36$.

EXERCISES 3 (c)

1. In how many ways can a team of eleven be picked from fifteen possible players?

2. How many different arrangements can be made by taking (i) five, (ii) all the letters of the word *special*?

3. How many numbers between 2000 and 3000 can be made from the digits 7, 3, 2, 5?

4. In how many ways can five books be distributed to four readers when each reader can have all the books?

5. How many numbers each of four digits can be formed from the digits 1, 2, 3, 4 when each digit can be repeated four times? Calculate the sum of all these numbers. (L.U.)

6. There are 10 articles, 2 of which are alike and the rest all different. In how many ways can a selection of 5 articles be made? (L.U.)

7. A signaller has six flags, of which one is blue, two are white and three are red. He sends messages by hoisting flags on a flagpole, the message being conveyed by the order in which the colours are arranged. Find how many different messages he can send (i) by using exactly six flags, (ii) by using exactly five flags. (L.U.)

8. What are the odds against drawing three black balls from a bag containing four white and five black balls?

9. Find the chance of throwing head and tail alternately with three successive tosses of a coin.

3.9. The binomial theorem for a positive integral index

By actual multiplication, we can show that

$$(1 + x)^2 = 1 + 2x + x^2,$$
$$(1 + x)^3 = 1 + 3x + 3x^2 + x^3,$$
$$(1 + x)^4 = 1 + 4x + 6x^2 + 4x^3 + x^4.$$

In the expressions for these powers of $(1 + x)$ we observe that:—

(i) the indices of x increase by unity as we go term by term from left to right, the index of the last term being the same as the power to which $(1 + x)$ is raised,

(ii) the first term and the coefficient of x in the last term are both unity and those of the other terms are $^2C_1$ in the expression for $(1 + x)^2$, $^3C_1$ and $^3C_2$ in that for $(1 + x)^3$ and $^4C_1$, $^4C_2$, $^4C_3$ in that for $(1 + x)^4$.

This suggests that the result for any positive integral power n of $(1 + x)$ will be

$$(1 + x)^n = 1 + {}^nC_1x + {}^nC_2x^2 + \ldots + {}^nC_rx^r + \ldots + x^n. \quad (3.22)$$

Assuming that this result is valid, multiplication by $(1 + x)$ and collection of the terms in like powers of x gives

$$(1 + x)^{n+1} = 1 + ({}^nC_1 + 1)x + ({}^nC_2 + {}^nC_1)x^2 + \ldots$$
$$+ ({}^nC_r + {}^nC_{r-1})x^r + \ldots + x^{n+1}.$$

Since $^nC_1 + 1 = n + 1 = {}^{n+1}C_1$ and $^{n+1}C_r = {}^nC_r + {}^nC_{r-1}$, a result already established in (3.19), this can be written

$$(1 + x)^{n+1} = 1 + {}^{n+1}C_1x + {}^{n+1}C_2x^2 + \ldots + {}^{n+1}C_rx^r$$
$$+ \ldots + x^{n+1}. \quad (3.23)$$

Hence if the assumption made in (3.22) is true for a positive integral index n, (3.23) shows that it is also true when n is increased to $n + 1$. But we know the assumption to be true for $n = 2, 3$ and 4, so that we infer that it is also true for $n = 5$ and therefore for $n = 6$ and so on. Hence the result is true for any positive integer n.

Often the coefficients in (3.22) are abbreviated by omitting the n; with this notation we should have

$$(1 + x)^n = 1 + C_1x + C_2x^2 + \ldots + C_rx^r + \ldots + x^n, \quad (3.24)$$

where the coefficient of x^r is given by

$$C_r = {}^nC_r = \frac{n(n-1)(n-2)\ldots(n-r+1)}{(r)!}. \tag{3.25}$$

The first and last terms are sometimes written as C_0 and C_n, so that C_0 and C_n are both unity and C_r is then always the coefficient of x^r in the expansion. The result established here is known as the *binomial theorem* as it gives the expansion (a series of $n + 1$ terms) for the nth power of the binomial expression $(1 + x)$. The method used here in establishing the theorem is known as a proof by *induction* and is a very powerful method in many branches of pure mathematics.

If we require the expansion of $(a + x)^n$, we have

$$(a + x)^n = a^n \left(1 + \frac{x}{a}\right)^n,$$

and, writing x/a in place of x in (3.24),

$$(a + x)^n = a^n \left(1 + C_1\frac{x}{a} + C_2\frac{x^2}{a^2} + \ldots + C_r\frac{x^r}{a^r} + \ldots + \frac{x^n}{a^n}\right)$$
$$= a^n + C_1 a^{n-1} x + C_2 a^{n-2} x^2 + \ldots$$
$$+ C_r a^{n-r} x^r + \ldots + x^n. \tag{3.26}$$

The numerical coefficients in the binomial expansion are given by the following table (*Pascal's arithmetical triangle*):—

Power			Coefficients				
1	1	1					
2	1	2	1				
3	1	3	3	1			
4	1	4	6	4	1		
5	1	5	10	10	5	1	
6	1	6	15	20	15	6	1

Apart from the first and last coefficients which are unity, any entry in the table is given by adding together the one immediately above it and the next entry on the left. Thus the entry 15 in the sixth line is the sum of 10 and 5 and so on. The coefficients in the expansion of $(1 + x)^7$ are therefore immediately obtained from the last line of the table as 1, 7, 21, 35, 35, 21, 7 and 1 and once these are available, those for $(1 + x)^8$ can be obtained similarly.

Example 17. *Expand $(x + 3y)^6$ by the binomial theorem and apply the expansion to evaluate $(1\cdot03)^6$ correct to five places of decimals.* (L.U.)

Writing x for a, $3y$ for x and taking $n = 6$, (3.26) gives

$$(x + 3y)^6 = x^6 + 6x^5(3y) + 15x^4(3y)^2 + 20x^3(3y)^3 + 15x^2(3y)^4$$
$$+ 6x(3y)^5 + (3y)^6,$$

the coefficients being taken from Pascal's triangle [or calculated from (3.25)]. This gives

$$(x + 3y)^6 = x^6 + 18x^5y + 135x^4y^2 + 540x^3y^3 + 1215x^2y^4$$
$$+ 1458xy^5 + 729y^6.$$

Taking $x = 1, y = 10^{-2}, x + 3y = 1\cdot03$ and hence

$$(1\cdot03)^6 = 1 + 18 \times 10^{-2} + 135 \times 10^{-4} + 540 \times 10^{-6} + 1215 \times 10^{-8}$$
$$+ 1458 \times 10^{-10} + 729 \times 10^{-12}$$

$$= 1\cdot0000000 + 0\cdot1800000 + 0\cdot0135000 + 0\cdot0005400$$
$$+ 0\cdot0000122 + 0\cdot0000001 + \ldots$$

$$= 1\cdot1940523 = 1\cdot19405 \text{ (to five places)}.$$

Example 18. *Expand* $(1 - \frac{3}{2}x - x^2)^5$ *in ascending powers of* x *as far as the term in* x^4.

Writing $-\frac{3}{2}x(1 + \frac{2}{3}x)$ in place of x in (3.24) and taking the coefficients from Pascal's triangle,

$$(1 - \tfrac{3}{2}x - x^2)^5 = \{1 - \tfrac{3}{2}x(1 + \tfrac{2}{3}x)\}^5$$
$$= 1 + 5(-\tfrac{3}{2}x)(1 + \tfrac{2}{3}x) + 10(-\tfrac{3}{2}x)^2(1 + \tfrac{2}{3}x)^2 + 10(-\tfrac{3}{2}x)^3(1 + \tfrac{2}{3}x)^3$$
$$+ 5(-\tfrac{3}{2}x)^4(1 + \tfrac{2}{3}x)^4 + \ldots$$

The last term in the expansion need not be included as it involves x^5 and higher powers only. Simplifying and retaining only terms which involve x^4 and lower powers we have

$$(1 - \tfrac{3}{2}x - x^2)^5 = 1 - \tfrac{15}{2}x(1 + \tfrac{2}{3}x) + \tfrac{45}{2}x^2(1 + \tfrac{4}{3}x + \tfrac{4}{9}x^2)$$
$$- \tfrac{135}{4}x^3(1 + 2x + \ldots) + \tfrac{405}{16}x^4(1 + \ldots) + \ldots$$

$$= 1 - \tfrac{15}{2}x + (-5 + \tfrac{45}{2})x^2 + (30 - \tfrac{135}{4})x^3$$
$$+ (10 - \tfrac{135}{2} + \tfrac{405}{16})x^4 + \ldots$$

$$= 1 - \tfrac{15}{2}x + \tfrac{35}{2}x^2 - \tfrac{15}{4}x^3 - \tfrac{515}{16}x^4 + \ldots$$

Example 19. *Find the term independent of* x *in the expansion of* $(2x + 1/x^2)^{12}$ *in descending powers of* x *and find the greatest term in the expansion when* $x = \frac{2}{3}$.
(L.U.)

We can write $(2x + 1/x^2)^{12}$ as $2^{12}x^{12}\left(1 + \dfrac{1}{2x^3}\right)^{12}$, so the term independent of

x is $2^{12}x^{12}$ times the term in x^{-12} in the expansion of $\left(1 + \dfrac{1}{2x^3}\right)^{12}$. This is

$${}^{12}C_4\left(\dfrac{1}{2x^3}\right)^4$$ or

$$\frac{12 \times 11 \times 10 \times 9}{1 \times 2 \times 3 \times 4} \cdot \frac{1}{2^4x^{12}}$$

i.e. $\dfrac{495}{2^4x^{12}}$, and multiplying by $2^{12}x^{12}$, the required term is 495×2^8 or 126720.

Let T_r be the rth term in the expansion of $\left(1 + \dfrac{1}{2x^3}\right)^{12}$. Then

$$T_r = {}^{12}C_{r-1}\left(\frac{1}{2x^3}\right)^{r-1} \text{ and } T_{r+1} = {}^{12}C_r\left(\frac{1}{2x^3}\right)^r. \text{ Hence}$$

$$\frac{T_{r+1}}{T_r} = \frac{{}^{12}C_r}{{}^{12}C_{r-1}} \cdot \frac{1}{2x^3} = \frac{12 - r + 1}{r} \cdot \frac{1}{2(2/3)^3}$$
$$= \frac{27}{16} \cdot \left(\frac{13 - r}{r}\right),$$

when $x = 2/3$ all the factors in the expressions for $^{12}C_r$ and $^{12}C_{r-1}$ cancelling except the last in numerator and denominator.

Thus the $(r + 1)$th term is greater than the rth so long as

$$\frac{27}{16}\left(\frac{13 - r}{r}\right) > 1,$$

i.e., so long as $351 > 43r$. The largest value of r consistent with this inequality is eight so the greatest term is the ninth.

Example 20. *If C_r denotes the coefficient of x^r in the expansion of $(1 + x)^n$, n being a positive integer, prove that*

$$C_0 - C_1 + C_2 - \ldots + (-1)^n C_n = 0,$$
$$C_0 + C_1 + C_2 + \ldots + C_n = 2^n,$$

and deduce that, if n is even

$$2C_0 + C_1 + 2C_2 + C_3 + 2C_4 + C_5 + \ldots + 2C_n = 3.2^{n-1}.$$

We have

$$C_0 + C_1 x + C_2 x^2 + \ldots + C_n x^n = (1 + x)^n. \tag{3.27}$$

Putting $x = -1$,

$$C_0 - C_1 + C_2 + \ldots + (-1)^n C_n = (1 - 1)^n = 0, \tag{3.28}$$

the coefficients of the various C's being ± 1 according as their suffices are even or odd.

Putting $x = 1$ in (3.27),

$$C_0 + C_1 + C_2 + \ldots + C_n = (1 + 1)^n = 2^n. \tag{3.29}$$

Multiplying (3.29) by 3/2, (3.28) by 1/2 and adding

$$2C_0 + C_1 + 2C_2 + \ldots + 2C_n = \tfrac{3}{2}.2^n = 3.2^{n-1},$$

the coefficient of C_n in (3.28) being $+1$, since n is even.

3.10. The binomial theorem when n is not a positive integer

When n is fractional or negative, it can be shown (but the proof is outside the range of the present book) that the series

$$1 + nx + \frac{n(n - 1)}{(2)!}x^2 + \frac{n(n - 1)(n - 2)}{(3)!}x^3 + \ldots \tag{3.30}$$

is convergent if $-1 < x < 1$ and that the limit of its sum is $(1 + x)^n$. This result is known as the binomial theorem for a fractional or negative index. The points of difference between it and the theorem for a positive integral index are:—

(i) for positive integral n, the series $1 + nx + \dfrac{n(n - 1)}{(2)!}x^2 + \ldots$ terminates at the term in x^n and its sum is $(1 + x)^n$ for all values of x,

(ii) for fractional or negative n, the series does not terminate, it is convergent and has $(1 + x)^n$ as the limit of its sum only when $-1 < x < 1$.

The following particular cases are worth noting. Putting $n = -1$ in (3.30) we deduce that the series

$$1 - x + x^2 - x^3 + \dots$$

converges for $-1 < x < 1$ and the limit of its sum is $\dfrac{1}{1 + x}$. Changing x to $-x$ we see that the series

$$1 + x + x^2 + x^3 + \dots$$

converges for $-1 < x < 1$ and that the limit of its sum is $\dfrac{1}{1 - x}$. These two series are geometrical progressions with first terms unity and common ratios $\mp x$ respectively and the limits of their sums as given here agree with those obtained from (3.13). Putting $n = -2$ and changing x to $-x$ in (3.30) we find that the series

$$1 + 2x + 3x^2 + 4x^3 + \dots$$

converges for $-1 < x < 1$ and the limit of its sum is $\dfrac{1}{(1 - x)^2}$. Finally writing $n = \frac{1}{2}$ in (3.30), the series

$$1 + \tfrac{1}{2}x - \tfrac{1}{8}x^2 + \tfrac{1}{16}x^3 - \dots$$

converges for $-1 < x < 1$ and the limit of its sum is $\sqrt{(1 + x)}$.

Example 21. *Expand* $\dfrac{7 + x}{(1 + x)(1 + x^2)}$ *in ascending powers of* x *as far as the term in* x^4.

Resolving the given expression into partial fractions, we find

$$\frac{7 + x}{(1 + x)(1 + x^2)} \equiv \frac{3}{1 + x} + \frac{4 - 3x}{1 + x^2}$$

$$= 3(1 + x)^{-1} + (4 - 3x)(1 + x^2)^{-1}$$

$$= 3(1 - x + x^2 - x^3 + x^4 - \dots)$$

$$\qquad\qquad + (4 - 3x)(1 - x^2 + x^4 - x^6 + \cdots)$$

$$= 7 - 6x - x^2 + 7x^4 + \dots,$$

the series being convergent if $-1 < x < 1$.

Example 22. *Use the binomial theorem to find* $\sqrt{(1 \cdot 05)}$ *to four places of decimals.*

$$\sqrt{(1 \cdot 05)} = (1 + 0 \cdot 05)^{1/2} = 1 + \tfrac{1}{2} \times 0 \cdot 05 - \tfrac{1}{8} \times (0 \cdot 05)^2 + \tfrac{1}{16} \times (0 \cdot 05)^3 - \dots$$

$$= 1 + 0 \cdot 025 - 0 \cdot 000313 + 0 \cdot 000008 - \dots$$

$$= 1 \cdot 0247 \text{ (to four places).}$$

EXERCISES 3 (d)

1. If x is so small that x^3 and higher powers can be neglected, show that

$$(1 - \tfrac{3}{2}x)^5(2 + 3x)^6 = 64 + 96x - 720x^2. \qquad\qquad \text{(L.U.)}$$

2. Find, by the binomial theorem, the coefficient of x^8 in the expansion of $(3 - 5x^2)^{\frac{1}{2}}$ in ascending powers of x. (Q.E.)

3. Write down and simplify the term independent of x in the expansion of $\left(3x^2 - \dfrac{1}{2x}\right)^9$. Which is the numerically greatest term in this expansion when $x = \frac{1}{2}$? (L.U.)

4. In the binomial expansion of $(1 + x)^{n+1}$, n being an integer greater than two, the coefficient of x^4 is six times the coefficient of x^2 in the expansion of $(1 + x)^{n-1}$. Determine the value of n. (L.U.)

5. If ^nC_r denotes the coefficient of x^r in the expansion of $(1 + x)^n$, prove that $^nC_r + 2(^nC_{r+1}) + {}^nC_{r+2} = {}^{n+2}C_{r+2}$. (L.U.)

6. Find the value of n for which the coefficients of x, x^2 and x^3 in the expansion of $(1 + x)^n$ are in arithmetical progression. (L.U.)

7. Express $2x^3/(1 + x^2)(1 - x)^2$ as a sum of three partial fractions; and obtain an expansion, in ascending powers of x, of this expression as far as the term involving x^7. (Q.E.)

8. Use the binomial theorem to evaluate $0.90^{2.2}$ correct to four significant figures. Check the result by using logarithms. (Q.E.)

EXERCISES 3 (e)

1. Find the sum of the terms from the $(n + 1)$th to the mth term inclusive of an arithmetical progression whose first term is a and whose second term is b. If $m = 13$, $n = 3$ and the sum is $12a$, find the ratio $b:a$. (L.U.)

2. If a^{-1}, b^{-1}, c^{-1}, d^{-1} are in arithmetical progression, prove that $b = 2ac/(a + c)$ and find b/d in terms of a and c. (L.U.)

3. Three unequal numbers a, b, c are such that $1/a$, $1/b$, $1/c$ are in arithmetical progression and a, c, b are in geometrical progression. Prove that b, a, c are in arithmetical progression. (L.U.)

4. S is the sum of n terms of a geometrical progression, P is the product of the n terms and R is the sum of the reciprocals of the terms. Prove that $(S/R)^n = P^2$.

5. The amplitude of the first oscillation of a pendulum is $15°$. If the amplitude of each succeeding oscillation is 0.89 of the amplitude of the preceding oscillation, find after how many oscillations the amplitude will first be less than $1°$. (L.U.)

6. £1000 is borrowed at 5 per cent. compound interest. Two-thirds of the amount then owing is paid back at the end of each year. How much will have been paid back at the end of five years?

7. The sum of £2000 is borrowed from a building society at 4 per cent. per annum compound interest, and the capital and interest is repaid by 20 annual payments of £A. Find A and the sum which would pay off the balance after 10 such payments. (L.U.)

8. If a geometrical progression with common ratio $(1 + c)/(1 - c)$ is convergent, prove that c must be negative, but that it is otherwise unrestricted.

Two such progressions, one with $c = c_1$ and the other with $c = c_2$, each have their first terms unity. If S_1 and S_2 are the corresponding limits to their sums, show that

$$S_1 - S_2 = (c_1 - c_2)/2c_1c_2.$$

Hence deduce that $S_1 > S_2$ when c_1 is less than c_2 numerically.   (L.U.)

9. The first term of a geometrical progression is 7 and its common ratio is $\frac{1}{2}$. Find how many terms of the progression must be taken in order that the sum may differ from the limit of its sum by less than 0·01.

10. Show by induction that the sum of the cubes of the first n positive integers is $\frac{1}{4}n^2(n + 1)^2$ and deduce that the sum of the cubes of the $n + 1$ odd integers from 1 to $(2n + 1)$ inclusive is $(n + 1)^2(2n^2 + 4n + 1)$.   (L.U.)

11. If n consecutive terms are taken from an arithmetical progression of common difference 2, show that

$$3(n S_n - s_n{}^2) = n^2(n^2 - 1),$$

where s_n is the sum of the n terms and S_n is the sum of their squares.   (L.U.)

12. A typical car registration number contains three letters of the alphabet and three of the digits 0, 1, 2, ..., 9. How many such numbers can be formed?   (L.U.)

13. In how many ways can a party of five people be selected from six men and four women so that there are always more men than women in the party?

14. Using all the digits 1, 2, 3, 4, 5, 6 how many arrangements can be made (i) beginning with an even digit, (ii) beginning and ending with an even digit?

15. Two straight lines intersect at O. Points A_1, A_2, ..., A_n are taken on one line and points B_1, B_2, ..., B_n on the other. Prove that the number of triangles that can be drawn with three of the points for vertices is
     (i) $n^2(n - 1)$, if the point O is not to be used,
     (ii) n^3, if the point O may be used.   (L.U.)

16. In a hand of twelve cards, five are red and seven black. If two cards are selected at random, find the odds against them both being black.

17. The odds against a student solving a certain problem are 4 to 3 and the odds in favour of a second student solving the same problem are 7 to 5. Find the chance that the problem will be solved if both students attempt it.

18. Prove that in the binomial expansion of $(1 + 0·03)^{12}$ the rth term is less than one-tenth of the $(r - 1)$th term if $r > 4$.   (L.U.)

19. If the coefficients of x^{r-1}, x^r, x^{r+1} in the binomial expansion of $(1 + x)^n$ are in arithmetical progression, prove that

$$n^2 - n(4r + 1) + 4r^2 - 2 = 0.$$

Find three consecutive coefficients of the expansion of $(1 + x)^{14}$ which form an arithmetical progression.   (L.U.)

20. In the expansion of $(1 + ax + 2x^2)^6$ in powers of x, the coefficients of x^2 and x^{11} are 27 and -192 respectively. Find a and the coefficients of x^3 and x^{10}.   (L.U.)

21. If C_r is the coefficient of x^r in the expansion of $(1 + x)^n$, n being a positive

integer, prove that
$$C_0 + C_1 + C_2 + \ldots + C_n = 2^n,$$

and that
$$C_0 + 2C_1 + 3C_2 + \ldots + (n + 1)C_n = (n + 2)2^{n-1}. \qquad \text{(L.U.)}$$

22. By using the identity $(1 - x^2)^n \equiv (1 - x)^n(1 + x)^n$, or otherwise, prove that, if n is a positive integer and C_r is the coefficient of x_r in the expansion of $(1 + x)^n$, then
$$1 - C_1{}^2 + C_2{}^2 - \ldots + (-1)^n C_n{}^2 = 0$$
if n is odd but is equal to $(-1)^{n/2} \dfrac{(n + 2)(n + 4) \ldots 2n}{2 . 4 \ldots n}$ if n is even. (L.U.)

23. Expand $\dfrac{\sqrt{(1 + 3x - 4x^2)}}{(1 - 2x)^2}$ in ascending powers of x as far as the term in x^3, assuming that the value of x is such that the expansion converges.

24. Express
$$\frac{5x + x^2}{(1 - x)(1 - x^2)}$$
in the form of partial fractions.

   Prove that, when the function is expanded as a series in *descending* powers of x, the coefficient of x^{-n} is $3n - 1 + (-1)^n$. (Q.E.)

25. Show that, if x is so small that x^4 and higher powers of x can be neglected, then
$$\left\{ \left(1 + \frac{x}{2}\right)^3 - (1 + 3x)^{\frac{1}{2}} \right\} \div \left(1 - \frac{5x}{6}\right) = \frac{15x^2}{8}. \qquad \text{(N.U.)}$$

CHAPTER 4

TRIGONOMETRICAL RATIOS FOR ANY ANGLE. GRAPHS OF TRIGONOMETRICAL RATIOS. TRIGONOMETRICAL EQUATIONS

4.1. Introduction

The student is assumed to be familiar with the idea of circular measure and the definitions of the trigonometrical ratios for acute angles. Thus we take for granted that

$$x \text{ degrees} = \pi x/180 \text{ radians.} \tag{4.1}$$

Also if x, y and r are respectively the base, height and hypotenuse of the right-angled triangle ABC in which the angle BCA is a right angle and the angle ABC is denoted by θ (Fig. 1), then

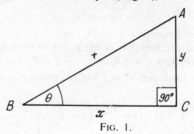

FIG. 1.

$$\sin \theta = y/r, \quad \cos \theta = x/r, \quad \tan \theta = y/x, \tag{4.2}$$

and

$$\operatorname{cosec} \theta = \frac{1}{\sin \theta}, \quad \sec \theta = \frac{1}{\cos \theta}, \quad \cot \theta = \frac{1}{\tan \theta}. \tag{4.3}$$

Immediate and important consequences of these definitions are

$$\tan \theta = \frac{y}{x} = \frac{y/r}{x/r} = \frac{\sin \theta}{\cos \theta}, \tag{4.4}$$

and, since by Pythagoras' theorem, $x^2 + y^2 = r^2$,

$$\sin^2 \theta + \cos^2 \theta = \frac{y^2 + x^2}{r^2} = 1. \tag{4.5}$$

Also,

$$1 + \tan^2 \theta = 1 + \frac{y^2}{x^2} = \frac{x^2 + y^2}{x^2} = \frac{r^2}{x^2} = \frac{1}{\cos^2 \theta} = \sec^2 \theta, \tag{4.6}$$

and

$$1 + \cot^2 \theta = 1 + \frac{x^2}{y^2} = \frac{y^2 + x^2}{y^2} = \frac{r^2}{y^2} = \frac{1}{\sin^2 \theta} = \operatorname{cosec}^2 \theta. \tag{4.7}$$

58

These relations enable all the trigonometrical ratios to be found when one is given and are often very useful in casting expressions involving the trigonometrical ratios into alternative forms.

Example 1. *If sin θ = 1/3 find the values of all the other trigonometrical ratios.*

The relation (4.5) gives $\cos^2 \theta = 1 - (1/3)^2 = 8/9$, so that

$$\cos \theta = 2\sqrt{2}/3.$$

From (4.4),　　　　　　　　$\tan \theta = \dfrac{1/3}{2\sqrt{2}/3} = \dfrac{1}{2\sqrt{2}}.$

Formulae (4.3) then give $\operatorname{cosec} \theta = 3$, $\sec \theta = 3/2\sqrt{2}$, $\cot \theta = 2\sqrt{2}$.

Example 2. *Show that* $\sin^3 \theta - \cos^3 \theta = (\sin \theta - \cos \theta)(1 + \sin\theta \cos \theta).$

$$\sin^3 \theta - \cos^3 \theta = (\sin \theta - \cos \theta)(\sin^2 \theta + \sin \theta \cos \theta + \cos^2 \theta)$$
$$= (\sin \theta - \cos \theta)(1 + \sin \theta \cos \theta),$$

using relation (4.5) in the second factor on the right.

4.2. The trigonometrical ratios for the general angle

Suppose a radius OP, starting from a standard initial position OX, is rotated in an *anti-clockwise* direction. It sweeps out angles which are conventionally termed *positive* angles and these may be of any magnitude. Thus the angle shown in Fig. 2 is 240° (240π/180 or 4π/3 radians). The same position will be reached by OP after rotations 240° + 360°, 240° + 720° or 240° + any integral multiple of 360°. We should speak of these angles as being of magnitudes 600° (or 10π/3 radians), 960° (or 16π/3 radians) and 240° + n × 360° [or (6n + 4)π/3 radians] respectively.

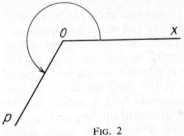

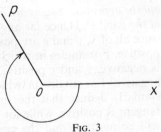

FIG. 2　　　　　　　　　　　　　　　　FIG. 3

Angles generated when OP rotates in a *clockwise* direction are called *negative* angles: that shown in Fig. 3 is an angle of −240°.

Taking O as origin, two perpendicular lines $X'OX$, $Y'OY$ as axes and OP of length r, let the abscissa and ordinate of the point P with respect to these axes be x and y. The axes divide the diagram into four quadrants XOY, YOX', $X'OY'$ and $Y'OX$: these are referred to as the first, second, third and fourth quadrants respectively. *The usual* *sign conventions used in elementary graphical work* are taken to apply to the coordinates x and y of the point P.* Thus when P is in the first quadrant x and y are both positive, when P is in the second quadrant

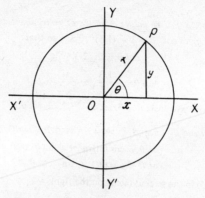

FIG. 4

x is negative and y positive, when P is in the third quadrant x and y are both negative and when P is in the fourth quadrant x is positive and y negative. r is taken to be positive for all positions of the line OP.

The trigonometrical ratios for angles *XOP* of any magnitude are defined in precisely the same way as for acute angles: thus

$$\sin \theta = y/r, \quad \cos \theta = x/r, \quad \tan \theta = y/x, \qquad (4.8)$$

and

$$\operatorname{cosec} \theta = \frac{1}{\sin \theta}, \quad \sec \theta = \frac{1}{\cos \theta}, \quad \cot \theta = \frac{1}{\tan \theta}, \qquad (4.9)$$

but the appropriate signs are attached to x and y according to the position of the point P. Hence for angles in which OP lies in the first quadrant, since all of x, y and r are positive, the sine, cosine and tangent will be positive. For angles in which OP lies in the second quadrant, since x is negative, y and r positive, the sine is positive, cosine and tangent negative. For angles in which OP is in the third quadrant we can similarly deduce that the sine and cosine are both negative but the tangent is positive, while for the fourth quadrant the sine and tangent will be negative and the cosine positive. The diagram below shows which of the ratios are positive in each quadrant and may be useful as an aid to memory.

Sine	All
Tangent	Cosine

* See § 14.1.

4.3. The trigonometrical ratios of some related angles

Some useful relations connecting the trigonometrical ratios of certain related angles can be obtained as follows.

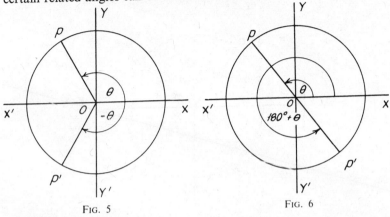

FIG. 5   FIG. 6

In Fig. 5 the radii OP, OP' correspond respectively to angles θ and $-\theta$. It is clear that the abscissae of P and P' are the same and that their ordinates are the same in magnitude but opposite in sign. In other words, changing from $-\theta$ to $+\theta$ is equivalent to reflecting OP' in the axis $X'OX$ and this changes the sign of y but leaves x unaltered. Hence, using (4.8),

$$\sin(-\theta) = -\sin\theta, \quad \cos(-\theta) = \cos\theta, \qquad (4.10)$$

and, by division,

$$\tan(-\theta) = -\tan\theta. \qquad (4.11)$$

In Fig. 6, POP' is a diameter of the circle centre O. The angles XOP, XOP' are respectively θ and $180° + \theta$. The addition of $180°$ to θ is equivalent to reflecting OP through O. This changes the signs of both x and y. Hence

$$\sin(180° + \theta) = -\sin\theta, \quad \cos(180° + \theta) = -\cos\theta, \qquad (4.12)$$

and, by division,

$$\tan(180° + \theta) = \tan\theta. \qquad (4.13)$$

Changing the sign of θ in (4.12), (4.13) and using the results of (4.10), (4.11) we have

$$\left.\begin{array}{l}\sin(180° - \theta) = -\sin(-\theta) = \sin\theta, \\ \cos(180° - \theta) = -\cos(-\theta) = -\cos\theta, \\ \tan(180° - \theta) = \tan(-\theta) = -\tan\theta.\end{array}\right\} \qquad (4.14)$$

In Fig. 7, the angle XOP is θ and XOP' is $90° + \theta$. The addition of $90°$ to the angle is equivalent to measuring θ from the axis $Y'OY$

instead of from $X'OX$. The ordinate of P' is therefore the same as the abscissa of P and the abscissa of P' is minus the ordinate of P. Hence, if x, y are the coordinates of P,

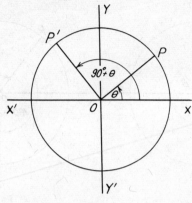

FIG. 7

$$\sin(90° + \theta) = x/r = \cos\theta, \quad \cos(90° + \theta) = -y/r = -\sin\theta, \quad (4.15)$$

and, by division,

$$\tan(90° + \theta) = -\cot\theta. \qquad (4.16)$$

Changing the sign of θ in (4.15), (4.16) and using the results of (4.10), (4.11),

$$\left.\begin{array}{l} \sin(90° - \theta) = \cos(-\theta) = \cos\theta, \\ \cos(90° - \theta) = -\sin(-\theta) = \sin\theta, \\ \tan(90° - \theta) = \cot\theta. \end{array}\right\} \qquad (4.17)$$

The trigonometrical ratios for angles $270° \pm \theta$ are obtained by the addition of $180°$ in those for $90° \pm \theta$. Thus, using (4.15) and (4.12) we have

$$\left.\begin{array}{l} \sin(270° + \theta) = \sin\{180° + (90° + \theta)\} \\ \qquad\qquad = -\sin(90° + \theta) = -\cos\theta, \\ \cos(270° + \theta) = \cos\{180° + (90° + \theta)\} \\ \qquad\qquad = -\cos(90° + \theta) = \sin\theta, \end{array}\right\} \qquad (4.18)$$

and, by division,

$$\tan(270° + \theta) = -\cot\theta. \qquad (4.19)$$

Changing the sign of θ and using (4.10), (4.11)

$$\left.\begin{array}{l} \sin(270° - \theta) = -\cos(-\theta) = -\cos\theta, \\ \cos(270° - \theta) = \sin(-\theta) = -\sin\theta, \\ \tan(270° - \theta) = -\cot(-\theta) = \cot\theta. \end{array}\right\} \qquad (4.20)$$

Finally, the addition of $360°$ (or any integral multiple thereof) does not alter the position of P. Hence we can drop the $360°$ and

$$\sin(360° + \theta) = \sin\theta, \quad \cos(360° + \theta) = \cos\theta, \qquad (4.21)$$

$$\left.\begin{array}{l} \sin(360° - \theta) = \sin(-\theta) = -\sin\theta, \\ \cos(360° - \theta) = \cos(-\theta) = \cos\theta, \end{array}\right\} \qquad (4.22)$$

and the results for $\tan(360° \pm \theta)$ can be obtained by division.

It should be noted that for angles $-\theta$, $180° \pm \theta$, $360° \pm \theta$, sines remain sines and cosines remain cosines. For $90° \pm \theta$, $270° \pm \theta$, sines become cosines and cosines become sines.

Example 3. *Express* $\sin 135°$, $\tan 140°$, $\sin 1220°$ *and* $\cos(-840°)$ *in terms of the trigonometrical ratios of positive acute angles.*

$$\sin 135° = \sin(180° - 45°) = \sin 45°.$$
$$\tan 140° = \tan(180° - 40°) = -\tan 40°.$$
$$\sin 1220° = \sin(3 \times 360° + 140°) = \sin 140°$$
$$= \sin(180° - 40°) = \sin 40°.$$
$$\cos(-840°) = \cos 840° = \cos(2 \times 360° + 120°) = \cos 120°$$
$$= \cos(180° - 60°) = -\cos 60°.$$

EXERCISES 4 (a)

1. If θ is an acute angle and $\sin\theta = 1/4$, find the values of the other trigonometrical ratios.

2. If $\tan\theta = 3/4$, find possible values for $\sin\theta$ and $\cos\theta$.

3. If $\cos\theta = -3/5$, find the values of $\sin\theta$ and $\tan\theta$ when θ is in (i) the second and (ii) the third quadrant.

4. Show that $\tan\theta + \cot\theta = \sec\theta \operatorname{cosec}\theta$ and that
    $$(\sin\theta + \cos\theta)(\cot\theta + \tan\theta) = \sec\theta + \operatorname{cosec}\theta.$$

5. Prove that $\tan^2 A - \sin^2 A = \sin^4 A \sec^2 A$.

6. Prove that
    $$\frac{\sin\theta}{1 + \cos\theta} + \frac{1 + \cos\theta}{\sin\theta} = \frac{2}{\sin\theta}.$$

7. Prove that $\sin 330° \cos 390° - \cos 570° \sin 510° = 0$.

8. Show that $\sin(270° - \theta) - \sin(270° + \theta) = \cos\theta + \cos(180° + \theta)$.

4.4. The graphs of the trigonometrical ratios for acute angles

The graph of $\sin\theta$ can be constructed as follows. Take points $P_1, P_2, P_3, \ldots$ on a circle of unit radius and draw a line marked off in degrees (or radians) in prolongation of the initial line CX (Fig. 8). If we then plot a point Q_1 such that its abscissa ON_1 is equal to the number of degrees (or radians) in the angle XCP_1 and its ordinate Q_1N_1 is equal to the height of P_1 above CX, Q_1 will lie on the graph of $\sin\theta$. Other points $Q_2, Q_3, \ldots$ can be plotted similarly. The graph commences at zero and rises to unity when θ reaches $90°$. A rough table of values of $\sin \theta$ could be made by measuring the heights of $P_1, P_2, P_3, \ldots$ above CX and recording these heights against the corresponding angles $XCP_1, XCP_2, XCP_3, \ldots$

Since, from (4.17), $\cos \theta = \sin(90° - \theta)$, the values taken by the cosine as the angle increases from $0°$ to $90°$ will be the same as those

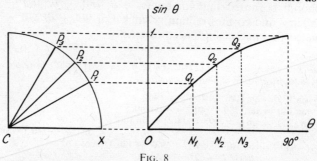

FIG. 8

taken by the sine as the angle decreases from $90°$ to $0°$. The graph of $\cos \theta$ for acute angles is shown in Fig. 9.

To construct the graph of $\tan \theta$ we draw a base line CX of unit length and mark off a line in degrees (or radians) in prolongation of CX. Points $P_1, P_2, P_3, \ldots$ are taken on the line XY (perpendicular

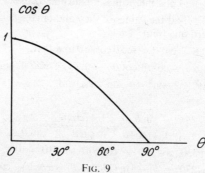

FIG. 9

to CX) as shown in Fig. 10. If we then plot a point Q_1 such that its abscissa ON_1 is equal to the number of degrees (or radians) in the angle XCP_1 and its ordinate is equal to P_1X, the point Q_1 will lie on the graph of $\tan \theta$. Other points $Q_2, Q_3, \ldots$ can be plotted similarly. The graph commences at zero and rises faster and faster as the angle approaches $90°$. Again, a rough table of values of $\tan \theta$ could be made by measuring $P_1X, P_2X, P_3X, \ldots$ and recording these lengths against the corresponding angles $XCP_1, XCP_2, XCP_3, \ldots$

Tables of the trigonometrical ratios for angles between $0°$ and $90°$ more accurate than could be obtained by measurement as indicated above are available. Details of their construction are beyond the scope of the present book but the student will be expected to be able to use the information contained in such tables.

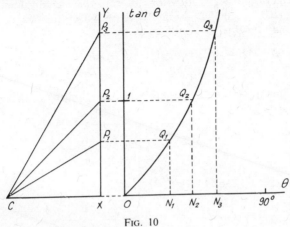

FIG. 10

Accurate values of the ratios for certain angles such as 45° and 60° can be obtained from the isosceles right-angled triangle and the equilateral triangle as follows. In Fig. 11, ABC is an isosceles triangle right-angled at C: the angle at B (and at A) is 45°. If CB, CA are each

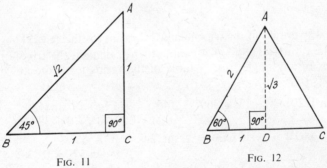

FIG. 11   FIG. 12

taken to be of unit length, the hypotenuse AB will, by Pythagoras' theorem, be of length $\sqrt{2}$. Hence

$$\sin 45° = AC/AB = 1/\sqrt{2}, \quad \cos 45° = BC/AB = 1/\sqrt{2}, \\ \tan 45° = AC/BC = 1. \qquad\qquad\qquad (4.23)$$

In Fig. 12, ABC is an equilateral triangle each of whose sides is taken to be of length 2. The perpendicular from A to BC will bisect it at D. Hence ABD is a right-angled triangle of hypotenuse 2, base 1 and height, by Pythagoras' theorem, $\sqrt{(2^2 - 1^2)}$ or $\sqrt{3}$. Therefore, since the angle ABC is 60°,

$$\sin 60^\circ = AD/AB = \sqrt{3}/2, \quad \left.\begin{array}{l} \cos 60^\circ = BD/AB = 1/2, \\ \tan 60^\circ = AD/BD = \sqrt{3}. \end{array}\right\} \quad (4.24)$$

The trigonometrical ratios for 30° can also be obtained from the triangle ABD of Fig. 12, using AD as its base and BD as its height, for the angle DAB is 30°. In this way we find

$$\sin 30^\circ = BD/AB = 1/2, \quad \left.\begin{array}{l} \cos 30^\circ = AD/AB = \sqrt{3}/2, \\ \tan 30^\circ = BD/AD = 1/\sqrt{3}. \end{array}\right\} \quad (4.25)$$

The following table may be a useful aid to memory for these results :—

θ	$\sin^2 \theta$	$\cos^2 \theta$	$\tan^2 \theta$
0°	0	1	0
30°	1/4	3/4	1/3
45°	1/2	1/2	1
60°	3/4	1/4	3
90°	1	0	∞

The cosecant, secant and cotangent of these angles follow directly from (4.3). For example,

$$\sec 60^\circ = \frac{1}{\cos 60^\circ} = 2.$$

4.5. The graphs of the trigonometrical ratios for the general angle

The results of §§ 4.3, 4.4 enable us to calculate the trigonometrical ratios of any angle from a table of the ratios for acute angles. Thus

$$\cos 170^\circ = \cos(180^\circ - 10^\circ) = -\cos 10^\circ = -0.9848,$$
$$\sin 1220^\circ = \sin(3 \times 360^\circ + 140^\circ) = \sin 140^\circ = \sin(180^\circ - 40^\circ)$$
$$= \sin 40^\circ = 0.6428,$$
$$\sin(-663^\circ) = -\sin 663^\circ = -\sin(2 \times 360^\circ - 57^\circ)$$
$$= \sin 57^\circ = 0.8387.$$

To draw the graph of $\sin\theta$ for any angle we have (4.15), $\sin(90^\circ + \theta) = \cos\theta$ so that the graph for $\sin\theta$ for values of θ between 90° and 180° is the same as that of $\cos\theta$ for θ between 0° and 90°. Also from (4.12), $\sin(180^\circ + \theta) = -\sin\theta$ so the graph for θ between 180° and 360° is the same as that for θ between 0° and 180° but is on the other side of the θ-axis. Since $\sin(360^\circ + \theta) = \sin\theta$ the graph repeats itself after 360° and again after 720° and so on. Finally, since, (4.10), $\sin(-\theta) = -\sin\theta$, the graph for negative θ can be obtained from that for positive θ by reflection in the origin. The graph is shown by the full curve of Fig. 13.

Since, from (4.17), $\cos\theta = \sin(90° - \theta)$, the graph of $\cos\theta$ can be obtained by displacing the graph of $\sin\theta$ to the left by 90°. This is shown dotted in Fig. 13.

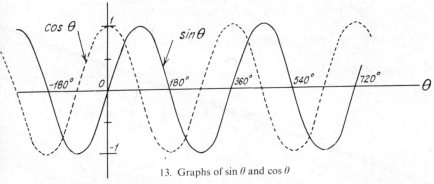

13. Graphs of $\sin\theta$ and $\cos\theta$

For the graph of $\tan\theta$, we have by (4.11), $\tan(-\theta) = -\tan\theta$, so the graph for $-90° < \theta < 0$ is obtained from Fig. 10 by reflection in the origin. Also from (4.13), $\tan(180° + \theta) = \tan\theta$, so the graph for θ between 90° and 270° is the same as that for θ between $-90°$ and 90° and so on. The full graph is shown in Fig. 14.

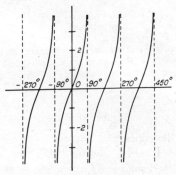

14. Graph of $\tan\theta$

The graphs of the trigonometrical ratios show their periodic nature. Each ratio repeats itself after a certain interval (called the *period*). The trigonometrical ratios are examples of *periodic* functions: the periods of $\sin\theta$ and $\cos\theta$ are both 360° (or 2π radians) while that of $\tan\theta$ is 180° (or π radians). The magnitude of $\sin\theta$ and $\cos\theta$ is always between ±1: half of this range of variation (i.e., unity) is called the *amplitude*.

The fundamental relation, $\sin^2\theta + \cos^2\theta = 1$, proved in § 4.1 for acute angles remains true for all values of θ. For, as we have seen in § 4.3, the addition or subtraction of any multiple of $90°$ can at worst change a sine into a cosine or vice versa and lead to an alteration in sign. But $\sin^2 \theta + \cos^2 \theta$ only involves a sine and a cosine and variations in sign are unimportant as both terms are squared. Hence the relation remains valid for all values of θ. Similar arguments can be applied to show that the relations $1 + \tan^2 \theta = \sec^2 \theta$ and $1 + \cot^2 \theta = \csc^2 \theta$ remain true for all values of θ.

Example 4. *Draw the graph of* $y = \sin(2x + 2\pi/3)$ *from* $x = 0$ *to* $x = 2\pi$. *Use your graph to find the positive values of* x *which satisfy the equation* $x = 5\sin(2x + 2\pi/3)$. (L.U.)

Working in radians, using tables and the relations of § 4.3, we can plot the graph shown in Fig. 15.

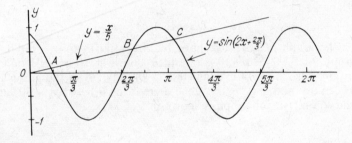

FIG. 15

Plotting the graph of $y = x/5$ on the same diagram, we find that the graphs intersect at three points A, B, C given by $x = 0.48$, 2.34 and 3.30 (approx.). These are the required positive values of x which satisfy the equation $x = 5\sin(2x + 2\pi/3)$.

EXERCISES 4 (b)

1. Use tables to find the values of
   (i) $\sin 212°$, (ii) $\cos(-110°)$, (iii) $\tan 1145°$, (iv) $\sec 1327°$.

2. If $90° < A < 180°$ and $\sin A = 0.6$, use tables to find the values of $\sin 2A$ and $\cos 2A$. (O.C.)

3. Find the pairs of angles between $0°$ and $180°$ which satisfy the equations
   $$\sin(x + y) = 0.5, \quad \sin(x - y) = -0.5.$$ (L.U.)

4. If x is in degrees, draw on the same diagram the graphs of $\sin 2x$ and $1 - \cos x$ for values of x between $0°$ and $180°$. Hence find an acute angle which satisfies the equation $\sin 2x = 1 - \cos x$. (O.C.)

5. By plotting $\tan x$ between 0 and $\pi/2$, show that the equation $\tan x = 2x$ has a positive root less than $\pi/2$ and find this root. Without any further *exact* plotting, show that the equation has roots lying just short of $3\pi/2$, $5\pi/2$, etc. (O.C.)

6. Draw the graph of $y = \sin x + \frac{1}{3}\sin 3x$ for values of x between $0°$ and $180°$. Hence find positive values of x less than $180°$ which satisfy the equation $\sin x + \frac{1}{3}\sin 3x - \frac{2}{3} = 0$. (O.C.)

4.6. The solution of trigonometrical equations

Trigonometrical equations differ from algebraic equations in that they often have an unlimited number of solutions. Some equations can only be solved by graphical methods (see, for instance, Example 4, page 68). In such cases it is usually best to rearrange the equation so that the simplest trigonometrical graph is drawn. Thus in solving the equation $x = 5\sin(2x + 2\pi/3)$ it is preferable to plot graphs of $y = \sin(2x + 2\pi/3)$ and $y = x/5$ rather than $y = 5\sin(2x + 2\pi/3)$ and $y = x$.

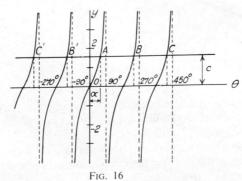

Fig. 16

When a theoretical solution of a trigonometrical equation can be obtained, the equation can often be reduced to one of the forms $\sin\theta = c$, $\cos\theta = c$ or $\tan\theta = c$, where c is a numerical quantity. We now consider the solution of these three equations.

The simplest to deal with is $\tan\theta = c$ and we commence with this equation. Angles satisfying this equation are given by the abscissae of the points of intersection of the graph of $y = \tan\theta$ with a line parallel to the θ-axis and at distance c from it. Suppose that (Fig. 16) $A, B, C, \ldots, B', C', \ldots$ are the points of intersection of the graph $y = \tan\theta$ with this line. The abscissae of all such points satisfy the equation $\tan\theta = c$. Suppose one of these abscissae (for convenience the smallest numerical one is usually selected) is $\alpha°$. This then is the abscissa of A. The abscissa of B is $180° + \alpha°$, that of C is $360° + \alpha°$, and so on for points to the right, while the abscissae of $B', C', \ldots$ are $-180° + \alpha°$, $-360° + \alpha°$ and so on. All these results can be included in the formula $(n \times 180° + \alpha°)$ where n is any positive or negative integer or zero. If α were quoted in radians instead of degrees we should write the formula $(n\pi + \alpha)$. Hence the general solution of the equation $\tan \theta = c$ is

$$(n \times 180 + \alpha) \text{ degrees or } (n\pi + \alpha) \text{ radians,}$$

where α is any solution of the equation, but generally taken for convenience to be the smallest numerical solution. As an example the general solution of the equation $\tan \theta = 1$ is, since $\tan 45° = 1$,

$$(n \times 180 + 45) \text{ degrees or } (n + \tfrac{1}{4})\pi \text{ radians.}$$

If we deal with the solution of $\cos \theta = c$ in the same way, Fig. 17 applies. It should first be noted that since $\cos \theta$ always lies between ± 1, there will be no solutions to an equation of this type for which

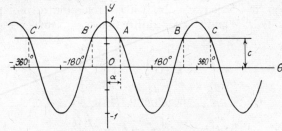

FIG. 17

c is numerically greater than unity. Taking the smallest solution $\alpha°$ as the abscissa of the point A, other solutions given by the abscissae of points B, C, ... and B', C', ... are $360° - \alpha°$, $360° + \alpha°$, ... and $-\alpha°$, $-360° + \alpha°$, ... These are all included in the formula $(n \times 360° \pm \alpha°)$. Thus the general solution of the equation $\cos \theta = c$ when $-1 < c < 1$ is

$$(n \times 360 \pm \alpha) \text{ degrees or } (2n\pi \pm \alpha) \text{ radians,}$$

where n is any positive or negative integer or zero and α is any solution of the equation. As an example the general solution of the equation $\cos \theta = 1/2$, since $\cos 60° = 1/2$, is

$$(n \times 360 \pm 60) \text{ degrees or } (2n \pm \tfrac{1}{3})\pi \text{ radians.}$$

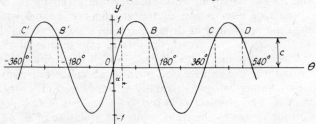

FIG. 18

For the equation $\sin \theta = c$, we again suppose that $-1 < c < 1$ and Fig. 18 applies. Taking the smallest solution $\alpha°$ as the abscissa of the point A, other solutions, given by the abscissae of points $B, C, D, \ldots$ and $B', C', \ldots$ are $180° - \alpha°, 360° + \alpha°, 540° - \alpha°, \ldots$ and $-180° - \alpha°$, $-360° + \alpha°, \ldots$ The general formula including all these is $n \times 180° + (-1)^n \alpha°$. Hence the general solution of the equation $\sin \theta = c$ when $-1 < c < 1$ is

$$(n \times 180 + (-1)^n \alpha) \text{ degrees or } (n\pi + (-1)^n \alpha) \text{ radians,}$$

where again n is any positive or negative integer or zero and α is any solution of the equation. For example, the general solution of the equation $\sin \theta = 1/\sqrt{2}$ is, since $\sin 45° = 1/\sqrt{2}$,

$$(n \times 180 + (-1)^n \times 45) \text{ degrees or } \left(n + \frac{(-1)^n}{4}\right)\pi \text{ radians.}$$

Taking $n = 0, 1, 2, 3, \ldots$ solutions are $45°, 135°, 405°, 495°, \ldots$ while $n = -1, -2, -3, \ldots$ gives the solutions $-225°, -315°, -585°, \ldots$

To sum up, if α radians is one solution of the equation

(i) $\sin \theta = c, (-1 < c < 1)$, the general solution is $n\pi + (-1)^n \alpha$,
(ii) $\cos \theta = c, (-1 < c < 1)$, the general solution is $2n\pi \pm \alpha$, $\quad$ (4.26)
(iii) $\tan \theta = c$, the general solution is $n\pi + \alpha$,

where n is any positive or negative integer or zero, and corresponding expressions for the solutions in degrees.

Many trigonometrical equations can be reduced to one of these forms and the general solution can then be written down. Examples are given below and further examples will be found in Chapter 5, page 83.

Example 5. *Find all the angles less than four right angles which satisfy the equation* $2 \cos^2 \theta = 1 + \sin \theta.$ (L.U.)

Since $\cos^2 \theta = 1 - \sin^2 \theta$, the equation can be written

$$2 - 2\sin^2 \theta = 1 + \sin \theta, \quad \text{or} \quad 2\sin^2 \theta + \sin \theta - 1 = 0.$$

This is equivalent to $(2 \sin \theta - 1)(\sin \theta + 1) = 0$. Hence either $\sin \theta = 1/2$ giving $\theta = n \times 180° + (-1)^n \times 30°$; angles between $0°$ and $360°$ included in this are those for $n = 0$, $n = 1$, i.e. $30°$ and $150°$; or $\sin \theta = -1$ giving $\theta = n \times 180° + (-1)^n(-90°)$; the only angle between $0°$ and $360°$ in this is that for $n = 1$, i.e. $270°$. Hence the required angles are $30°, 150°$ and $270°$.

Example 6. *Find all the angles which satisfy the equation* $4 \sec^2 \theta = 3 \tan \theta + 5.$ (L.U.)

Since $\sec^2 \theta = 1 + \tan^2 \theta$, the equation can be written

$$4 + 4\tan^2 \theta = 3 \tan \theta + 5, \quad \text{or} \quad 4\tan^2 \theta - 3 \tan \theta - 1 = 0.$$

This is equivalent to $(\tan \theta - 1)(4 \tan \theta + 1) = 0$. One set of solutions corresponds to $\tan \theta = 1$ and is $\theta = n \times 180° + 45°$. The other corresponds to $\tan \theta = -1/4$. From tables, the angle whose tangent is -0.25 is $-14° 2'$, so that another set of solutions is given by

$$\theta = n \times 180° - 14° 2'.$$

Example 7. *Find all the angles between* $0°$ *and* $360°$ *which satisfy the equation* $\sin 2\theta = \cos 3\theta$. (L.U.)

Since $\cos 3\theta = \sin(90° - 3\theta)$, we have $\sin 2\theta = \sin(90° - 3\theta)$ so that the general solution is $2\theta = n \times 180° + (-1)^n \times (90° - 3\theta)$. Putting $n = 0, 2, 4, 6$ and 8 respectively gives

> (i) $2\theta = 90° - 3\theta$ or $\theta = 18°$,
> (ii) $2\theta = 360° + 90° - 3\theta$ leading to $\theta = 90°$,
> (iii) $2\theta = 720° + 90° - 3\theta$ leading to $\theta = 162°$,
> (iv) $2\theta = 1080° + 90° - 3\theta$ giving $\theta = 234°$,

and (v) $2\theta = 1440° + 90° - 3\theta$ giving $\theta = 306°$.

Higher even values of n lead to values of θ greater than $360°$ while odd positive values and negative even values of n all lead to negative values of θ. The value $n = -1$, gives $2\theta = -180° - 90° + 3\theta$ leading to $\theta = 270°$ and values like $n = -3, -5$, etc., give angles in excess of $360°$. Hence the required angles are $18°, 90°, 162°, 234°, 270°$ and $306°$.

Example 8. *Find the general value of* θ, *in degrees, which satisfies simultaneously the equations* $\tan \theta = \sqrt{3}$, $\sec \theta = -2$. (L.U.)

If $\tan \theta = \sqrt{3}$, the general solution is $\theta = n \times 180° + 60°$; if $\sec \theta = -2$, $\cos \theta = -1/2$ and $\theta = 2n \times 180° \pm 120°$. Solutions of the first equation are therefore $-480°, -300°, -120°, 60°, 240°, 420°, 600°$, etc., while those of the second are $-480°, -240°, -120°, 120°, 240°, 480°, 600°, 840°$, etc. Values simultaneously satisfying the two equations are therefore $-480°, -120°, 240°, 600°$, etc., all of which are included in the formula $n \times 360° + 240°$.

EXERCISES 4 (c)

1. Find all the values of θ which satisfy the equation
$$2 \tan \theta + 3 \sec \theta = 4 \cos \theta. \qquad \text{(L.U.)}$$

2. Find the values of x between $0°$ and $360°$ satisfying the equation $10 \sin^2 x + 10 \sin x \cos x - \cos^2 x = 2$. (L.U.)

3. Find the values of A and B between $0°$ and $180°$ which satisfy the equations $A - B = 12° \, 18'$, $\cos(A + B) = 0.4457$. (L.U.)

4. Give the general solution (in radians) of the equation
$$\cos(\theta - \pi/4) = \sin 2\theta.$$

5. Find the general solution of the equation $10 \sec^2 \theta - 3 = 17 \tan \theta$.

6. What is the most general value of θ which satisfies both the equations $\tan \theta = 1/\sqrt{3}$ and $\sin \theta = -1/2$?

EXERCISES 4 (d)

1. If $\sin \theta = \dfrac{a^2 - b^2}{a^2 + b^2}$, find the values of $\cos \theta$ and $\tan \theta$.

2. Find the value of
$$\sin^2 A \operatorname{cosec}\left(\frac{\pi}{2} - A\right) - \cot^2\left(\frac{\pi}{2} - A\right) \cos A. \qquad \text{(L.U.)}$$

3. If θ is an angle in the first quadrant and $\tan \theta = t$, express all the other trigonometrical ratios in terms of t.

4. Find the values of $\cos 3360°$, $\operatorname{cosec}(-840°)$. (L.U.)

5. Prove the identity
$$(1 + \sec x + \tan x)(1 + \operatorname{cosec} x + \cot x)$$
$$= 2(1 + \tan x + \cot x + \sec x + \operatorname{cosec} x),$$
and verify this result when $x = \pi/4$. (L.U.)

6. If $\tan^2 \alpha - 2 \tan^2 \beta = 1$, find the possible ratios of $\cos \alpha$ to $\cos \beta$. (L.U.)

7. If $\sec \theta - \cos \theta = a$ and $\operatorname{cosec} \theta - \sin \theta = b$, prove that
$$a^2 b^2 (a^2 + b^2 + 3) = 1.$$

8. Prove that $\dfrac{\cot \alpha + \tan \beta}{\cot \beta + \tan \alpha} = \cot \alpha \tan \beta.$

9. Show that
$$\frac{\cos \theta - 1}{\sec \theta + \tan \theta} + \frac{\cos \theta + 1}{\sec \theta - \tan \theta} = 2(1 + \tan \theta).$$

10. Prove that
$$\frac{1 + \sin \theta}{\cos \theta} = \frac{\cos \theta}{1 - \sin \theta} = \sec \theta + \tan \theta.$$

11. If $x \cos \theta + y \sin \theta = a$ and $x \sin \theta - y \cos \theta = b$, prove that
$$\tan \theta = \frac{bx + ay}{ax - by} \quad \text{and} \quad x^2 + y^2 = a^2 + b^2. \quad \text{(L.U.)}$$

12. If $\tan \theta + \sin \theta = x$ and $\tan \theta - \sin \theta = y$, prove that
$$(x^2 - y^2)^2 = 16xy. \quad \text{(L.U.)}$$

13. Plot on the same diagram the graphs of $\cos 2\theta$ and $\tan(40° - \theta)$ between $\theta = -20°$ and $\theta = 60°$. Hence find two approximate solutions of the equation $\cos 2\theta \cot(40° - \theta) = 1$.

14. Sketch the graph of $y = (\pi/2) \sin^2 x$ and use your graph to solve the equation $2x = \pi \sin^2 x$. (O.C.)

15. Find graphically the values of x between $0°$ and $180°$ which satisfy the equation $\sin x = 3 \cos^2 x$. (O.C.)

16. Draw on the same diagram the graphs of $4 \sin(x + 30°)$ and $2 + \tan x$ for values of x from $0°$ to $360°$. From your graphs obtain the solutions, within this range, of the equation $4 \sin(x + 30°) - \tan x = 2$. (L.U.)

17. Draw the graph of $y = 2 \sin(x + \pi/4)$ between $x = -\pi$ and $x = \pi$. By drawing another graph, using the same scales and axes, solve approximately the equation $4x^2 + 16 \sin^2(x + \pi/4) = \pi^2$. (L.U.)

18. Find all the angles between $0°$ and $360°$ which satisfy the equation $6 \sin^2 x + 5 \cos x = 7$.

19. Find the general solution of the equation $\tan 3\theta = \cot 2\theta$.

20. Find the general values of x satisfying the equation
$$4 \cos x + 5 = 6 \sin^2 x.$$

21. Assuming r is positive, find r and a value of θ between $-180°$ and $180°$ to satisfy the equations $r \cos \theta = -4$, $r \sin \theta = 2\cdot5$.

22. Find all the angles between $0°$ and $360°$ which satisfy the equation
$$3 \tan^3 \theta - 3 \tan^2 \theta = \tan \theta - 1.$$

23. Find the general solution of the equation
$$(2 \tan x - 1)^2 = 3(\sec^2 x - 2).$$

24. Find the general solution of the equation $\tan x \tan 4x = 1$.

25. Find the values of x, in radians, between 0 and 2π, which satisfy the equation $6 \tan^2 x - 4 \sin^2 x = 1$.   (L.U.)

CHAPTER 5

ADDITION THEOREMS. MULTIPLE AND SUB-MULTIPLE ANGLES. FURTHER TRIGONOMETRICAL EQUATIONS. THE INVERSE NOTATION. SMALL ANGLES

5.1. The addition theorems for the sine and cosine

We now consider formulae expressing the trigonometrical ratios of the sum of two angles in terms of the trigonometrical ratios of the separate angles. Such formulae are known as *addition theorems* and we start by deriving them for a restricted range of angles. Generalisation of the results to cover all angles can be made but the process is rather troublesome: a compact general method of derivation of the formulae making use of a result in coordinate geometry is available and this is given in Chapter 14, page 262.

Formulae for sin $(A + B)$ and cos $(A + B)$ for cases in which the *angle $(A + B)$ is acute* can be obtained from Fig. 19. Here the angle AOB is A, the angle BOC is B, P is any point on OC and PM, PN are perpendicular to the lines OA, OB respectively. NH and NK are

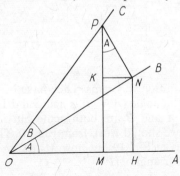

FIG. 19

perpendiculars from the point N on to OA, PM respectively. Since PM, PN are perpendicular to the arms OA, OB respectively of the angle AOB, the angle MPN is A. From the right-angled triangle MPO,

$$OP \sin (A + B) = MP = MK + KP$$
$$= HN + KP, \qquad (5.1)$$

since $MHNK$ is a rectangle by construction. The right-angled triangle OHN gives $HN = ON \sin A$, and the right-angled triangle ONP gives $ON = OP \cos B$. Hence $HN = OP \sin A \cos B$. Also, from the right-

75

angled triangle PKN, $KP = PN \cos A$, while the triangle ONP gives $PN = OP \sin B$. Thus $KP = OP \cos A \sin B$. Substituting for HN, KP in (5.1) and dividing both sides by OP, we have

$$\sin(A + B) = \sin A \cos B + \cos A \sin B. \tag{5.2}$$

The corresponding formula for $\cos(A + B)$ is obtained similarly; thus, from Fig. 19,

$$OP \cos(A + B) = OM = OH - HM$$
$$= OH - NK.$$

Also $OH = ON \cos A$ and $ON = OP \cos B$ so that $OH = OP \cos A \cos B$. $NK = PN \sin A$ and $PN = OP \sin B$, giving $NK = OP \sin A \sin B$. Substitution and division by OP leads to

$$\cos(A + B) = \cos A \cos B - \sin A \sin B. \tag{5.3}$$

The formulae for $\sin(A + B)$, $\cos(A + B)$ given in (5.2), (5.3) are

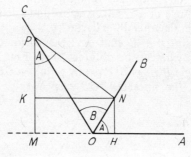

FIG. 20

the fundamental addition theorems. They have been derived only for the case in which the angle $(A + B)$ is acute and Fig. 19 applies. For the case in which A and B are both acute but in which their sum $(A + B)$ is obtuse, we should work from Fig. 20. The lettering has the same significance as in Fig. 19 but now M lies on AO produced. We now have, from the triangle MPO

$$MP = OP \sin MOP = OP \sin(180° - A - B).$$

Since $\sin(180° - \theta) = \sin \theta$, this can still be written

$$OP \sin(A + B) = MP = MK + KP = HN + KP, \tag{5.4}$$

and we obtain (5.2) in exactly the same way as before. The formula for $\cos(A + B)$ can be similarly extended to cases in which $(A + B)$ is obtuse.

The theorems can be extended to cases in which one of the angles, say A', lies between $90°$ and $180°$ as follows. Let $A' = 90° + A$, so that A is acute. Then since $\sin A' = \sin(90° + A) = \cos A$ and $\cos A' = \cos(90° + A) = -\sin A$,

$$\sin (A' + B) = \sin (90° + A + B)$$
$$= \cos (A + B)$$
$$= \cos A \cos B - \sin A \sin B,$$

since A and B are both acute,

$$= \sin A' \cos B + \cos A' \sin B.$$

Similarly,

$$\cos (A' + B) = \cos (90° + A + B)$$
$$= -\sin (A + B)$$
$$= -\sin A \cos B - \cos A \sin B$$
$$= \cos A' \cos B - \sin A' \sin B.$$

Thus both the addition theorems are true when A' lies between one and two right angles. A similar argument holds if B is increased by 90°. Hence the theorems are true for any angles between 0° and 180°. The argument can be extended to show that the theorems are valid for angles of any magnitude but the full proof becomes rather long. As stated previously, a much shorter proof for general angles can be given when an elementary result in coordinate geometry is available (see Chapter 14, page 262): for the present we shall take the theorems as applying to angles of any size.

By writing $-B$ for B in the theorems, we have, using (4.10), the two results,

$$\sin (A - B) = \sin A \cos (-B) + \cos A \sin (-B)$$
$$= \sin A \cos B - \cos A \sin B, \qquad (5.5)$$

and

$$\cos (A - B) = \cos A \cos (-B) - \sin A \sin (-B)$$
$$= \cos A \cos B + \sin A \sin B. \qquad (5.6)$$

Example 1. *Show that* $\cos 15° = \dfrac{\sqrt{3} + 1}{2\sqrt{2}}.$

$$\cos 15° = \cos (45° - 30°)$$
$$= \cos 45° \cos 30° + \sin 45° \sin 30°$$
$$= \frac{1}{\sqrt{2}} \cdot \frac{\sqrt{3}}{2} + \frac{1}{\sqrt{2}} \cdot \frac{1}{2} = \frac{\sqrt{3} + 1}{2\sqrt{2}}.$$

Example 2. *Use the addition formula to show that* $\cos (90° + A) = -\sin A.$
$$\cos (90° + A) = \cos 90° \cos A - \sin 90° \sin A$$
$$= -\sin A,$$
since $\cos 90° = 0$, $\sin 90° = 1$.

5.2. The addition theorem for the tangent

By division, the addition theorems for the sine and cosine give

$$\tan (A + B) = \frac{\sin (A + B)}{\cos (A + B)}$$

$$= \frac{\sin A \cos B + \cos A \sin B}{\cos A \cos B - \sin A \sin B}$$

$$= \frac{\tan A + \tan B}{1 - \tan A \tan B}, \qquad (5.7)$$

dividing numerator and denominator by $\cos A \cos B$.

Writing $-B$ in place of B,

$$\tan (A - B) = \frac{\tan A + \tan (-B)}{1 - \tan A \tan (-B)}$$

$$= \frac{\tan A - \tan B}{1 + \tan A \tan B}. \qquad (5.8)$$

Example 3. *Show that* $\tan (45° + A) = \dfrac{1 + \tan A}{1 - \tan A}$.

$$\tan (45° + A) = \frac{\tan 45° + \tan A}{1 - \tan 45° \tan A} = \frac{1 + \tan A}{1 - \tan A},$$

since $\tan 45° = 1$.

EXERCISES 5 (a)

1. Show that $\cos (\alpha + \beta) \cos (\alpha - \beta) = \cos^2 \alpha - \sin^2 \beta$.

2. Prove that $\cot (A + B) = \dfrac{\cot A \cot B - 1}{\cot A + \cot B}$.

3. Show that
$$\sin (A + B + C) = \cos A \cos B \cos C(\tan A + \tan B$$
$$+ \tan C - \tan A \tan B \tan C)$$
    and deduce that, if A, B, C are the angles of a triangle, then
$$\cot A \cot B + \cot B \cot C + \cot C \cot A = 1. \qquad \text{(L.U.)}$$

4. Show that $a \sin x + b \cos x = \sqrt{(a^2 + b^2)} \sin (x + \alpha)$ where $\tan \alpha = b/a$.

5. If $k \cos \theta = \cos (\theta - \alpha)$, show that $\tan \theta = k \operatorname{cosec} \alpha - \cot \alpha$.

6. Prove that
$$\frac{\sin (A - B)}{\cos A \cos B} + \frac{\sin (B - C)}{\cos B \cos C} + \frac{\sin (C - A)}{\cos C \cos A} = 0.$$

5.3. Multiple angles

By writing $B = A$ in the three addition formulae (5.2), (5.3) and (5.7) we obtain expressions for the sine, cosine and tangent of $2A$ in terms of the trigonometrical ratios of A. Thus, from (5.2)

$$\sin (A + A) = \sin A \cos A + \cos A \sin A,$$

or, $\qquad \sin 2A = 2 \sin A \cos A. \qquad (5.9)$

The addition formula for the cosine, (5.3), gives similarly

$$\cos 2A = \cos^2 A - \sin^2 A. \tag{5.10}$$

By writing $\sin^2 A = 1 - \cos^2 A$, this can be written in the alternative form

$$\cos 2A = 2\cos^2 A - 1, \tag{5.11}$$

and by writing $\cos^2 A = 1 - \sin^2 A$ in (5.10), yet another equivalent form is

$$\cos 2A = 1 - 2\sin^2 A. \tag{5.12}$$

Sometimes, particularly in the integral calculus, it is necessary to express $\sin^2 A$ and $\cos^2 A$ in terms of $\cos 2A$. This can be done by rearranging the last two formulae to give

$$\cos^2 A = \tfrac{1}{2}(1 + \cos 2A) \text{ and } \sin^2 A = \tfrac{1}{2}(1 - \cos 2A). \tag{5.13}$$

By writing $B = A$ in the addition formula for the tangent (5.7) we have

$$\tan 2A = \frac{2\tan A}{1 - \tan^2 A}. \tag{5.14}$$

Expressions for the trigonometrical ratios of $3A$ can be obtained as follows. By writing $B = 2A$ in (5.2) we have

$$\sin 3A = \sin(A + 2A)$$
$$= \sin A \cos 2A + \cos A \sin 2A.$$

Substituting $\cos 2A = 1 - 2\sin^2 A$, $\sin 2A = 2\sin A \cos A$ gives

$$\sin 3A = \sin A(1 - 2\sin^2 A) + 2\sin A \cos^2 A,$$

and, writing $\cos^2 A = 1 - \sin^2 A$ we have, after slight reduction,

$$\sin 3A = 3\sin A - 4\sin^3 A. \tag{5.15}$$

A similar process applied to the addition formula for the cosine, (5.3), gives

$$\cos 3A = \cos(A + 2A)$$
$$= \cos A \cos 2A - \sin A \sin 2A$$
$$= \cos A (2\cos^2 A - 1) - 2\sin^2 A \cos A$$
$$= \cos A (2\cos^2 A - 1) - 2(1 - \cos^2 A)\cos A$$
$$= 4\cos^3 A - 3\cos A. \tag{5.16}$$

Proceeding similarly from (5.7), we have

$$\tan 3A = \tan(A + 2A)$$
$$= \frac{\tan A + \tan 2A}{1 - \tan A \tan 2A}$$
$$= \frac{\tan A + \dfrac{2\tan A}{1 - \tan^2 A}}{1 - \tan A\left(\dfrac{2\tan A}{1 - \tan^2 A}\right)},$$

using (5.14). After reduction this gives

$$\tan 3A = \frac{3 \tan A - \tan^3 A}{1 - 3 \tan^2 A}. \tag{5.17}$$

5.4. Submultiple angles

By writing $A = x/2$ in the formulae of the last section we have, from (5.9),

$$\sin x = 2 \sin \tfrac{1}{2}x \cos \tfrac{1}{2}x, \tag{5.18}$$

from (5.10), (5.11) and (5.12),

$$\left. \begin{aligned} \cos x &= \cos^2 \tfrac{1}{2}x - \sin^2 \tfrac{1}{2}x \\ &= 2 \cos^2 \tfrac{1}{2}x - 1 \\ &= 1 - 2 \sin^2 \tfrac{1}{2}x, \end{aligned} \right\} \tag{5.19}$$

and from (5.14),

$$\tan x = \frac{2 \tan \tfrac{1}{2}x}{1 - \tan^2 \tfrac{1}{2}x}. \tag{5.20}$$

These formulae enable us to express the sine, cosine and tangent of an angle in terms of the tangent of the half angle. If we write

$$t = \tan \tfrac{1}{2}x, \tag{5.21}$$

(5.20) gives immediately

$$\tan x = 2t/(1 - t^2). \tag{5.22}$$

Formula (5.18) can be written

$$\begin{aligned} \sin x &= 2 \tan \tfrac{1}{2}x \cos^2 \tfrac{1}{2}x \\ &= \frac{2 \tan \tfrac{1}{2}x}{\sec^2 \tfrac{1}{2}x} \\ &= \frac{2 \tan \tfrac{1}{2}x}{1 + \tan^2 \tfrac{1}{2}x} = \frac{2t}{1 + t^2}. \end{aligned} \tag{5.23}$$

Also, from (5.19),

$$\begin{aligned} \cos x &= \cos^2 \tfrac{1}{2}x(1 - \tan^2 \tfrac{1}{2}x) \\ &= \frac{1 - \tan^2 \tfrac{1}{2}x}{\sec^2 \tfrac{1}{2}x} \\ &= \frac{1 - \tan^2 \tfrac{1}{2}x}{1 + \tan^2 \tfrac{1}{2}x} = \frac{1 - t^2}{1 + t^2}. \end{aligned} \tag{5.24}$$

The three formulae

$$\sin x = \frac{2t}{1 + t^2}, \quad \cos x = \frac{1 - t^2}{1 + t^2}, \quad \tan x = \frac{2t}{1 - t^2}, \tag{5.25}$$

where $t = \tan \tfrac{1}{2}x$ are useful in the solution of a certain type of trigonometrical equation (see p. 84). They also have other important applications.

Example 4. *Prove that* $\dfrac{\cos 3\theta}{\cos \theta} - \dfrac{\cos 6\theta}{\cos 2\theta} = 2\,(\cos 2\theta - \cos 4\theta)$.   (L.U.)

$$\frac{\cos 3\theta}{\cos \theta} - \frac{\cos 6\theta}{\cos 2\theta} = \frac{4\cos^3 \theta - 3\cos \theta}{\cos \theta} - \frac{4\cos^3 2\theta - 3\cos 2\theta}{\cos 2\theta}$$

$$= 4\cos^2 \theta - 3 - (4\cos^2 2\theta - 3)$$
$$= 4\,(\cos^2 \theta - \cos^2 2\theta)$$
$$= 2\{(1 + \cos 2\theta) - (1 + \cos 4\theta)\} = 2\,(\cos 2\theta - \cos 4\theta).$$

Example 5. *If* $\tan \theta = 4/3$ *and if* $0° < \theta < 360°$, *find, without tables, the possible values of* $\tan \tfrac{1}{2}\theta$ *and* $\sin \tfrac{1}{2}\theta$.

Let $t = \tan \tfrac{1}{2}\theta$, then $4/3 = \tan \theta = 2t/(1 - t^2)$, giving $4 - 4t^2 = 6t$ or $2t^2 + 3t - 2 = 0$. This gives $(2t - 1)(t + 2) = 0$ leading to $t = 1/2$ or -2, and these are the required values of $\tan \tfrac{1}{2}\theta$.

To find $\sin \tfrac{1}{2}\theta$, we have

$$t = \tan \tfrac{1}{2}\theta = \sin \tfrac{1}{2}\theta \sec \tfrac{1}{2}\theta = \sin \tfrac{1}{2}\theta(1 + \tan^2 \tfrac{1}{2}\theta)^{1/2},$$

so that $\sin \tfrac{1}{2}\theta = t/\sqrt{(1 + t^2)}$. With $t = 1/2$ this gives

$$\sin \tfrac{1}{2}\theta = (1/2)/\sqrt{(1 + 1/4)} = 1/\sqrt{5};$$

and with $t = -2$, $\sin \tfrac{1}{2}\theta = -2/\sqrt{(1 + 4)} = -2/(\pm\sqrt{5}) = 2/\sqrt{5}$ if θ is to be less than 360° and therefore $\tfrac{1}{2}\theta$ less than 180°.

EXERCISES 5 (b)

1. If $\tan^2 \alpha - 2\tan^2 \beta = 1$, prove that $\cos 2\alpha + \sin^2 \beta = 0$.   (L.U.)
2. If $t = \tan \tfrac{1}{2}\theta$, express the square root of

$$(1 + \sin \theta)(3\sin \theta + 4\cos \theta + 5)$$

 in terms of t.   (L.U.)
3. If $\sin 3\theta = p$ and $\sin^2 \theta = \tfrac{3}{4} - q$, prove that $p^2 + 16q^3 = 12q^2$.   (L.U.)
4. Prove that $2\cot \tfrac{1}{2}A + \tan A = \tan A \cot^2 \tfrac{1}{2}A$.   (L.U.)
5. If $2\cos \theta = x + 1/x$, show that $2\cos 3\theta = x^3 + 1/x^3$.   (L.U.)
6. If $\sec A - \tan A = x$, prove that $\tan \tfrac{1}{2}A = \dfrac{1 - x}{1 + x}$.   (L.U.)

5.5. The factor formulae

The addition formulae for the sine and cosine can be used to express sums and differences of sines and cosines as products. Starting from the addition formulae

$$\sin (A + B) = \sin A \cos B + \cos A \sin B,$$
$$\sin (A - B) = \sin A \cos B - \cos A \sin B,$$

addition leads to

$$2\sin A \cos B = \sin (A + B) + \sin (A - B),$$ (5.26)

and subtraction gives

$$2\cos A \sin B = \sin (A + B) - \sin (A - B).$$ (5.27)

Similarly the addition formulae for the cosine

$$\cos(A + B) = \cos A \cos B - \sin A \sin B,$$
$$\cos(A - B) = \cos A \cos B + \sin A \sin B,$$

give $\qquad 2 \cos A \cos B = \cos(A + B) + \cos(A - B),$ $\qquad$ (5.28)

and $\qquad -2 \sin A \sin B = \cos(A + B) - \cos(A - B).$ $\qquad$ (5.29)

These four formulae express products as sums.

By writing $A + B = C$, $A - B = D$, so that $A = \frac{1}{2}(C + D)$ and $B = \frac{1}{2}(C - D)$ these formulae become

$$\left.\begin{array}{l}
\sin C + \sin D = 2 \sin \frac{1}{2}(C + D) \cos \frac{1}{2}(C - D), \\
\sin C - \sin D = 2 \cos \frac{1}{2}(C + D) \sin \frac{1}{2}(C - D), \\
\cos C + \cos D = 2 \cos \frac{1}{2}(C + D) \cos \frac{1}{2}(C - D), \\
\cos C - \cos D = -2 \sin \frac{1}{2}(C + D) \sin \frac{1}{2}(C - D).
\end{array}\right\} \qquad (5.30)$$

These formulae, which express sums as products, are of great use and are often called the *factor formulae*. It is useful to remember the results in words; e.g., the sum of the sines of two angles is equal to twice the sine of half the sum of the angles multiplied by the cosine of half the difference of the angles and so on. *The minus sign in the last formula should be noted.*

Example 6. *Prove that* $\dfrac{\sin 3A \sin 6A + \sin A \sin 2A}{\sin 3A \cos 6A + \sin A \cos 2A} = \tan 5A.$

Using (5.29) and (5.26),

$$\frac{\sin 3A \sin 6A + \sin A \sin 2A}{\sin 3A \cos 6A + \sin A \cos 2A}$$

$$= \frac{-\frac{1}{2} \cos 9A + \frac{1}{2} \cos(-3A) - \frac{1}{2} \cos 3A + \frac{1}{2} \cos(-A)}{\frac{1}{2} \sin 9A + \frac{1}{2} \sin(-3A) + \frac{1}{2} \sin 3A + \frac{1}{2} \sin(-A)}$$

$$= \frac{\cos A - \cos 9A}{-\sin A + \sin 9A}$$

$$= \frac{-2 \sin \frac{1}{2}(A + 9A) \sin \frac{1}{2}(A - 9A)}{2 \cos \frac{1}{2}(A + 9A) \sin \frac{1}{2}(9A - A)}, \text{ using (5.30)},$$

$$= \frac{\sin 5A \sin 4A}{\cos 5A \sin 4A} = \tan 5A.$$

Example 7. *Without using tables, prove that* $\cos 165° + \sin 165° = \cos 135°.$

By the last of the factor formulae (5.30),

$$\cos 165° - \cos 135° = -2 \sin \tfrac{1}{2}(165° + 135°) \sin \tfrac{1}{2}(165° - 135°)$$
$$= -2 \sin 150° \sin 15°$$
$$= -2 \sin(180° - 150°) \sin(180° - 15°)$$
$$= -2 \sin 30° \sin 165°$$
$$= -\sin 165°,$$

since $\sin 30° = 1/2$. Hence $\cos 165° + \sin 165° = \cos 135°.$

5.6. Further trigonometrical equations

The formulae of the last section enable certain trigonometrical equations to be reduced to equations of the form $\sin \theta = c$, $\cos \theta = c$ or $\tan \theta = c$, the general solution of which is known (Chapter 4, page 71). The following examples illustrate the methods.

Example 8. *Find all the values of θ which satisfy the equation*
$$\cos p\theta + \cos (p + 2)\theta = \cos \theta.$$

Using the third of the factor formulae, the equation can be written
$$2 \cos (p + 1)\theta \cos \theta = \cos \theta,$$

or,
$$\cos \theta \{2 \cos (p + 1)\theta - 1\} = 0.$$

Hence either $\cos \theta = 0$, giving $\theta = \left(2n \pm \dfrac{1}{2}\right)\pi$ or $\cos (p + 1)\theta = \dfrac{1}{2}$

giving $(p + 1)\theta = \left(2n \pm \dfrac{1}{3}\right)\pi$, i.e., $\theta = \dfrac{1}{p + 1}\left(2n \pm \dfrac{1}{3}\right)\pi$.

Example 9. *Find the general solution of the equation $2 \sin 3x \sin x = 1$.*

By formula (5.29), the equation can be written
$$-\cos 4x + \cos 2x = 1,$$

and using the formula $2 \cos^2 2x - 1 = \cos 4x$, we have
$$-2 \cos^2 2x + \cos 2x = 0 \quad \text{or} \quad \cos 2x(2 \cos 2x - 1) = 0,$$

so that either $\cos 2x = 0$ giving $2x = (2n \pm \frac{1}{2})\pi$, i.e., $x = (n \pm \frac{1}{4})\pi$; or $\cos 2x = \frac{1}{2}$ giving $2x = (2n \pm \frac{1}{3})\pi$, i.e., $x = (n \pm \frac{1}{6})\pi$.

5.7. The equation $a \cos \theta + b \sin \theta = c$

The equation $a \cos \theta + b \sin \theta = c$ in which a, b and c are supposed known numerical quantities, often occurs in practical applications. There are various methods of solution: here we shall consider two.

In the first method, we divide by $\sqrt{(a^2 + b^2)}$ and obtain

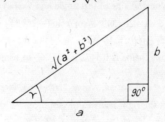

FIG. 21

$$\frac{a}{\sqrt{(a^2 + b^2)}} \cos \theta + \frac{b}{\sqrt{(a^2 + b^2)}} \sin \theta = \frac{c}{\sqrt{(a^2 + b^2)}}.$$

If we introduce an angle γ whose tangent is b/a, a glance at Fig. 21 shows that $a/\sqrt{(a^2 + b^2)}$ is $\cos \gamma$ and $b/\sqrt{(a^2 + b^2)}$ is $\sin \gamma$. Hence the equation can be written

$$\cos \theta \cos \gamma + \sin \theta \sin \gamma = \frac{c}{\sqrt{(a^2 + b^2)}},$$

or, $$\cos(\theta - \gamma) = \frac{c}{\sqrt{(a^2 + b^2)}}.$$

The equation has now been reduced to one of the standard forms whose general solution is known. Hence a general value of $\theta - \gamma$ can be written down and, since γ is a known angle, θ can be found. For real solutions to exist it is necessary for c to be numerically less than $\sqrt{(a^2 + b^2)}$. More precise details of the method of solution can be obtained from Example 10 below.

The second method makes use of the formulae (5.25), i.e., if $t = \tan\frac{1}{2}\theta$ then

$$\sin\theta = \frac{2t}{1 + t^2}, \quad \cos\theta = \frac{1 - t^2}{1 + t^2}.$$

Substitution in the given equation $a\cos\theta + b\sin\theta = c$ and multiplication throughout by $1 + t^2$, gives

$$a(1 - t^2) + 2bt = c(1 + t^2),$$

or, $$(a + c)t^2 - 2bt - (a - c) = 0.$$

This quadratic equation gives two values of t (or $\tan\frac{1}{2}\theta$) from which general values of θ can be derived. Again precise details will be found in Example 10 where the equation is solved by both methods.

Example 10. *Find the general solution of the equation $2\cos\theta - \sin\theta = 1$.*

Method (i) Dividing by $\sqrt{\{2^2 + (-1)^2\}}$ or $\sqrt{5}$ we have

$$\frac{2}{\sqrt{5}}\cos\theta - \frac{1}{\sqrt{5}}\sin\theta = \frac{1}{\sqrt{5}}.$$

Taking $\tan\gamma = -1/2$ so that, from tables, $\gamma = -26°\ 34'$, $2/\sqrt{5}$, $-1/\sqrt{5}$ are respectively the cosine and sine of this angle and we have

$$\cos\theta\cos(-26°\ 34') + \sin\theta\sin(-26°\ 34') = 1/\sqrt{5},$$

or, $$\cos(\theta + 26°\ 34') = 1/\sqrt{5} = 0·4472.$$

Now the angle whose cosine is $0·4472$ is $63°\ 26'$, so using the general solution given in (4.26),

$$\theta + 26°\ 34' = n \times 360° \pm 63°\ 26'.$$

The positive sign on the right leads to the solutions

$$\theta = n \times 360° + 36°\ 52',$$

while the negative sign gives $\theta = n \times 360° - 90°$.

Method (ii) Writing $\sin\theta = 2t(1 + t^2)$, $\cos\theta = (1 - t^2)/(1 + t^2)$, where $t = \tan\frac{1}{2}\theta$, and multiplying throughout by $1 + t^2$, we have

$$2(1 - t^2) - 2t = 1 + t^2,$$

or, $$3t^2 + 2t - 1 = 0.$$

This can be written $(t + 1)(3t - 1) = 0$. The root $t = -1$ gives $\tan\frac{1}{2}\theta = -1$, so that $\frac{1}{2}\theta = n \times 180° - 45°$ and $\theta = n \times 360° - 90°$. The root $t = 1/3$ leads to $\tan\frac{1}{2}\theta = \frac{1}{3}$, $\frac{1}{2}\theta = n \times 180° + 18°\ 26'$ and $\theta = n \times 360° + 36°\ 52'$ as before.

EXERCISES 5 (c)

1. Prove that $\sin A \sin (60° - A) \sin (60° + A) = \frac{1}{4} \sin 3A$.   (L.U.)

2. If $A + B + C = 180°$, prove that
$$\sin A + \sin B + \sin C = 4 \cos \tfrac{1}{2}A \cos \tfrac{1}{2}B \cos \tfrac{1}{2}C.$$ (Q.E.)

3. If $\cos A - \cos B = p$ and $\sin A - \sin B = q$, express $\cos (A - B)$ and $\sin (A + B)$ in terms of p and q.   (L.U.)

4. Find all the angles between $0°$ and $360°$ which satisfy the equation
$$\cos \theta + \sin 3\theta = \cos 2\theta.$$

5. Find the general solution of the equation
$$\sin \theta - \sin 2\theta = \sin 4\theta - \sin 3\theta.$$

6. Use the appropriate factor theorem to find the general value of x satisfying the equation $\cos px + \cos qx = 0$.

7. Find all the angles between $0°$ and $360°$ which satisfy the equation
$$\cos x + 7 \sin x = 5.$$

8. Find the value of θ less than $360°$ which satisfies the equation
$$3 \cos \theta - 4 \sin \theta = 5.$$

5.8. The inverse notation

If $\sin \theta = x$ where x is a given quantity numerically less than unity we know that θ can be any one of a whole series of angles. Thus if $\sin \theta = 1/2$, $\theta = n\pi + (-1)^n(\pi/6)$ and θ is "many-valued". The inverse notation $\theta = \sin^{-1} x$ is used to denote the angle whose sine is x and

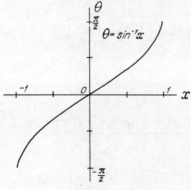

FIG. 22

the *numerically smallest* angle satisfying the relation $x = \sin \theta$ is chosen as the *principal value*. Here and in what follows we shall deal only with principal values and understand the statement $\theta = sin^{-1} x$ to mean that θ is the angle lying between $-\pi/2$ and $\pi/2$ radians whose sine is x. The statement $\theta = \sin^{-1} x$ is read as θ equals the inverse sine of

x (or sine minus one x) and an alternative notation, more commonly used on the Continent, is $\theta = \text{arc sin } x$.

The graph of $\theta = \sin^{-1} x$ is, on this understanding, easily seen to be that part of the graph $x = \sin \theta$ given by $-\pi/2 < \theta < \pi/2$ with the x-axis horizontal and the θ-axis vertical. This is shown in Fig. 22.

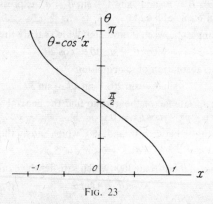

FIG. 23

In a similar way, $\theta = \cos^{-1} x$ will be taken to denote the smallest angle whose cosine is x. Since the cosine takes the same values for negative as for the corresponding positive angles and we require a notation which gives an unique value to θ when x is given, we conventionally take θ *to be the angle lying between* 0 *and* π *radians whose cosine is* x. For example, $\cos^{-1}\left(\dfrac{1}{2}\right) = \dfrac{\pi}{3}$ and $\cos^{-1}\left(-\dfrac{1}{2}\right) = \dfrac{2\pi}{3}$. The graph of $\theta = \cos^{-1} x$ is easily derived from that of $x = \cos \theta$ and is shown in Fig. 23.

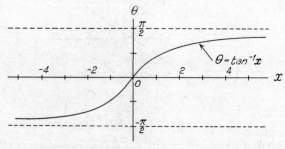

FIG. 24

The inverse tangent is similarly defined but since, unlike the sine and cosine, the tangent can take all values, x is quite unrestricted in value. $\theta = \tan^{-1} x$ is taken to mean that θ *is the smallest angle whose tangent is* x *and* θ *lies between* $-\pi/2$ *and* $\pi/2$ *radians.* Thus

$\tan^{-1}(1) = \pi/4$ and $\tan^{-1}(-1) = -\pi/4$. The graph of $\theta = \tan^{-1} x$ is given in Fig. 24.

The inverse cosecant, secant and cotangent are similarly defined. In order that they may be single-valued, we choose $\operatorname{cosec}^{-1} x$ to denote the angle between $-\pi/2$ and $\pi/2$ whose cosecant is x while $\sec^{-1} x$ and $\cot^{-1} x$ are taken to mean the angles lying between 0 and π radians whose secant and cotangent respectively are x.

It follows from these definitions that

$$\sin(\sin^{-1} x) = x, \quad \cos(\cos^{-1} x) = x, \quad \tan(\tan^{-1} x) = x, \text{etc.} \quad (5.31)$$

and these relations will be found useful in some of the examples and exercises given below. Care should be taken to avoid confusion between the inverse sine, cosine, etc. of x and the reciprocals of $\sin x$, $\cos x$, etc. The latter should always be written $\dfrac{1}{\sin x}$ or $\operatorname{cosec} x$, $\dfrac{1}{\cos x}$ or $\sec x$, etc.

The general solutions of the three equations $\sin \theta = c$, $\cos \theta = c$, and $\tan \theta = c$ given in (4.26) can be compactly expressed in this inverse notation. Thus if

$$\left.\begin{array}{ll}
\text{(i)} \ \sin \theta = c, (-1 < c < 1), & \theta = n\pi + (-1)^n \sin^{-1} c, \\
\text{(ii)} \ \cos \theta = c, (-1 < c < 1), & \theta = 2n\pi \pm \cos^{-1} c, \\
\text{(iii)} \ \tan \theta = c, & \theta = n\pi + \tan^{-1} c.
\end{array}\right\} \quad (5.32)$$

Example 11. *Show that*

$$(a) \ \cos^{-1}(-x) = \pi - \cos^{-1} x, \quad (b) \ \sin^{-1}(-x) = -\sin^{-1} x.$$

(a) (i) Suppose that x is positive and let $\theta = \cos^{-1} x$. Then θ lies between 0 and $\pi/2$ and $x = \cos \theta$. Hence

$$-x = -\cos \theta = \cos(\pi - \theta),$$

giving $\cos^{-1}(-x) = \pi - \theta = \pi - \cos^{-1} x.$

(ii) If x is negative, let $x = -y$ so that y is positive. Then $\cos^{-1}(-y) = \pi - \cos^{-1} y$ by (i) above. Hence

$$\cos^{-1} x = \pi - \cos^{-1}(-x)$$

and slight rearrangement gives

$$\cos^{-1}(-x) = \pi - \cos^{-1} x.$$

(iii) If x is neither positive nor negative, it must be zero. Hence $\cos^{-1}(-x) = \cos^{-1}(0) = \pi/2$ and $\cos^{-1} x = \cos^{-1}(0) = \pi/2$, so that

$$\pi - \cos^{-1} x = \pi - \tfrac{1}{2}\pi = \tfrac{1}{2}\pi = \cos^{-1}(-x).$$

Hence $\cos^{-1}(-x) = \pi - \cos^{-1} x$ for all values of x (provided, of course, that x is not numerically greater than unity).

(b) The identity $\sin^{-1}(-x) = -\sin^{-1} x$ can be established in a similar way and is left as an exercise for the student.

Example 12. *Show that* $\cos^{-1} x + \sin^{-1} x = \pi/2$.

(i) Let x be positive and let $\theta = \cos^{-1} x$. Then θ lies between 0 and $\pi/2$, and

$x = \cos\theta$. Hence

$$x = \cos\theta = \sin(\tfrac{1}{2}\pi - \theta),$$

giving $\qquad\qquad \sin^{-1}x = \tfrac{1}{2}\pi - \theta = \tfrac{1}{2}\pi - \cos^{-1}x,$

and the required result follows.

(ii) Let x be negative and let $x = -y$ so that y is positive. By (i) above, $\cos^{-1}y + \sin^{-1}y = \pi/2$ giving $\cos^{-1}(-x) + \sin^{-1}(-x) = \pi/2$. But, by Example 11,

$$\cos^{-1}(-x) = \pi - \cos^{-1}x \quad\text{and}\quad \sin^{-1}(-x) = -\sin^{-1}x,$$

so that

$$\pi - \cos^{-1}x - \sin^{-1}x = \pi/2,$$

giving $\qquad\qquad \cos^{-1}x + \sin^{-1}x = \pi/2.$

(iii) If x is neither positive nor negative, it must be zero. Hence

$$\cos^{-1}x = \cos^{-1}(0) = \pi/2 \quad\text{and}\quad \sin^{-1}x = \sin^{-1}(0) = 0,$$

giving $\cos^{-1}x + \sin^{-1}x = \pi/2$.

Hence the identity $\cos^{-1}x + \sin^{-1}x = \pi/2$ is valid for all values of $x(-1 < x < 1)$.

Example 13. *Show that* $\tan^{-1}(1/3) + \sin^{-1}(1/\sqrt5) = \pi/4$.

Let $\alpha = \tan^{-1}(1/3)$, $\beta = \sin^{-1}(1/\sqrt5)$, so that $\tan\alpha = 1/3$ and $\sin\beta = 1/\sqrt5$. Hence β is the angle shown in the right-angled triangle (Fig. 25) in which the

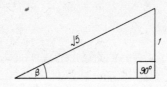

FIG. 25

height is 1 and the hypotenuse is $\sqrt5$. The base is $\sqrt{(5 - 1)}$ or 2 and we deduce that $\tan\beta = 1/2$. Hence

$$\tan^{-1}(1/3) + \sin^{-1}(1/\sqrt5) = \alpha + \beta$$
$$= \tan^{-1}\{\tan(\alpha + \beta)\}, \text{ by (5.31)},$$
$$= \tan^{-1}\left\{\frac{\tan\alpha + \tan\beta}{1 - \tan\alpha\tan\beta}\right\}$$
$$= \tan^{-1}\left\{\frac{\tfrac{1}{3} + \tfrac{1}{2}}{1 - \tfrac{1}{3} \times \tfrac{1}{2}}\right\}$$
$$= \tan^{-1}(1) = \pi/4.$$

EXERCISES 5 (*d*)

1. Evaluate $\sin^{-1}(1/\sqrt2)$, $\cos^{-1}(-\sqrt3/2)$, $\sec^{-1}2$ and $\cot^{-1}(\sqrt3)$.
2. Show that $\tan^{-1}(-x) = -\tan^{-1}x$.
3. Prove that $\tan^{-1}x + \cot^{-1}x = \pi/2$.

4. Prove that $2 \sin^{-1} \left(\frac{1}{2} \right) = \sin^{-1} \left(\frac{\sqrt{3}}{2} \right)$.

5. Show that $\cos^{-1} \left(\frac{63}{65} \right) + 2 \tan^{-1} \left(\frac{1}{5} \right) = \sin^{-1} \left(\frac{3}{5} \right)$.

6. If all the angles are acute, show that
$$\cos^{-1} x + \cos^{-1} y = \cos^{-1} \left[xy - \sqrt{\{(1 - x^2)(1 - y^2)\}} \right].$$

7. Show that there is a positive value of x which satisfies the equation $\tan^{-1} (2x + 1) + \tan^{-1} (2x - 1) = \tan^{-1} 2$, and find this value.

8. By drawing the graphs of $y = \tan x$ and $y = 2x$ show that the smallest positive root of the equation $\tan^{-1} 2x = x$ is the circular measure of an angle of about $67°$.

5.9. Small angles

If we plot on the same diagram the graphs of $\sin \theta$, θ and $\tan \theta$

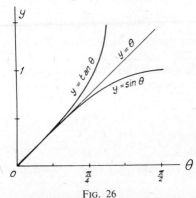

FIG. 26

for values of θ in radians a figure like Fig. 26 results. It is apparent that for $0 < \theta < \pi/2$,

$$\sin \theta < \theta < \tan \theta, \tag{5.33}$$

and that these three quantities are approximately equal to one another for small values of the angle θ.

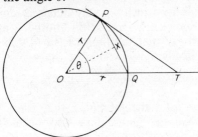

FIG. 27

These inequalities and approximations can also be inferred from Fig. 27 in which the chord PQ subtends an acute angle θ at the centre O of a circle of radius r. The tangent to the circle at P meets OQ produced at T. If OX is the bisector of the angle POQ it will bisect the base of the isosceles triangle POQ at right angles at X. Hence $OX = r \cos \frac{1}{2}\theta$, $PQ = 2PX = 2r \sin \frac{1}{2}\theta$ and the area of the triangle $POQ = \frac{1}{2}PQ.OX = \frac{1}{2}(2r \sin \frac{1}{2}\theta . r \cos \frac{1}{2}\theta) = \frac{1}{2}r^2 \sin \theta$. The area of the sector POQ is $\frac{1}{2}r^2 \theta$ (where θ is in radians) and since OPT is a right angle, $PT = r \tan \theta$ and the area of the triangle OTP

$$= \tfrac{1}{2}OP.PT = \tfrac{1}{2}r^2 \tan \theta.$$

From the figure,

area of triangle POQ < area of sector POQ < area of triangle POT,

or, $\qquad \tfrac{1}{2}r^2 \sin \theta < \tfrac{1}{2}r^2 \theta < \tfrac{1}{2}r^2 \tan \theta,$

which, on division by $\frac{1}{2}r^2$ give

$$\sin \theta < \theta < \tan \theta.$$

It is also clear that the areas of the three figures considered approach equality as the angle θ diminishes.

Dividing the inequalities (5.33) by $\sin \theta$ we find $1 < \theta/\sin \theta < 1/\cos \theta$, and this can be written $1 > (\sin \theta)/\theta > \cos \theta$. This in turn can be written

$$0 < 1 - \frac{\sin \theta}{\theta} < 1 - \cos \theta.$$

Since $1 - \cos \theta = 2 \sin^2 \frac{1}{2}\theta$ and since $\sin \theta < \theta$, $\sin \frac{1}{2}\theta$ will be less than $\frac{1}{2}\theta$ so we have $1 - \cos \theta < 2(\frac{1}{2}\theta)^2$ or $1 - \cos \theta < \frac{1}{2}\theta^2$. Hence

$$0 < 1 - \frac{\sin \theta}{\theta} < \tfrac{1}{2}\theta^2,$$

and $1 - (\sin \theta)/\theta$ can therefore be made as small as we please by making θ sufficiently small. Another way of expressing this is to write

$$\frac{\sin \theta}{\theta} = 1 - \epsilon,$$

where ϵ is a quantity which we may make as small as we please by taking θ to be sufficiently small. Still another way of expressing the same thing is to say that the limit of $(\sin \theta)/\theta$ as θ tends to zero is unity or, symbolically

$$\lim_{\theta \to 0} \left(\frac{\sin \theta}{\theta} \right) = 1. \qquad (5.34)$$

We also infer from the foregoing that *when θ is small, $\sin \theta$ is approximately equal to θ (in radians)*. Also, since $\cos \theta = 1 - 2 \sin^2 \frac{1}{2}\theta$ and since $\sin \frac{1}{2}\theta$ is approximately equal to $\frac{1}{2}\theta$, an approximate value of $\cos \theta$ when θ is small will be $1 - 2(\frac{1}{2}\theta)^2$ or $1 - \frac{1}{2}\theta^2$. A coarser approximation will be $\cos \theta = 1$ *for small* θ. As examples, consider the values given by these approximations when the angle is 4°. The radian measure of 4° is 0·0698 so that the approximation $\sin \theta = \theta$ gives $\sin 4° = 0·0698$, while the approximation $\cos \theta = 1 - \frac{1}{2}\theta^2$ gives $\cos 4° = 1 - \frac{1}{2}(0·00487) = 0·9976$. Both of these values are correct to four decimal places.

Example 14. *The elevations of the top Q of a flagstaff PQ from three distant points A, B, C which are in a horizontal line with P are* θ, 2θ *and* 3θ *respectively. Prove that AB = 3BC approximately.* (L.U.)

The three right-angled triangles CPQ, BPQ, APQ give

$$\frac{CP}{QP} = \cot 3\theta = \frac{\cos 3\theta}{\sin 3\theta},$$

$$\frac{BC + CP}{QP} = \cot 2\theta = \frac{\cos 2\theta}{\sin 2\theta},$$

$$\frac{AB + BC + CP}{QP} = \cot \theta = \frac{\cos \theta}{\sin \theta}.$$

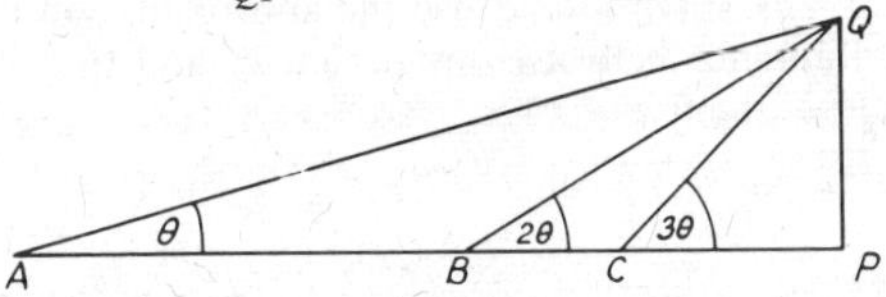

FIG. 28

Since the points A, B, C are all far from P, the angles θ, 2θ, 3θ will all be small. The cosines can therefore be replaced approximately by unity and the sines by the angles in radian measure. Hence we can write approximately

$$\frac{CP}{QP} = \frac{1}{3\theta}, \quad \frac{BC}{QP} + \frac{CP}{QP} = \frac{1}{2\theta} \quad \text{and} \quad \frac{AB}{QP} + \frac{BC}{QP} + \frac{CP}{QP} = \frac{1}{\theta}.$$

Subtracting the first of these relations from the second and the second from the third we shall have approximately

$$\frac{BC}{QP} = \frac{1}{2\theta} - \frac{1}{3\theta} = \frac{1}{6\theta} \quad \text{and} \quad \frac{AB}{QP} = \frac{1}{\theta} - \frac{1}{2\theta} = \frac{1}{2\theta}.$$

By division we then have $AB/BC = 3$, approximately.

Example 15. *Find an approximate value of the acute angle which satisfies the equation* $\sin \theta = 0·52$.

Since $\sin \theta$ is nearly equal to 0·5, θ must be nearly $\pi/6$ radians. Let $\theta = \dfrac{\pi}{6} + \epsilon$ where ϵ is therefore small. Then

$$0·52 = \sin\left(\frac{\pi}{6} + \epsilon\right) = \sin\frac{\pi}{6}\cos\epsilon + \cos\frac{\pi}{6}\sin\epsilon.$$

Since $\sin(\pi/6) = 1/2$, $\cos(\pi/6) = \sqrt{3}/2$, and, because ϵ is small, $\cos\epsilon = 1$, $\sin\epsilon = \epsilon$ approximately, we have

$$0·52 = \frac{1}{2} + \frac{\sqrt{3}}{2}\epsilon,$$

giving $\epsilon = \dfrac{2}{\sqrt{3}} \times 0\cdot02 = 0\cdot0231.$

Hence $\epsilon = 0\cdot0231$ radians or $1°\ 19'$ approximately and $\theta = 31°\ 19'$.

EXERCISES 5 (e)

1. Find an acute angle which approximately satisfies the equation $\sin \theta = 0\cdot48$.

2. The diameter of the bull's eye of a certain target is $0\cdot0254$ m. At what distance will it subtend an angle of 30 minutes?

3. In a right-angled triangle ABC, C is the right angle, the side $BC = a$ and the side $AC = b$. Show that the angle ABC lies between $b/\sqrt{(a^2 + b^2)}$ and b/a radians.

4. Assuming that $\sin \theta = \theta - k\theta^3$, where k is a numerical constant, is a sufficient approximation to the value of $\sin \theta$ when θ is a small angle, use the formula $\sin 3\theta = 3 \sin \theta - 4 \sin^3 \theta$ to show that $k = 1/6$. (This result gives, of course, a better approximation than $\sin \theta = \theta$.)

5. Prove that the perimeter of a regular polygon of n sides inscribed in a circle of radius R is $2nR \sin (\pi/n)$ and use the approximation of Exercise 4 above to find the difference between this perimeter and the circumference of the circle when $R = 30$ m, $n = 200$.

6. Prove that if $0 < \theta < \pi/2$, $\sin \theta > \theta - \frac{1}{4}\theta^3$.

EXERCISES 5 (f)

1. Prove that $\tan (45° + \theta) - \tan (45° - \theta) = 2 \tan 2\theta$.

2. If $\tan \alpha = a/(a + 1)$ and $\tan \beta = 1/(2a + 1)$ find the smallest value of the angle $\alpha + \beta$.

3. Show that $\sin (\alpha + \beta) \sin (\alpha - \beta) = \sin^2 \alpha - \sin^2 \beta$.

4. Express $\tan (A + B + C)$ in terms of $\tan A$, $\tan B$ and $\tan C$.

5. Prove that

$$\frac{\sin \alpha \sin \beta}{\cos \alpha + \cos \beta} = \frac{2 \tan \dfrac{\alpha}{2} \tan \dfrac{\beta}{2}}{1 - \tan^2 \dfrac{\alpha}{2} \tan^2 \dfrac{\beta}{2}} \qquad \text{(Q.E.)}$$

6. Prove the identity
$$(\sin 2\alpha - \sin 2\beta) \tan (\alpha + \beta) = 2(\sin^2 \alpha - \sin^2 \beta). \qquad \text{(L.U.)}$$

7. If $\tan \frac{1}{2}x = \operatorname{cosec} x - \sin x$, prove that $\tan^2 \frac{1}{2}x = -2 \pm \sqrt{5}$. (L.U.)

8. If $\sin \theta + \sin 2\theta = a$ and $\cos \theta + \cos 2\theta = b$, prove that
$$(a^2 + b^2)(a^2 + b^2 - 3) = 2b. \qquad \text{(L.U.)}$$

9. Prove that
$$\tan \alpha + \tan (60° + \alpha) + \tan (120° + \alpha) = 3 \tan 3\alpha. \qquad \text{(L.U.)}$$

10. Prove that
$$\sin^3 A + \sin^3 (120° + A) + \sin^3 (240° + A) = -\tfrac{3}{4} \sin 3A. \qquad \text{(L.U.)}$$

11. Establish the identity
$$\sin\theta + \sin(\theta + \alpha) + \sin(\theta + 2\alpha) + \sin(\theta + 3\alpha)$$
$$= 4\sin\left(\theta + \frac{3\alpha}{2}\right)\cos\alpha\cos\frac{\alpha}{2}.\quad \text{(L.U.)}$$

12. Prove that
$$\sin 3x + 2\sin 5x \sin^2 x + \sin 7x = \cos x(\sin 6x + \sin 4x).\quad \text{(L.U.)}$$

13. Prove that
$$2\cot 2A \cot \tfrac{1}{2}A = \frac{\cos A}{1 - \cos A} - \frac{1 + \cos A}{\cos A}.\quad \text{(L.U.)}$$

14. Find all the angles between $0°$ and $180°$ (inclusive) which satisfy the equation $\cos x - \cos 7x = \sin 4x$. (L.U.)

15. Solve the equation $10\sin^2\theta - 5\sin 2\theta = 4$, giving the values of θ between $0°$ and $360°$. (L.U.)

16. Solve the equation $2\sin\theta + 3\cos\theta = -1$, stating all solutions for the range $0 < \theta < 2\pi$. (Q.E.)

17. Solve completely the equation $\sin 3\theta \cos 3\theta - \cos^2 2\theta + \tfrac{1}{2} = 0$. (L.U.)

18. Find a pair of angles lying between $0°$ and $180°$ and satisfying the equations $\sin A + \sin B = 0.95$, $A - B = 120°$. (L.U.)

19. Find all the angles between $0°$ and $360°$ satisfying the equation $\sin 2\theta - 2\cos 2\theta = \sin\theta - 2\cos\theta + 2$. (L.U.)

20. If $\tan\theta = \lambda\tan(A - \theta)$, show that
$$(\lambda - 1)\sin A = (\lambda + 1)\sin(2\theta - A).$$
Hence, or otherwise, find the values of θ between $0°$ and $360°$ which satisfy the equation $\tan\theta = 2\tan(60° - \theta)$. (L.U.)

21. Find two values of θ less than $\pi/2$ satisfying the equation
$$8\sec\theta - 4\tan\theta = 7.$$
Write down the general solution of the equation. (L.U.)

22. Prove that
$$4\tan^{-1}\left(\tfrac{1}{5}\right) - \tan^{-1}\left(\tfrac{1}{239}\right) = \frac{\pi}{4}.$$

23. Prove that
$$\cot^{-1}\left(\tfrac{1}{3}\right) = \cot^{-1}(3) + \cos^{-1}\left(\tfrac{3}{5}\right).$$

24. Find x from the equation $\tan^{-1} 2x + \tan^{-1} 3x = \pi/4$.

25. If θ is an acute angle such that $\cos\theta = 1 - x$, where x is so small that x^2 is negligible compared with unity, prove that $\cos 2\theta = 1 - 4x$ and $\cos 3\theta = 1 - 9x$ approximately. (L.U.)

CHAPTER 6

RELATIONS BETWEEN THE SIDES AND ANGLES OF A TRIANGLE. THE SOLUTION OF TRIANGLES. HEIGHTS AND DISTANCES

6.1. Notation

A triangle has six parts or elements—three sides and three angles. If A, B and C are used to denote the angles of the triangle, it is conventional to denote the sides opposite these angles by the corresponding small letters a, b and c respectively (Fig. 29).

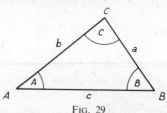

FIG. 29

The sides of a triangle are independent of one another except for the fact that the sum of any two of them must be greater than the third. The angles, however, are not independent. Since the sum of the angles of any triangle is $180°$, the third angle is known if two angles are given. There are thus five independent elements in a triangle, three sides and two angles. Later in this chapter we shall see that if three elements of a triangle, one at least of which is a side, are known, the other three can be found. The process of calculating the unknown elements of a triangle when three of its elements are given is termed the solution of the triangle.

The early part of this chapter is devoted to the derivation of relations between the sides and angles of a triangle. In the later sections, the arrangement of the computations involved in the numerical solution of triangles is discussed and examples are given of some practical applications to problems in heights and distances.

6.2. The sine formula

Let O be the centre of the circle circumscribing the triangle ABC. Join BO and produce it to meet the circle again at D. Join DC. Figs. 30, 31 apply respectively when the angle A of the triangle ABC is acute or obtuse. In both diagrams, the angle BCD, being the angle in a semi-circle, is a right angle. In Fig. 30, the angle BDC is equal to the angle BAC in the same segment, while in Fig. 31, the angle BDC is equal to the supplement of BAC, since the points B, A, C and D

are concyclic. If R is the radius of the circumcircle so that $BD = 2R$, the right-angled triangle BCD gives,

> in Fig. 30, $BC = 2R \sin A$,
> in Fig. 31, $BC = 2R \sin (180° - A) = 2R \sin A$.

Hence in both cases, since $BC = a$, $a = 2R \sin A$.

By joining AD instead of DC we could prove similarly that $c = 2R \sin C$. By starting our construction from C, instead of B, we

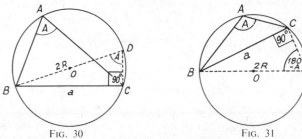

FIG. 30   FIG. 31

could show in the same way that $b = 2R \sin B$. These three results can be displayed in the formula

$$\frac{a}{\sin A} = \frac{b}{\sin B} = \frac{c}{\sin C} = 2R, \tag{6.1}$$

a result usually known as the *sine formula*. In certain cases (see for instance, page 105) this formula enables the solution of a triangle to be carried out and it also enables the radius R of the circumcircle of a given triangle to be found.

6.3. The cosine formula

In Figs. 32 and 33, BD is the perpendicular from B on to the base

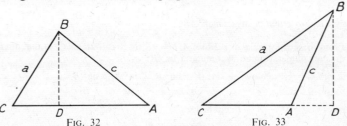

FIG. 32   FIG. 33

CA, or CA produced, of the triangle ABC. The first diagram applies when the angle A is acute, the second when A is obtuse. In each figure the right-angled triangle DAB gives

$$BD = c \sin A.$$

In Fig. 32, $DA = c \cos A$ and $CD = CA - DA = b - c \cos A$.
In Fig. 33, $AD = c \cos D\hat{A}B = c \cos (180° - A) = - c \cos A$

and $$CD = CA + AD = b - c \cos A.$$

Applying Pythagoras' theorem to the right-angled triangle DCB in either figure,

$$CB^2 = CD^2 + BD^2,$$

or, $$a^2 = (b - c \cos A)^2 + c^2 \sin^2 A$$
$$= b^2 - 2bc \cos A + c^2 (\cos^2 A + \sin^2 A),$$

which, using the identity $\sin^2 A + \cos^2 A = 1$, gives

$$a^2 = b^2 + c^2 - 2bc \cos A. \tag{6.2}$$

The two similar formulae

$$b^2 = c^2 + a^2 - 2ca \cos B,$$
$$c^2 = a^2 + b^2 - 2ab \cos C,$$

can be similarly derived. These are the *cosine formulae* and are useful in the solution of triangles when at least two sides are given (see page 109).

Example 1. *In a triangle ABC, prove that $a^2 = (b - c)^2 + 4bc \sin^2 \frac{1}{2}A$ and hence that $a = (b - c) \sec \phi$ where $\tan \phi = \dfrac{2\sqrt{(bc)} \sin \frac{1}{2}A}{b - c}$.* (L.U.)

Since $\cos A = 1 - 2 \sin^2 \frac{1}{2}A$, the cosine formula

$$a^2 = b^2 + c^2 - 2bc \cos A$$

gives

$$a^2 = b^2 + c^2 - 2bc(1 - 2 \sin^2 \frac{1}{2}A)$$
$$= (b - c)^2 + 4bc \sin^2 \frac{1}{2}A.$$

Using the given expression for $\tan \phi$, this can be written

$$a^2 = (b - c)^2 + (b - c)^2 \tan^2 \phi$$
$$= (b - c)^2 (1 + \tan^2 \phi)$$
$$= (b - c)^2 \sec^2 \phi,$$

leading to $\qquad a = (b - c) \sec \phi.$

Example 2. *Prove that in any triangle ABC, $\dfrac{a^2 - b^2}{c^2} = \dfrac{\sin (A - B)}{\sin (A + B)}$.* (L.U.)

From the sine formula, $a = 2R \sin A, b = 2R \sin B, c = 2R \sin C$, so that after dividing numerator and denominator by $4R^2$,

$$\frac{a^2 - b^2}{c^2} = \frac{\sin^2 A - \sin^2 B}{\sin^2 C} = \frac{\sin^2 A - \sin^2 B}{\sin^2 (A + B)},$$

since $C = 180° - A - B$ and hence

$$\sin C = \sin (180° - A - B) = \sin (A + B).$$

This can be written

$$\frac{a^2 - b^2}{c^2} = \frac{(\sin A + \sin B)(\sin A - \sin B)}{\sin^2 (A + B)}$$

$$= \frac{2 \sin \frac{1}{2}(A + B) \cos \frac{1}{2}(A - B) . 2 \cos \frac{1}{2}(A + B) \sin \frac{1}{2}(A - B)}{2 \sin \frac{1}{2}(A + B) \cos \frac{1}{2}(A + B) \sin (A + B)}$$

using (5.30) and (5.9).

This simplifies to
$$\frac{a^2 - b^2}{c^2} = \frac{2 \sin \frac{1}{2}(A - B) \cos \frac{1}{2}(A - B)}{\sin (A + B)} = \frac{\sin (A - B)}{\sin (A + B)}.$$

EXERCISES 6 (a)

1. With the usual notation for a triangle ABC, prove that
$$a \cos A + b \cos B = c \cos (A - B). \qquad \text{(L.U.)}$$

2. With the usual notation for a triangle ABC, show that c^2 can be expressed in the form
$$(a + b)^2(1 - k^2 \cos^2 \tfrac{1}{2}C),$$
and obtain the value of k^2.

3. Prove that in a triangle ABC with circumcircle of radius R,
$$a \cos A + b \cos B + c \cos C = 4R \sin A \sin B \sin C.$$

4. If, in a triangle ABC, $ab = c^2$, prove that
$$\cos (A - B) + \cos C + \cos 2C = 1.$$

5. With the usual notation for a triangle ABC, prove that
$$(b + c) \tan \tfrac{1}{2}A - (b - c) \cot \tfrac{1}{2}A = 2b \cot B.$$

6. Prove, in the usual notation for a triangle, that if
$$\frac{b + c}{11} = \frac{c + a}{12} = \frac{a + b}{13},$$
then,
$$\frac{\sin A}{7} = \frac{\sin B}{6} = \frac{\sin C}{5} \quad \text{and} \quad \frac{\cos A}{7} = \frac{\cos B}{19} = \frac{\cos C}{25}.$$
$$\text{(L.U.)}$$

6.4. The area of a triangle

Let Δ denote the area of the triangle ABC and, in Fig. 34, let BD

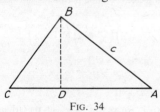

FIG. 34

be the perpendicular from B on to AC. Then since $BD = c \sin A$
$$\Delta = \tfrac{1}{2}CA \cdot BD = \tfrac{1}{2}bc \sin A. \qquad (6.3)$$
In the same way we could show that $\Delta = \tfrac{1}{2}ca \sin B$ or $\tfrac{1}{2}ab \sin C$.

By writing (6.3) in the form $\dfrac{a}{\sin A} = \dfrac{\tfrac{1}{2}abc}{\Delta}$, we could rewrite the sine formula (6.1) as
$$\frac{a}{\sin A} = \frac{b}{\sin B} = \frac{c}{\sin C} = 2R = \frac{abc}{2\Delta}. \qquad (6.4)$$

To find an expression for the area of a triangle in terms of the sides alone, we have from (6.3),

$$2bc \sin A = 4\Delta,$$

and from the cosine formula (6.2),

$$2bc \cos A = b^2 + c^2 - a^2.$$

Squaring and adding, and using $\sin^2 A + \cos^2 A = 1$, these give

$$4b^2 c^2 = 16\Delta^2 + (b^2 + c^2 - a^2)^2,$$

so that
$$\begin{aligned}
\Delta^2 &= \tfrac{1}{16}\{(4b^2 c^2 - (b^2 + c^2 - a^2)^2\} \\
&= \tfrac{1}{16}(2bc + b^2 + c^2 - a^2)(2bc - b^2 - c^2 + a^2) \\
&= \tfrac{1}{16}\{(b + c)^2 - a^2\}\{a^2 - (b - c)^2\} \\
&= \tfrac{1}{16}(b + c + a)(b + c - a)(a - b + c)(a + b - c).
\end{aligned}$$

If we write

$$2s = a + b + c, \tag{6.5}$$

so that s is half the perimeter of the triangle, this can be written

$$\Delta^2 = \tfrac{1}{16} \cdot 2s \cdot (2s - 2a)(2s - 2b)(2s - 2c),$$

or,

$$\Delta = \sqrt{\{s(s - a)(s - b)(s - c)\}}. \tag{6.6}$$

Example 3. *The sides of a triangle are in arithmetical progression and its area is 3/5ths that of an equilateral triangle of the same perimeter. Prove that its sides are in the ratio* 3:5:7.

Let the sides of the triangle be $x - d$, x, $x + d$. If $2s$ is its perimeter, $2s = 3x$. From (6.6) the square of its area

$$= \frac{3x}{2}\left(\frac{3x}{2} - x + d\right)\left(\frac{3x}{2} - x\right)\left(\frac{3x}{2} - x - d\right) = \frac{3}{4}x^2\left(\frac{1}{4}x^2 - d^2\right).$$

For the equilateral triangle of the same perimeter each side will be x and the square of its area

$$= \frac{3x}{2}\left(\frac{3x}{2} - x\right)\left(\frac{3x}{2} - x\right)\left(\frac{3x}{2} - x\right) = \frac{3x^4}{16}.$$

Hence $\frac{3}{4}x^2(\frac{1}{4}x^2 - d^2) = \frac{9}{25} \times \frac{3}{16}x^4,$

giving $\frac{1}{4}x^2 - d^2 = \frac{9}{100}x^2.$

Thus $d^2 = (\frac{1}{4} - \frac{9}{100})x^2 = \frac{16}{100}x^2,$

so that $d = 4x/10 = 2x/5$. Hence the sides of the triangle are $(x - 2x/5)$, x and $(x + 2x/5)$ or $3x/5$, x, $7x/5$, which are in the ratio 3:5:7.

6.5. The radius of the inscribed circle

Let I be the centre of the inscribed circle and D, E, F the points of contact of the circle with the sides BC, CA, AB of the triangle ABC. Then

area triangle BIC + area triangle CIA + area triangle AIB

$$= \text{area triangle } ABC = \Delta.$$

If r is the radius of the inscribed circle, the heights of the triangles

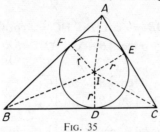

FIG. 35

BIC, CIA, AIB are each r and their bases are respectively a, b and c. Hence

$$\tfrac{1}{2}ra + \tfrac{1}{2}rb + \tfrac{1}{2}rc = \Delta,$$

or since $\tfrac{1}{2}(a + b + c) = s$,

$$rs = \Delta,$$

giving $$r = \Delta/s. \qquad (6.7)$$

This formula, together with (6.6) enables the radius of the inscribed circle to be found in terms of the three sides of the triangle.

   Alternative expressions for r can be found as follows. Since AI bisects the angle CAB, the angles AFI, IEA are right angles and AI is common, the triangles AFI, IEA are congruent and $AE = AF$. Hence

$$2AE = AE + AF$$

and similarly

$$2BD = BD + BF,$$
$$2CD = CD + CE.$$

By addition

$$2AE + 2(BD + CD) = AE + AF + BD + BF + CD + CE$$

or $$2AE + 2BC = \text{perimeter of the triangle } ABC.$$

Hence $$2AE + 2a = 2s,$$

giving $$AE = s - a.$$

In the right-angled triangle IEA, the angle EAI is $\tfrac{1}{2}A$ and since $IE = r$.

$$\frac{r}{AE} = \tan \tfrac{1}{2}A,$$

giving $$r = (s - a) \tan \tfrac{1}{2}A.$$

Similarly we can show that $r = (s - b) \tan \tfrac{1}{2}B$, $r = (s - c) \tan \tfrac{1}{2}C$, and, combining these with (6.7), we have

$$r = (s - a) \tan \tfrac{1}{2}A = (s - b) \tan \tfrac{1}{2}B = (s - c) \tan \tfrac{1}{2}C = \Delta/s. \qquad (6.8)$$

   Yet another expression for r can be obtained by starting from the relation (see Fig. 35), $BD + DC = BC = a$. Now the angle DBI is

$\frac{1}{2}B$ and the angle DCI is $\frac{1}{2}C$, so that the right-angled triangles DBI, DCI give

$$DB = r \cot \tfrac{1}{2}B, \quad DC = r \cot \tfrac{1}{2}C.$$

Hence

$$r\{\cot \tfrac{1}{2}B + \cot \tfrac{1}{2}C\} = a.$$

This can be written

$$r \left\{ \frac{\cos \tfrac{1}{2}B}{\sin \tfrac{1}{2}B} + \frac{\cos \tfrac{1}{2}C}{\sin \tfrac{1}{2}C} \right\} = a,$$

giving

$$r\{\sin \tfrac{1}{2}C \cos \tfrac{1}{2}B + \cos \tfrac{1}{2}C \sin \tfrac{1}{2}B\} = a \sin \tfrac{1}{2}B \sin \tfrac{1}{2}C,$$

or, $\qquad\qquad r \sin \tfrac{1}{2}(B + C) = a \sin \tfrac{1}{2}B \sin \tfrac{1}{2}C.$

Since $A + B + C = 180°$,

$$\tfrac{1}{2}(B + C) = 90° - \tfrac{1}{2}A,$$
$$\sin \tfrac{1}{2}(B + C) = \sin (90° - \tfrac{1}{2}A) = \cos \tfrac{1}{2}A.$$

Also, from the sine formula (6.1),

$$a = 2R \sin A = 4R \sin \tfrac{1}{2}A \cos \tfrac{1}{2}A.$$

Hence

$$r \cos \tfrac{1}{2}A = 4R \sin \tfrac{1}{2}A \cos \tfrac{1}{2}A \sin \tfrac{1}{2}B \sin \tfrac{1}{2}C,$$

giving, $\qquad\qquad r = 4R \sin \tfrac{1}{2}A \sin \tfrac{1}{2}B \sin \tfrac{1}{2}C. \qquad\qquad (6.9)$

6.6. The radii of the escribed circles

Let I_1 be the centre of the escribed circle touching BC internally at

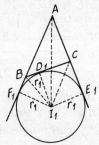

FIG. 36

D_1 and AB, AC externally at F_1, E_1 respectively. Let the radius of this circle be r_1. Then, Fig. 36,

area triangle BI_1A + area triangle CI_1A − area triangle BI_1C
$$= \text{area triangle } ABC = \Delta.$$

The heights of the triangles BI_1A, CI_1A, BI_1C are each r_1 and their bases are respectively c, b and a. Hence

$$\tfrac{1}{2}r_1c + \tfrac{1}{2}r_1b - \tfrac{1}{2}r_1a = \Delta,$$

or, $\qquad\qquad \tfrac{1}{2}r_1(c + b - a) = \Delta.$

Since $2s = a + b + c, c + b - a = 2s - 2a$, and therefore

$$r_1(s - a) = \Delta.$$

Hence

$$r_1 = \frac{\Delta}{s - a}, \tag{6.10}$$

and we can derive similar formulae $r_2 = \Delta/(s - b)$, $r_3 = \Delta/(s - c)$ for the radii of the escribed circles opposite B and C respectively.

Since, from (6.7), $\Delta = rs$, this can be written

$$r_1 = \frac{rs}{s - a},$$

and use of (6.8) for r then gives

$$r_1 = s \tan \tfrac{1}{2}A. \tag{6.11}$$

Similarly we can show that $r_2 = s \tan \tfrac{1}{2}B$ and $r_3 = s \tan \tfrac{1}{2}C$.

Example 4. *In a triangle ABC, r_1, r_2, r_3 respectively denote the radii of the three escribed circles. Prove that $r_2r_3 + r_3r_1 + r_1r_2 = s^2$, where $2s = a + b + c$.*
(L.U.)

Since $r_2 = \Delta/(s - b)$, $r_3 = \Delta/(s - c)$, $r_2r_3 = \Delta^2/(s - b)(s - c)$. But $\Delta^2 = s(s - a)(s - b)(s - c)$ so that $r_2r_3 = s(s - a)$. Similarly

$$r_3r_1 = s(s - b) \quad \text{and} \quad r_1r_2 = s(s - c).$$

Hence

$$\begin{aligned}
r_2r_3 + r_3r_1 + r_1r_2 &= s(s - a) + s(s - b) + s(s - c)\\
&= s(3s - a - b - c)\\
&= s(3s - 2s) = s^2.
\end{aligned}$$

EXERCISES 6 (b)

1. If E is the middle point of the side CA of the triangle ABC and if Δ is the area of the triangle, prove that

$$\cot AEB = \frac{BC^2 - BA^2}{4\Delta}. \tag{L.U.}$$

2. If Δ is the area and R the radius of the circumcircle of the triangle ABC, prove that $\cos A + \cos (B - C) = 2\Delta/aR$.

3. If Δ is the area and r, r_1, r_2, r_3 are respectively the radii of the inscribed and the three escribed circles of a triangle, prove that $\Delta = (rr_1r_2r_3)^{1/2}$.

4. If r, r_1, r_2, r_3 are respectively the radii of the inscribed and the three escribed circles of a triangle and if R is the radius of the circumcircle, prove that $r_1 + r_2 + r_3 - r = 4R$. (L.U.)

5. If r_1, r_2, r_3 are respectively the radii of the three escribed circles of a triangle ABC and if R is the radius of the circumcircle, prove that

$$r_1 = 4R \sin \tfrac{1}{2}A \cos \tfrac{1}{2}B \cos \tfrac{1}{2}C, \quad r_2 = 4R \cos \tfrac{1}{2}A \sin \tfrac{1}{2}B \cos \tfrac{1}{2}C,$$
$$r_3 = 4R \cos \tfrac{1}{2}A \cos \tfrac{1}{2}B \sin \tfrac{1}{2}C.$$

6. If r, r_1, r_2, r_3 are respectively the radii of the inscribed and the three escribed circles of a triangle ABC, prove that

$$r_1r_2r_3 = r^3 \cot^2 \tfrac{1}{2}A \cot^2 \tfrac{1}{2}B \cot^2 \tfrac{1}{2}C. \tag{L.U.}$$

6.7. Formulae for the angles of a triangle in terms of the sides

By rearranging the cosine formula

$$a^2 = b^2 + c^2 - 2bc \cos A$$

in the form

$$\cos A = (b^2 + c^2 - a^2)/2bc,$$

we can express the angle A in terms of the sides of a triangle. This formula is not, however, well suited to computation by logarithms and we develop alternative formulae as follows.

From the formula (6.8) we have

$$\tan \tfrac{1}{2}A = r/(s - a).$$

Using (6.7), $r = \Delta/s$, so that

$$\tan \tfrac{1}{2}A = \frac{\Delta}{s(s - a)}, \tag{6.12}$$

and using the expression (6.6) giving the area in terms of the sides, we have

$$\tan \tfrac{1}{2}A = \frac{\sqrt{\{s(s - a)(s - b)(s - c)\}}}{s(s - a)},$$

or,

$$\tan \tfrac{1}{2}A = \sqrt{\left\{\frac{(s - b)(s - c)}{s(s - a)}\right\}}. \tag{6.13}$$

The corresponding formulae

$$\tan \tfrac{1}{2}B = \sqrt{\left\{\frac{(s - c)(s - a)}{s(s - b)}\right\}},$$

$$\tan \tfrac{1}{2}C = \sqrt{\left\{\frac{(s - a)(s - b)}{s(s - c)}\right\}}$$

can be similarly derived.

If in (6.12) we use the formula $\Delta = \tfrac{1}{2}bc \sin A$ instead of (6.6) we have

$$\tan \tfrac{1}{2}A = \frac{bc \sin A}{2s(s - a)} = \frac{bc \sin \tfrac{1}{2}A \cos \tfrac{1}{2}A}{s(s - a)},$$

giving

$$\cos^2 \tfrac{1}{2}A = \frac{s(s - a)}{bc},$$

or,

$$\cos \tfrac{1}{2}A = \sqrt{\left\{\frac{s(s - a)}{bc}\right\}}. \tag{6.14}$$

Multiplication of (6.13) and (6.14) gives

$$\sin \tfrac{1}{2}A = \sqrt{\left\{\frac{(s - b)(s - c)}{bc}\right\}}, \tag{6.15}$$

and, of course, there are formulae corresponding to (6.14), (6.15) for the angles B and C.

Example 5. *Prove that in a triangle ABC,*

$$\frac{1}{a}\cos^2\frac{1}{2}A + \frac{1}{b}\cos^2\frac{1}{2}B + \frac{1}{c}\cos^2\frac{1}{2}C = \frac{(a+b+c)^2}{4abc}.$$

From (6.14) and the two similar formulae,

$$\frac{1}{a}\cos^2\tfrac{1}{2}A + \frac{1}{b}\cos^2\tfrac{1}{2}B + \frac{1}{c}\cos^2\tfrac{1}{2}C = \frac{s(s-a)+s(s-b)+s(s-c)}{abc}$$

$$= \frac{3s^2 - s(a+b+c)}{abc}$$

$$= \frac{3s^2 - 2s^2}{abc}$$

$$= \frac{s^2}{abc} = \frac{(a+b+c)^2}{4abc}.$$

6.8. The tangent formula

We now derive another set of formulae which are useful in the numerical solution of triangles. Starting from the sine formula (6.1), $b = 2R \sin B$, $c = 2R \sin C$, and after division of numerator and denominator by $2R$, we have

$$\frac{b-c}{b+c} = \frac{\sin B - \sin C}{\sin B + \sin C}$$

$$= \frac{2\cos\tfrac{1}{2}(B+C)\sin\tfrac{1}{2}(B-C)}{2\sin\tfrac{1}{2}(B+C)\cos\tfrac{1}{2}(B-C)}$$

$$= \cot\tfrac{1}{2}(B+C)\tan\tfrac{1}{2}(B-C).$$

Hence $$\tan\tfrac{1}{2}(B-C) = \left(\frac{b-c}{b+c}\right)\tan\tfrac{1}{2}(B+C),$$

and since $\tfrac{1}{2}(B+C) = 90° - \tfrac{1}{2}A$, this can be written

$$\tan\tfrac{1}{2}(B-C) = \left(\frac{b-c}{b+c}\right)\cot\tfrac{1}{2}A. \qquad (6.16)$$

The two corresponding formulae

$$\tan\tfrac{1}{2}(C-A) = \left(\frac{c-a}{c+a}\right)\cot\tfrac{1}{2}B, \quad \tan\tfrac{1}{2}(A-B) = \left(\frac{a-b}{a+b}\right)\cot\tfrac{1}{2}C,$$

can be similarly derived.

Example 6. *Prove that in any triangle ABC,*

$$\sin\tfrac{1}{2}(B-C) = \left(\frac{b-c}{a}\right)\cos\tfrac{1}{2}A.$$

Working as in § 6.8, we have

$$\frac{b-c}{a} = \frac{\sin B - \sin C}{\sin A}$$

$$= \frac{\sin B - \sin C}{\sin (B+C)},$$

since $A = 180° - B - C$. This can be written

$$\frac{b - c}{a} = \frac{2 \cos \frac{1}{2}(B + C) \sin \frac{1}{2}(B - C)}{2 \sin \frac{1}{2}(B + C) \cos \frac{1}{2}(B + C)}$$

$$= \frac{\sin \frac{1}{2}(B - C)}{\sin \frac{1}{2}(B + C)} = \frac{\sin \frac{1}{2}(B - C)}{\cos \frac{1}{2}A},$$

since $\frac{1}{2}(B + C) = 90° - \frac{1}{2}A$. The result follows immediately after multiplication by $\cos \frac{1}{2}A$.

EXERCISES 6 (c)

1. With the usual notation for a triangle ABC, show that

$$\sin A = \frac{2}{bc}\sqrt{\{s(s - a)(s - b)(s - c)\}}.$$

2. Prove that in any triangle ABC,

$$(a + b + c)(\tan \tfrac{1}{2}A + \tan \tfrac{1}{2}B) = 2c \cot \tfrac{1}{2}C.$$

3. Show that in a triangle ABC,

$$b + c = a \cos \tfrac{1}{2}(B - C) \operatorname{cosec} \tfrac{1}{2}A.$$

4. Show that in a triangle ABC, if $2s = a + b + c$,

$$1 - \tan \tfrac{1}{2}A \tan \tfrac{1}{2}B = \frac{c}{s}. \qquad \text{(L.U.)}$$

5. Assuming the cosine formula for a triangle ABC, prove that

$$\sin \tfrac{1}{2}A = \sqrt{\left\{\frac{(s - b)(s - c)}{bc}\right\}}. \qquad \text{(L.U.)}$$

6. If in a triangle ABC,

$$\tan \theta = \left(\frac{b + c}{b - c}\right) \tan \tfrac{1}{2}A,$$

prove that $a = (b - c) \cos \tfrac{1}{2}A \sec \theta$. $\qquad \text{(L.U.)}$

6.9. Summary of formulae for the triangle

The more important relations between the sides, angles and area of a triangle and the radii of its associated circles are here collected for reference.

If R is the radius of the circumcircle, r that of the inscribed circle, r_1, r_2 and r_3 the radii of the three escribed circles, Δ the area of the triangle and $2s = a + b + c$, then

$$\frac{a}{\sin A} = \frac{b}{\sin B} = \frac{c}{\sin C} = 2R = \frac{abc}{2\Delta}, \qquad (6.4)$$

$$a^2 = b^2 + c^2 - 2bc \cos A, \qquad (6.2)$$

$$\Delta = \sqrt{\{s(s - a)(s - b)(s - c)\}}, \qquad (6.6)$$

$$\tan \tfrac{1}{2}A = \sqrt{\left\{\frac{(s - b)(s - c)}{s(s - a)}\right\}}, \qquad (6.13)$$

$$\tan \tfrac{1}{2}(B - C) = \left(\frac{b - c}{b + c}\right) \cot \tfrac{1}{2}A, \qquad (6.16)$$

$$r = \Delta/s = (s-a)\tan\tfrac{1}{2}A = (s-b)\tan\tfrac{1}{2}B = (s-c)\tan\tfrac{1}{2}C, \quad (6.8)$$

$$r_1 = \frac{\Delta}{s-a} = s\tan\tfrac{1}{2}A, \quad (6.10, 6.11)$$

and similar relations for r_2, r_3.

6.10. The numerical solution of triangles

When three elements of a triangle, one at least of which is a side, are given the other three can be found. The sine and cosine formulae (6.4), (6.2) are all that are required to effect the calculation but some improvement in the numerical work involved can be made in certain cases by using the half angle formula (6.13) or the tangent formula (6.16). However, we shall start by giving examples in which only the sine and cosine formulae are employed.

It is advisable to set out the computational work in a systematic manner and to employ checks whenever possible. Some suggested lay-outs arc shown in the examples given in the following paragraphs.

6.11. Examples of the use of the sine formula

(i) *One side and two angles given*

In this case the remaining angle can be found immediately from the fact that the sum of the angles of a triangle is $180°$. Suppose the given side is c, then since the angle C is known, the diameter $2R$ of the circumcircle can be found from

$$2R = \frac{c}{\sin C}.$$

The remaining sides a and b can then be calculated from

$$a = 2R\sin A, \quad b = 2R\sin B,$$

the angles A and B being known. The adaptation of the method to the case where either of the sides a or b is given instead of c should cause no difficulty.

Example 7. *Solve the triangle in which*

$$c = 26·83 \ m, \quad A = 80° \ 30', \quad B = 40° \ 12'.$$

	No.	log.
	26·83	1·4286
	sin 50° 18′	$\bar{1}$·9344
	2R	1·4942
	sin 80° 30′	$\bar{1}$·9940
	a = 30·77	1·4882
	2R	1·4942
	sin 40° 12′	$\bar{1}$·8099
	b = 20·14	1·3041

$C = 180° - (A + B)$
$\quad = 180° - (80° \ 30' + 40° \ 12')$
$\quad = 180° - 120° \ 42' = 59° \ 18'$.

$2R = c/\sin C = 26·83/\sin 59° \ 18'$.
$a = 2R\sin A = 2R\sin 80° \ 30'$.
$b = 2R\sin B = 2R\sin 40° \ 12'$.

The required solution is therefore
$\qquad C = 59° \ 18',$
$\qquad a = 3·77 \ m,$
$\qquad b = 20·14 \ m.$

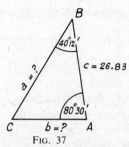

Fig. 37

(ii) *Two sides and the non-included angle given (the ambiguous case)*

To fix ideas, we shall take b, c as the given sides and B as the given angle. The angle C can be found from the sine formula in the form

$$\sin C = \frac{c \sin B}{b}. \qquad (6.17)$$

Various possibilities may arise.

(a) The side b may be sufficiently small for $b < c \sin B$. This would require $\sin C$ to be greater than unity and no triangle will exist with the given values of b, c and B. This is illustrated in Fig. 38.

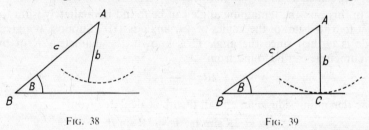

Fig. 38   Fig. 39

(b) The given values b, c and B may be such that $b = c \sin B$. In this case, $\sin C = 1$ and $C = 90°$. Here the triangle is right-angled at C (Fig. 39). Since, when one angle of a triangle is a right angle the other two are necessarily acute, this case can only occur when $B < 90°$.

(c) When b, c and B are such that $b > c \sin B$, $\sin C < 1$ and there will be two values of C (less than $180°$) which can satisfy equation (6.17). One of these values of C, say C_1, will be acute and the other, C_2, will be obtuse. We now have to enquire if both these values give possible solutions.

If B is obtuse, then it is the only obtuse angle of the triangle and the angle C_2 must be excluded as a possible solution.

If B is acute, values of C greater than $90°$ are not immediately excluded. If, however, $b > c$, such values are excluded on the grounds that the angle C would then be greater than the angle B and the greater angle

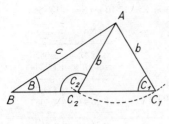

would be opposite the lesser side. For the case $B < 90°$, $b < c$, both values C_1, C_2 of C are possible (Fig. 40). This case gives rise to two possible triangles ABC_1, ABC_2 and is often called the *ambiguous* case.

To sum up, when the sides b, c and the angle B of a triangle are given, we have to consider the following cases.

(a) $b < c \sin B$. There is no solution.

(b) $b = c \sin B$. There is one solution and the triangle is right-angled at C.

(c) $b > c \sin B$. If $B \geqslant 90°$, there is one solution and C is the acute angle derived from equation (6.17). If $B < 90°$ and $b > c$, there is one solution and C is the acute angle derived from (6.17). If $B < 90°$ and $b < c$, there are two possible triangles ABC_1, ABC_2, the angles C_1, C_2 being respectively the acute and obtuse angles satisfying equation (6.17).

Once C has been found, the remaining angle and side are then found as in Example 7. In the ambiguous case there will be two values a_1, a_2 of the side a to be found corresponding to the two values A_1, A_2 deduced for the angle A. Some numerical examples follow.

Example 8. *Is there a triangle in which $b = 5$, $c = 7$ and $B = 48° 35'$? If so, solve the triangle.*

The value of C is given by

$$\sin C = \frac{c \sin B}{b} = \frac{7 \sin 48° 35'}{5}$$

This leads to a value of $\log \sin C$ of $0{\cdot}0212$ and hence $\sin C = 1{\cdot}05$. Since this is greater than unity, there is no possible triangle with the given sides and angle.

No.	log.
7	0·8451
sin 48° 35'	1̄·8751
5	0·7202
	0·6990
sin C	0·0212

Example 9. *Solve the triangle in which $b = 5{\cdot}6$, $c = 7{\cdot}0$ and $B = 53° 8'$.*

Here

$$\sin C = \frac{7 \sin 53° 8'}{5{\cdot}6}.$$

This gives $\log \sin C = 0$, $\sin C = 1$ and $C = 90°$.
The remaining angle A is given by

$$A = 180° - (B + C)$$
$$= 180° - (53° 8' + 90°)$$
$$= 36° 52'.$$

For the side a, we have

$$a = \frac{c}{\sin C} \cdot \sin A = \frac{7}{\sin 90°} \cdot \sin 36° 52'$$

$$= 7 \sin 36° 52' = 4{\cdot}2.$$

The solution is therefore $A = 36° 52'$, $C = 90°$, $a = 4{\cdot}2$.

No.	log.
7	0·8451
sin 53° 8'	1̄·9031
5·6	0·7482
	0·7482
sin C	0·0000
7	0·8451
sin 36° 52'	1̄·7781
a = 4·200	0·6232

Example 10. *Solve the triangle in which* $b = 24.93$ *m,* $c = 12.10$ *m,* $B = 122° 51'$.

Here

$$\sin C = \frac{12.1 \sin 122° 51'}{24.93} = \frac{12.1 \sin 57° 9'}{24.93},$$

since $\sin 122° 51' = \sin(180° - 122° 51')$.

Hence $\log \sin C = \bar{1}.6105$, giving $C = 24° 4'$, an obtuse value for C being impossible as B is greater than 90°.

$$\begin{aligned} A &= 180° - (B + C) \\ &= 180° - (122° 51' + 24° 4') \\ &= 33° 5'. \end{aligned}$$

$$\begin{aligned} a &= \frac{c}{\sin C} . \sin A \\ &= \frac{12.10}{\sin 24° 4'} . \sin 33° 5' = 16.20. \end{aligned}$$

No.	log
12·10	1·0828
sin 57° 9'	$\bar{1}$·9244
	1·0072
24·93	1·3967
sin C	$\bar{1}$·6105
12·10	1·0828
sin 33° 5'	$\bar{1}$·7371
	0·8199
sin 24° 4'	$\bar{1}$·6105
a = 16·20	1·2094

The solution is therefore

$$A = 33° 5', \quad C = 24° 4', \quad a = 16.20 \text{ m}.$$

Example 11. *Solve completely the triangle in which* $b = 2.718, c = 3.142, B = 54° 18'$.
(L.U.)

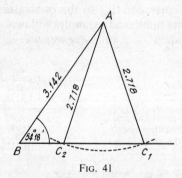

Fig. 41

No.	log.
3·142	0·4972
sin 54° 18'	$\bar{1}$·9096
	0·4068
2·718	0·4343
sin C	$\bar{1}$·9725
3·142	0·4972
sin 55° 51'	$\bar{1}$·9178
	0·4150
sin 69° 51'	$\bar{1}$·9725
$a_1 = 2.770$	0·4425
3·142	0·4972
sin 15° 33'	$\bar{1}$·4283
	$\bar{1}$·9255
sin 110° 9'	$\bar{1}$·9725
$a_2 = 0.8974$	$\bar{1}$·9530

Here

$$\sin C = \frac{3.142 \sin 54° 18'}{2.718}.$$

Hence $\log \sin C = \bar{1}.9725$ giving $C_1 = 69° 51'$ and $C_2 = 180° - 69° 51' = 110° 9'$, since this is the ambiguous case in which $B < 90°$ and $b < c$. If we denote the angles BAC_1, BAC_2 respectively by A_1 and A_2 we have

(i) $A_1 = 180° - (54° 18' + 69° 51') = 180° - 124° 9' = 55° 51'$.

By the sine rule, if $BC_1 = a_1$,

$$a_1 = \frac{c}{\sin C_1} \sin A_1 = \frac{3.142 \sin 55° 51'}{\sin 69° 51'} = 2.770,$$

and

(ii) $A_2 = 180° - (54° 18' + 110° 9') = 180° - 164° 27' = 15° 33'$, so that if $BC_2 = a_2$,

$$a_2 = \frac{c}{\sin C_2} \sin A_2 = \frac{3 \cdot 142 \sin 15° 33'}{\sin 110° 9'} = 0 \cdot 897.$$

The required solution is therefore,

$$A = 55° 51', \quad C = 69° 51', \quad a = 2 \cdot 770,$$

or, $\qquad A = 15° 33', \quad C = 110° 9', \quad a = 0 \cdot 897.$

6.12. Examples of the use of the cosine formula

(i) *Two sides and the included angle given*

Suppose for example that the two sides b, c and the included angle A are given. The side a can be calculated from the cosine formula

$$a^2 = b^2 + c^2 - 2bc \cos A,$$

tables of squares and square roots being useful in the arithmetical work. The remaining angles B and C can then be found from the sine rule arranged as

$$\sin B = \frac{b \sin A}{a} \quad \text{and} \quad \sin C = \frac{c \sin A}{a}.$$

Since all the sides are known, any question of the values of the angles B and C is settled by taking angles which are in the same order of magnitude as the sides opposite them. Alternatively, once one of the angles B or C has been found from the sine formula, since A is given, the other can be found from the fact that the sum of the angles is $180°$.

Example 12. *Solve the triangle in which $b = 10 \cdot 67\,m$, $c = 21 \cdot 7\,m$, $A = 44° 46'$.*

$a^2 = b^2 + c^2 - 2bc \cos A$
$\quad = (10 \cdot 67)^2 + (21 \cdot 7)^2 - 2(10 \cdot 67)(21 \cdot 7) \cos 44° 46'$
$\quad = 584 \cdot 8 - 328 \cdot 8 = 256 \cdot 0.$

$a = 16 \cdot 0\,m.$

$\sin B = \dfrac{b \sin A}{a} = \dfrac{10 \cdot 67 \sin 44° 46'}{16 \cdot 0}$

Hence, $\log \sin B = \bar{1} \cdot 6718$. Since the angle B, being opposite the smallest side, is necessarily acute, this gives $B = 28° 1'$.
$C = 180° - (A + B)$
$\quad = 180° - (44° 46' + 28° 1')$
$\quad = 180° - 72° 47' = 107° 13'.$

The required solution is therefore,
$B = 28° 1'$, $C = 107° 13'$, $a = 16 \cdot 0\,m.$

	No.	log.
$b^2 = (10 \cdot 67)^2$	$113 \cdot 9$	
$c^2 = (21 \cdot 7)^2$	$470 \cdot 9$	
$b^2 + c^2$	$548 \cdot 8$	
$2b = 2 \times 10 \cdot 67$	$21 \cdot 34$	$1 \cdot 3292$
c	$21 \cdot 7$	$1 \cdot 3365$
$\cos A$	$\cos 44° 46'$	$\bar{1} \cdot 8512$
$2bc \cos A$	$328 \cdot 8$	$2 \cdot 5169$
b	$10 \cdot 67$	$1 \cdot 0282$
$\sin A$	$\sin 44° 46'$	$\bar{1} \cdot 8477$
		$0 \cdot 8759$
a	$16 \cdot 0$	$1 \cdot 2041$
$\sin B$		$\bar{1} \cdot 6718$

(ii) *Three sides given*

When all three sides are given, the angles can be found from the cosine formula arranged in the form

$$\cos A = \frac{b^2 + c^2 - a^2}{2bc},$$

with corresponding formulae for $\cos B$ and $\cos C$. Alternatively, one of the angles can be found in this way and the other two found from the sine formula.

Example 13. *Solve the triangle in which* $a = 16$m, $b = 10\cdot67$ m, $c = 21\cdot7$ m.

$$\cos A = \frac{b^2 + c^2 - a^2}{2bc}$$

$$= \frac{(10\cdot67)^2 + (21\cdot7)^2 - (16\cdot0)^2}{2\,(10\cdot67)\,(21\cdot7)}$$

$$A = 44° \; 46'.$$

B and C then follow as in Example 12.

	No.	log.
$b^2 = (10\cdot67)^2$	113·9	
$c^2 = (21\cdot7)^2$	470·9	
$b^2 + c^2$	584·8	
$a^2 = (16\cdot0)^2$	256·0	
$b^2 + c^2 - a^2$	328·8	
$2b = 2 \times 10\cdot67$	21·34	1·3292
c	21·7	1·3365
$2bc$		2·6657
$b^2 + c^2 - a^2$	328·8	2·5169
$2bc$		2·6657
$\cos A$		$\bar{1}\cdot8512$

EXERCISES 6 (d)

Solve the triangles in which:—

1. $c = 15\cdot6$ m, $B = 34° \; 20'$, $C = 62° \; 9'$.

2. $a = 0\cdot5$ m, $b = 0\cdot7$ m $A = 62°$.

3. $a = 17\cdot00$ m, $b = 21\cdot42$ m, $B = 51° \; 34'$.

4. $b = 107\cdot2$ m, $c = 76\cdot69$ m, $B = 102° \; 25'$.

5. $b = 15\cdot6$ m, $c = 12\cdot3$ m, $C = 34° \; 20'$.

6. $a = 7\cdot00$ m, $b = 3\cdot59$ m, $C = 47°$.

7. $a = 0\cdot17$ m, $b = 0\cdot11$ m, $c = 0\cdot10$ m.

8. Find the third side and the radius of the circumcircle for the triangle in which $a = 6\cdot324, b = 8\cdot222, C = 64° \; 32'$.

6.13. Example of the use of the tangent rule

As shown in the last paragraph, when two sides and the included angle are given a triangle can be solved by using the cosine formula to find the remaining side and then the sine formula for the remaining angles. The cosine formula is not, however, well suited to work with logarithms. Unless the given sides are two-figure numbers and therefore easy to square, it is best to use the tangent formula to deal with this case.

Suppose the sides b, c and the angle A are given. Then the tangent formula (6.16)

$$\tan \tfrac{1}{2}(B - C) = \left(\frac{b - c}{b + c}\right)\cot \tfrac{1}{2}A,$$

enables $\tfrac{1}{2}(B - C)$ to be found. Since $\tfrac{1}{2}(B + C) = 90° - \tfrac{1}{2}A$, we can then find B and C by addition and subtraction. The remaining side a is then found from the sine formula.

Example 14. *Solve the triangle in which $b = 10\cdot67\ m$, $c = 21\cdot7\ m$, $A = 44°\ 46'$.*

Here $b < c$ and it is best to rewrite the tangent formula as

$$\tan \frac{1}{2}(C - B) = \left(\frac{c - b}{c + b}\right)\cot \frac{1}{2}A.$$

Hence $$\tan \frac{1}{2}(C - B) = \frac{21\cdot7 - 10\cdot67}{21\cdot7 + 10\cdot67}\cot 22°\ 23'$$

$$= \frac{11\cdot03}{32\cdot37}\cot 22°\ 23'.$$

Hence

$$\tfrac{1}{2}(C - B) = 39°\ 37'.$$

Also,

$$\tfrac{1}{2}(C + B) = 90° - \tfrac{1}{2}A$$
$$= 90° - 22°\ 23' = 67°\ 37'.$$

By addition and subtraction,

$$C = 107°\ 14', \quad B = 28°.$$

$$a = \frac{c}{\sin C}.\sin A = \frac{21\cdot7 \sin 44°\ 46'}{\sin 107°\ 14'}$$

$$= \frac{21\cdot7 \sin 44°\ 46'}{\sin 72°\ 46'} = 16\cdot0.$$

Hence the required solution is

$B = 28°, \quad C = 107°\ 14', \quad a = 16\cdot0$ m.

No.	log.
11·03	1·0426
cot 22° 23′	0·3853
	1·4279
32·37	1·5101
tan ½(C − B)	1̄·9178
21·7	1·3365
sin 44° 46′	1̄·8477
	1·1842
sin 72° 46′	1̄·9801
a	1·2041

6.14. Alternative methods of solution of triangle with three sides given

The cosine formula, rearranged so that the angles are expressed in terms of the sides, is again not well suited to work with logarithms. If all three angles are to be found it is probably best to use formula (6.13),

$$\tan \tfrac{1}{2}A = \sqrt{\left\{\frac{(s - b)(s - c)}{s(s - a)}\right\}},$$

and the two similar formulae for B and C. To save repetition in the logarithmic work, this can be written

$$\tan \tfrac{1}{2}A = \frac{1}{s-a} \sqrt{\left\{ \frac{(s-a)(s-b)(s-c)}{s} \right\}}, \qquad (6.18)$$

with corresponding expressions for B and C. The logarithm of the expression under the square root sign has only to be worked out once and the log tangents of the half angles then follow by subtracting $\log(s-a)$, $\log(s-b)$ and $\log(s-c)$.

Alternatively, the angles could be calculated from (6.14) or (6.15),

$$\cos\tfrac{1}{2}A = \sqrt{\left\{\frac{s(s-a)}{bc}\right\}}, \quad \sin\tfrac{1}{2}A = \sqrt{\left\{\frac{(s-b)(s-c)}{bc}\right\}},$$

or from

$$\sin A = 2\Delta/bc$$
$$= \frac{2}{bc}\sqrt{\{s(s-a)(s-b)(s-c)\}},$$

a formula which is easily deduced from (6.4) and (6.6).

Example 15. *Solve the triangle in which $a = 16$, $b = 10\cdot67$, $c = 21\cdot7$.*

We first find s from the formula

$$2s = a + b + c.$$

Then form $(s-a)$, $(s-b)$, $(s-c)$.
A check is provided from $(s-a) + (s-b) + (s-c) = s$.
We then find
$\log\{(s-a)(s-b)(s-c)/s\}$
by adding the logarithms of $(s-a)$, $(s-b)$, $(s-c)$ and subtracting that of s. The logarithm of the square root of this quantity follows by division by two, and the angles are found from (6.18) and two similar formulae. From the working on the right we find

	No.		log.
a	16		
b	10·67		
c	21·7		
$2s =$ sum	48·37		
s	24·19		
$s - a$	8·19		0·9133
$s - b$	13·52		1·1310
$s - c$	2·49		0·3962
$(s-a)(s-b)(s-c)$			2·4405
s	24·19		1·3836
$(s-a)(s-b)(s-c)/s$			1·0569
$\sqrt{\{(s-a)(s-b)(s-c)/s\}}$			0·5285
$s - a$			0·9133
$\tan\tfrac{1}{2}A$			$\bar{1}$·6152
$\sqrt{\{(s-a)(s-b)(s-c)/s\}}$			0·5285
$s - b$			1·1310
$\tan\tfrac{1}{2}B$			$\bar{1}$·3975
$\sqrt{\{(s-a)(s-b)(s-c)/s\}}$			0·5285
$s - c$			0·3962
$\tan\tfrac{1}{2}C$			0·1323

$\tfrac{1}{2}A = 22° 24'$, $\quad A = 44° 48'$,
$\tfrac{1}{2}B = 14° 1'$, $\quad B = 28° 2'$,
$\tfrac{1}{2}C = 53° 36'$, $\quad C = 107° 12'$.

There is a final check that the sum of the angles should be 180°. As there is a possible error of at least half a minute in each angle due to the use of four-figure tables, the slight differences between the angles found here and those in Examples 12 and 13 and the slight difference of $A + B + C$ from 180° is not surprising.

EXERCISES 6 (e)

1. Calculate the remaining angles of a triangle in which two sides are 13·45 m, 54·31 m and the included angle is 67° 24′.

2. Find the angles of a triangle whose sides are 10·4, 12·8 and 17·6 m.

3. In a triangle $a = 4·832$ m, $b = 2·186$ m, $A - B = 34°$ 16′. Solve the triangle.

4. In a triangle ABC, $b = 6·27$, $c = 4·32$, $A = 51°$. Calculate the angles B and C and the length of the internal bisector BD of the angle ABC.   (L.U.)

5. The sides a, b, c of a triangle are respectively $(k^2 + k + 1)$, $(2k + 1)$ and $(k^2 - 1)$ where $k > 1$. Show that the angle A is 120°.

6. Find the area of a triangle having sides of length 322·2 m, 644·7 m and 432·1 m.   (Q.E.)

6.15. Heights and distances

In the work of the surveyor and in navigation, elementary trigonometry finds an important application. It is often possible by measuring certain distances and angles to calculate other distances and angles which cannot be measured directly. Such calculations are usually simply practical applications of the formulae relating to the sides and angles of a triangle. We give below a few typical examples explaining the few technical terms used in this type of work as they occur.

Example 16. *The angle of elevation of the top of a vertical tower from a point A is* α. *From a point B, in a direct line between A and the foot of the tower and at distance d from A, the angle of elevation to the top of the tower is* β. *Find the height of the tower.*

The *angle of elevation* of the top T of the tower from A is the angle the line AT makes with the horizontal through A and the foot of the tower. If the level of T had been below that of A we should speak of this angle as the *angle of depression.*

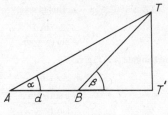

FIG. 42

In the diagram, TT' is the tower, T being the top and T' the foot. ABT' is horizontal, TT' vertical, the angle $AT'T$ is a right angle, the distance AB is d and the angles of elevation α, β of T from A and B are as shown.
Applying the sine formula to the triangle ABT, since the external angle $T'BT$

is equal to the sum of the angles TAB, BTA and therefore the angle BTA is $\beta - \alpha$, we have

$$\frac{TB}{\sin \alpha} = \frac{d}{\sin (\beta - \alpha)},$$

giving

$$TB = \frac{d \sin \alpha}{\sin (\beta - \alpha)}.$$

The right-angled triangle TBT' now gives for the height of the tower

$$TT' = TB \sin \beta$$

$$= \frac{d \sin \alpha \sin \beta}{\sin (\beta - \alpha)}.$$

Thus, by measuring two angles and one distance, this formula will enable the height of an inaccessible object to be found.

Example 17. *The angle of elevation of the top T of a vertical tower from a point A is α. B is a second point on the same level as A and the foot of T' of the tower. B is not in the straight line joining A and the foot of the tower and the distance AB is c. The angles TAB, TBA are measured and found to be γ and δ respectively. Derive a formula giving the height of the tower in terms of α, γ, δ and c.*

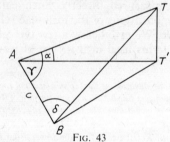

Fig. 43

From the triangle ABT,

$$\frac{AT}{\sin \delta} = \frac{c}{\sin A\hat{T}B}.$$

Since the angle ATB is $180° - \gamma - \delta$, this gives

$$AT = \frac{c \sin \delta}{\sin (\gamma + \delta)}.$$

The right-angled triangle $AT'T$, gives

$$TT' = AT \sin \alpha,$$

$$= \frac{c \sin \alpha \sin \delta}{\sin (\gamma + \delta)},$$

when the expression for AT is substituted.

Example 18. *A and B are two posts on one edge of a straight canal and C is a post on the opposite edge; $AB = 110$ m, the angle $CAB = 43° 20'$ and the angle $CBA = 65° 52'$. Find the width of the canal to the nearest metre.*

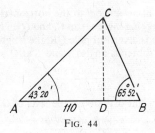

Fig. 44

The angle $ACB = 180° - 43° 20' - 65° 52'$
$= 180° - 109° 12' = 70° 48'$.

From the triangle ABC

$$AC = \frac{110 \sin 65° 52'}{\sin 70° 48'}.$$

If CD is perpendicular to AB, the right-angled triangle ADC gives

$$CD = AC \sin 43° 20'$$
$$= \frac{110 \sin 65° 52' \sin 43° 20'}{\sin 70° 48'}$$
$$= 73 \text{ m}.$$

No.	log.
110	2·0414
sin 65° 52'	1̄·9603
sin 43° 20'	1̄·8365
	1̄·8382
sin 70° 48'	1̄·9751
$CD = 72·96$	1̄·8631

EXERCISES 6 (f)

1. An aeroplane is observed at the same instant from three stations A, B, C in a horizontal straight line but not in a vertical plane through the aeroplane. If $AB = BC = c$ and the angles of elevation from A, B, C are respectively α, β, γ prove that the height of the aeroplane is

$$\frac{c\sqrt{2}}{(\cot^2 \alpha + \cot^2 \gamma - 2 \cot^2 \beta)^{1/2}}.$$ (L.U.)

2. An aeroplane is observed simultaneously from two points A and B, at the same level, A being at a distance c due north of B. From A the bearing of the aeroplane is θ east of south at an elevation α and from B the bearing is ϕ east of north. Show that the aeroplane is at a height $\dfrac{c \tan \alpha \sin \phi}{\sin (\theta + \phi)}$ and find its elevation from B.   (L.U.)

3. A vertical rectangular hoarding 2·4 m long and 3 m high, is held in position by four equal stay ropes, two attached to each top corner, the bottom end of each rope being fixed to a point in the ground. If these four points are the vertices of a square of side 3·6 m, calculate the length of each stay and its inclination to the horizontal.   (L.U.)

4. The base A of a vertical tower AB, of height h, and the base X of a vertical flagpole XY, of height H, are on level ground, and $H > h$. The angles of elevation of Y from A and B are α and β respectively. Prove that

$$H = h \sin \alpha \cos \beta \operatorname{cosec} (\alpha - \beta).$$

Taking $h = 40$ m, $\alpha = 30°$ and $\beta = 10°$, find the angle of elevation of B from X, giving your answer to the nearest tenth of a degree.   (O.C.)

5. A straight path rises at an angle θ to the horizontal; O, P and Q are three points on the path, P being higher than O, and Q higher than P; the distance OP is x ft. At Q there is a vertical pole QR, the height of R above Q being h ft. Prove that, if QR subtends angles α and β at O and P respectively, then

$$h = \frac{x \sin \alpha \sin \beta}{\sin (\beta - \alpha) \cos \theta}.$$

Prove also that the height of R above the level of O is

$$\frac{x \sin \beta \sin (\alpha + \theta)}{\sin (\beta - \alpha)}. \qquad \text{(O.C.)}$$

6. A tripod consists of three rods AB, AC and AD which are 100 m, 100 m and 125 m long respectively. The ends B, C and D of the three legs stand on a horizontal plane; D is equidistant from B and C and 50 m from the line BC. Find the height of the apex A above the ground, given that the angle BAC is 25°. (Q.E.)

7. A, B, C are three towns: B is 10 kilometres from A in a direction 47° E. of N.; C is 17 kilometres from B in a direction 20° N. of W. Calculate the distance and direction of A from C. (O.C.)

8. A horizontal tunnel AB is bored through a ridge in a direction perpendicular to the line of the ridge, and a path goes from A to B over the ridge. Show that if l is the length of the tunnel, and α, β are the inclination of the two portions of the path to the horizontal, the height of the ridge above the tunnel is

$$\frac{l \sin \alpha \sin \beta}{\sin (\alpha + \beta)}.$$

What is the length of the path if $l = 1000$ m, $\alpha = 10°$ and $\beta = 7\frac{1}{2}°$? (O.C.)

EXERCISES 6 (g)

1. Assuming that, in any triangle ABC,

$$\frac{\sin A}{a} = \frac{\sin B}{b} = \frac{\sin C}{c},$$

prove that $\dfrac{a + b - c}{a + b + c} = \tan \frac{1}{2}A \tan \frac{1}{2}B.$

Calculate the value of c for the triangle in which

$$a + b = 18{\cdot}5 \text{ m}, \quad A = 72° \ 14', \quad B = 45° \ 42'. \qquad \text{(O.C.)}$$

2. If in any triangle ABC

$$\sin \theta = \frac{2\sqrt{(bc)}}{b + c} \cos \frac{1}{2}A,$$

prove that $(b + c) \cos \theta = a.$

For the case $b = 123$, $c = 41{\cdot}2$, $A = 40° \ 50'$, find the value of $\sin \theta$ and hence the value of a. (L.U.)

3. In a triangle ABC, perpendiculars from the vertices to the opposite sides are AD, BE, CF. Find the angles of the triangle DEF in terms of those of

the triangle ABC and prove that the perimeter of the triangle DEF is
$$a \cos A + b \cos B + c \cos C. \qquad \text{(L.U.)}$$

4. Prove that in a triangle ABC,
$$a^2 + b^2 + c^2 = 2(bc \cos A + ca \cos B + ab \cos C).$$

5. In a triangle ABC, X is the mid-point of the side BC. Prove that
$$\sin A\hat{X}B = \frac{2b \sin C}{\sqrt{(2b^2 + 2c^2 - a^2)}},$$
and that
$$\sin X\hat{A}C = \frac{a \sin C}{\sqrt{(2b^2 + 2c^2 - a^2)}}.$$

6. The area of a triangle is 336 m², the sum of the three sides is 84 m and one side is 28 m. Calculate the lengths of the other two sides.   (L.U.)

7. The angle A of a triangle ABC is $60°$ and the area of the triangle is equal to that of an equilateral triangle with sides of length x. Show that $b^2 + c^2 - a^2 = x^2$.

8. In the triangle ABC the incentre and circumcentre are at the same distance from the side BC. Prove that
$$4 \sin \tfrac{1}{2}A \sin \tfrac{1}{2}B \sin \tfrac{1}{2}C = \cos A,$$
and deduce that $\cos B + \cos C = 1$.   (L.U.)

9. The radii of the incircle and circumcircle of the triangle ABC are r, R respectively. Prove that the area of the triangle ABC is
$$r^2 \cot \tfrac{1}{2}A \cot \tfrac{1}{2}B \cot \tfrac{1}{2}C.$$
If the tangents at A, B, C to the circumcircle meet in D, E, F, prove that the area of the triangle DEF is
$$R^2 \tan A \tan B \tan C. \qquad \text{(L.U.)}$$

10. Prove that the area of a triangle ABC is $2R^2 \sin A \sin B \sin C$ where R is the radius of the circumcircle.
    If r is the radius of the circle inscribed in the triangle ABC, and if DEF is the triangle formed by joining the points of contact of the inscribed circle with the sides, prove that
    (i) the area of the triangle DEF is $2r^2 \cos \tfrac{1}{2}A \cos \tfrac{1}{2}B \cos \tfrac{1}{2}C$;
    (ii) the radius of the circle inscribed in the triangle DEF is
$$\frac{2r \cos \tfrac{1}{2}A \cos \tfrac{1}{2}B \cos \tfrac{1}{2}C}{\cos \tfrac{1}{2}A + \cos \tfrac{1}{2}B + \cos \tfrac{1}{2}C}. \qquad \text{(L.U.)}$$

11. O is the centre and R the radius of the circle circumscribing the triangle ABC. AO, BO and CO meet the opposite sides in L, M, N respectively. Show that
    (i) $AL = \dfrac{b \sin C}{\cos (B - C)}$;
    (ii) $\dfrac{1}{AL} + \dfrac{1}{BM} + \dfrac{1}{CN} = \dfrac{2}{R}.$ (L.U.)

12. In any triangle ABC prove that
$$\frac{b-c}{a} = \frac{\sin\frac{1}{2}(B-C)}{\cos\frac{1}{2}A}.$$
The inscribed circle of the triangle ABC touches BC at E and one of the escribed circles touches BC at F. If $b > c$, prove that $BE = s - b$ and $EF = b - c$, where $2s = a + b + c$.

If $A = 36°\ 42'$, $a = 4·32$ and $EF = 1·67$, calculate the lengths of b and c.
(L.U.)

13. In the triangle ABC, prove that $c\cos\frac{1}{2}(A - B) = (a + b)\sin\frac{1}{2}C$.

If the sum of the lengths of two sides of a triangle is 21, the length of the third side is 15 and the angle opposite the third side is 52°, solve the triangle completely.
(L.U.)

14. If $2s = a+b+c$, show that the area of the triangle ABC is given by $s^2\tan\frac{1}{2}A\tan\frac{1}{2}B\tan\frac{1}{2}C$.

Find the angles and area of a triangle whose sides are 4·13, 3·28 and 1·67 metres.

15. In a triangle ABC, prove that
$$\tan B\cot C = \frac{a^2 + b^2 - c^2}{a^2 - b^2 + c^2}.$$

16. Find the difference between the areas of the two possible triangles ABC in which $A = 29°\ 13'$, $a = 51·47$ m, $b = 102·3$ m.
(Q.E.)

17. In a triangle ABC, $a = 18·9$ m, $b = 12·2$ m, $A - B = 37°$. Find the values of A and c.
(L.U.)

18. The sum of two sides of a triangle is 0·337 m and the included angle is $56°\ 24'$. Calculate the remaining angles when the third side is 0·163 m.
(L.U.)

19. Find all the sides and angles of a triangle ABC of which the area is 1008 m^2 and in which $a = 65$ m, $b + c = 97$ m.

20. The sides AB, AC of a triangle ABC are equal to one another. The perpendicular from A on to BC is 4 m and that from B on to AC is 2 m. Find the angle ABC and the area of the triangle.

21. A, B, C and D are four landmarks on the same horizontal level. B is 4 km N. 31° E. from A; C is 6 km S. 10° 15′ E. from B; D is 3 km E. from C. Calculate the distance and bearing of D from A.
(L.U.)

22. From a point A, due south-east of a tower, the elevation of the top of the tower is observed to be 32°. From A the observer walks in a straight line 100° west of north for 320 m and finds that he is then due south-west of the tower. Assuming that the foot of the tower and the line of the observer's walk are in the same horizontal plane, find the height of the tower.

23. From a point A a beacon on a mountain is observed on a bearing 64° 6′ east of north and at an angle of elevation of 10° 12′. From another point, at a distance of exactly 5 kilometres from A and at the same level, the bearing of the beacon is 30° 45′ east of north and its elevation is 6° 18′. Find the height of the beacon above A.
(Q.E.)

24. Two towers A and B on a level plain subtend an angle of $90°$ at an observer's eye. The observer walks directly towards B a distance of 630 m and then finds the angle subtended to be $143°\ 24'$. Find the distance of the tower A from each position of the observer.

25. At a point in the horizontal plane through the base of a circular tower, the elevation of the top of the tower is β and the elevation of the highest point of a flagstaff which stands on the top of the tower in the centre is γ. At a point a ft nearer the tower, the highest point of the flagstaff is just visible above the edge of the tower at an elevation α. Prove that the height of the flagstaff is

$$a \sin^2 \alpha \sin (\gamma - \beta) \operatorname{cosec} (\alpha - \beta) \operatorname{cosec} (\alpha - \gamma).$$

CHAPTER 7

INTRODUCTION TO THE IDEAS OF THE DIFFERENTIAL CALCULUS

7.1. Functions and functional notation

When two variable quantities x and y are so related that the value of one quantity y depends on the value of the other x, then y is said to be a *function* of x. The relation containing the two variable quantities may be a simple formula such as $y = 4x^2 + x + 2$ or $y = \sin x$, or the relation may be expressed by means of a graph. Sometimes the graph relating two variable quantities is available but it may be impossible (or very difficult) to express the relationship by a formula. For example, a recording barometer gives a record showing the atmospheric pressure p plotted against the time t—the record displays p as a function of t but it is not generally possible to express p in terms of t by a formula.

A means of expressing that y is a function of x when the formula connecting the two variable quantities is not known or when we are dealing with a general rather than a particular function of x is to write $y = f(x)$, $y = \phi(x)$ or $y = F(x)$. This notation does not mean that y is f multiplied by x but is simply an abbreviation for the words "function of x". The different letters f, ϕ, F, ... are used to denote different functions.

If $y = f(x)$ and we wish to specify y for a given value of x, say $x = 3$, we write the result as $f(3)$. For example, if $f(x) = 4x^2 + x + 2$, $f(3) = 4 \times 3^2 + 3 + 2 = 41$ and this is the value of y when $x = 3$. Similarly, if $\phi(x) = \sin x$, $\phi(\pi/4) = \sin \pi/4 = 1/\sqrt{2}$.

Functions such as we have considered above are termed *explicit* functions. The variable quantity x is called the *independent* variable. The second variable y, whose value depends on that given to x, is referred to as the *dependent* variable.

Sometimes the relation between two variable quantities x and y is given in a form such as

$$x^2 + y^2 = 2x, \quad \text{or} \quad x + y + \cos y = 4;$$

these are called *implicit* functions. In the first example given above we could solve for y and obtain

$$y = \sqrt{(2x - x^2)},$$

and in this form, y is an explicit function of x. In the second example it is not possible to find y explicitly as a function of x. We shall be concerned in this book chiefly with explicit functions.

A function like $y = x^2$ is a *single-valued* function of x—since, to a given value of x, say $x = 4$, there corresponds one and only one value of y ($y = 4^2 = 16$). On the other hand, a function like $y = \sqrt{x}$ is a two-valued function—for a given value of x, say $x = 4$, there are two values $y = \pm 2$ which satisfy the given relation. In general, if to a given value of x there corresponds more than one value of y, y is said to be a *multi-* or *many-valued* function of x.

7.2. The gradient of a curve

It is often helpful to depict the variation of a function by means of its graph. Suppose, to fix ideas, we consider the simple explicit function $y = x^2$. The graph is easily drawn and is shown in Fig. 45.

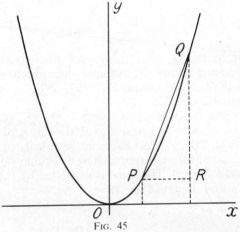

FIG. 45

We notice that as x increases from zero through positive values, y increases. The reverse is true when x increases to zero from negative values. We could describe the changes in y as x increases by saying that y decreases so long as $x < 0$ and that y increases when $x > 0$.

If x_1, y_1 are the abscissa and ordinate of a point P and x_2, y_2 those of a point Q, the change in the value of y as x changes from x_1 to x_2 is $y_2 - y_1$. The average rate of change of y as x changes from x_1 to x_2 is defined as

$$\frac{y_2 - y_1}{x_2 - x_1},$$

and from the figure this is seen to be the tangent of the angle QPR, PR being parallel to the x-axis. This quantity is referred to as the *slope* of the chord PQ. It may help to make matters clearer to take variables s and t instead of x and y, s being the distance moved by a body in time t. Instead of the graph $y = x^2$, we should have the "space-time" graph $s = t^2$. The average rate of change of distance as t changes from t_1 to t_2 would be $(s_2 - s_1)/(t_2 - t_1)$ and this is simply the average speed of the body in the time interval $(t_2 - t_1)$.

Let us now consider a numerical example. Take the simple function $y = x^2$ whose graph is shown in Fig. 45 and take the abscissa of the point P to be $x_1 = 2$. The ordinate of P is $y_1 = 2^2 = 4$. If we take the abscissa of the point Q to be $x_2 = 3$, its ordinate $y_2 = 3^2 = 9$. The tangent of the angle QPR, or the slope of the chord PQ, is

$$\frac{9 - 4}{3 - 2} = 5.$$

If we take another point Q_1, nearer to P than Q, say one whose abscissa is 2·4, its ordinate will be $2·4^2$ or 5·76 and the slope of the chord PQ_1 will be

$$\frac{5·76 - 4}{2·4 - 2} = 4·4.$$

Taking other points Q_2, Q_3, Q_4, ... each one nearer to P than the preceding one and taking, for example, their abscissae to be 2·2, 2·1, 2·05, ... the slopes of the chords PQ_2, PQ_3, PQ_4, ..., calculated in the same way, are 4·2, 4·1, 4·05, ...

When we join the point P to points Q, Q_1, Q_2, Q_3, Q_4, ... as the latter approach nearer and nearer to P, the slopes of the chords PQ, PQ_1, PQ_2, PQ_3, PQ_4, ... decrease and the calculated values of these slopes suggest that they are approaching a limiting value which might well be of magnitude 4. The same is suggested by Fig. 46 which is a magnified version of part of the curve of Fig. 45.

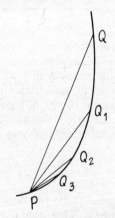

FIG. 46

To decide if the slopes of the various chords do in fact approach a limit as the point Q approaches P and to find the value of this limit, let the abscissa of Q be $2 + h$. The ordinate of Q is $(2 + h)^2$ and the slope of the chord PQ is

$$\frac{(2 + h)^2 - 2^2}{(2 + h) - 2} = \frac{h(4 + h)}{h} = 4 + h.$$

As Q approaches P, the value of h becomes smaller and smaller and it is apparent that the limiting value, as h decreases to zero, of the slope of the chord PQ is 4. In other words, we can find the abscissa $(2 + h)$ of a point Q so that the slope of the chord PQ differs from 4 by as little as we please. For example, if we wish to find a point Q such that the slope of the chord PQ is 4·0001, we have to take the abscissa of Q to be 2·0001. The line through P with slope 4 is the *limiting position* of the chord PQ as Q approaches P and is the *tangent* to the curve at P. The slope of this tangent line is defined as the *gradient* of the curve at the point P.

7.3. The increment notation

A convenient notation for a small increase, or increment, in the value of a variable x is the symbol δx called "delta x". This notation does not mean δ multiplied by x but $\delta x = x_1 - x$ where x_1 differs from x by a small quantity. When y is a function of x, the symbol δy is used to denote the change in the value of y corresponding to a change δx in the value of the independent variable x. In Fig. 47,

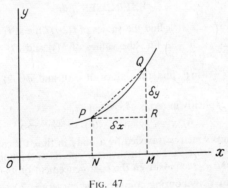

FIG. 47

P is a point on the graph of a function $y = f(x)$ whose abscissa and ordinate are x and y respectively and Q is a neighbouring point on the graph with coordinates $x + \delta x$ and $y + \delta y$. N and M are the projections of the points P and Q on the x-axis, and PR is drawn parallel to NM. Then $ON = x$, $PN = y$, $OM = x + \delta x$, $QM = y + \delta y$. It follows that $QR = QM - RM = QM - PN = y + \delta y - y = \delta y$ and that $PR = NM = OM - ON = x + \delta x - x = \delta x$. The average rate of change of y as x changes to $x + \delta x$ is measured by the tangent of the angle QPR and this is clearly equal to the ratio $\delta y / \delta x$. The gradient of the curve at the point P, or the slope of the tangent to the curve at the point P is the limiting value of the ratio $\delta y/\delta x$ as δx approaches zero.

As an example of the use of this notation, let us find the gradient of the curve $y = x^2$ at the point P whose coordinates are (x, y). At the point P we have

$$y = x^2,$$

and at the neighbouring point Q whose coordinates are $(x + \delta x, y + \delta y)$,

$$y + \delta y = (x + \delta x)^2.$$

By subtraction,

$$\delta y = (x + \delta x)^2 - x^2 = (2x + \delta x)\delta x,$$

and

$$\frac{\delta y}{\delta x} = 2x + \delta x.$$

As δx approaches zero, a convenient notation for which is "as $\delta x \to 0$", it is clear that this ratio approaches the limiting value $2x$. The gradient of the curve $y = x^2$ at the point (x, y), or the slope of the tangent to the curve at this point, is therefore $2x$. At the point where $x = 2$, this result gives the gradient of the curve to be 4 as found in the last section.

EXERCISES 7 (a)

1. If $f(x) = 2x^2 + x - 1$, find the values of $f(2), f(0)$ and $f(-1)$.

2. If $F(t) = 2t^2 + 3t - 2$, find the values of $F(\frac{1}{2})$ and $F(-\frac{1}{2})$. What values of t make $F(t) = 0$?

3. If $\phi(\theta) = \frac{1}{2} - \sin \theta$, find the values of $\phi(0)$ and $\phi(\pi/2)$. What values of θ make $\phi(\theta) = 0$?

4. Express y explicitly in terms of x when

$$\text{(i)}\ x^2 + 4y^2 = x, \quad \text{(ii)}\ xy + y^2 = x^2.$$

5. The distance s metres travelled by a body in time t seconds is given by the formula $s = 2t^2 + 2t$. Find the average speeds of the body in time intervals of $\frac{1}{2}, \frac{1}{4}, \frac{1}{8}$ and $\frac{1}{16}$ seconds from the commencement of its motion.

6. P and Q are neighbouring points on the curve $y = 2(x - x^2)$. P is the point (x, y) and Q the point $(x + \delta x, y + \delta y)$. Find the value of the ratio $\delta y/\delta x$ and determine the gradient of the curve at the point P.

7. Find the gradient of the curve $y = x^3$ at the point whose abscissa is x.

7.4. Differentiation from first principles. The differential coefficient

The process of calculating the ratio of the incremental change in a function y of x to the incremental change in x, that is of determining an expression for $\delta y/\delta x$, and then finding the limiting value of this ratio as δx approaches zero, is known *as differentiation from first principles.* The limit found in this way is generally denoted by the symbol $\dfrac{dy}{dx}$ and is called the *differential coefficient of y with respect to x.*

Alternative notations for the differential coefficient of a function $f(x)$ of x are

$$D_x f(x), \quad \frac{d}{dx} f(x), \quad \text{and} \quad f'(x),$$

and alternative terms for this quantity are the *derivative* or *derived function.* The process of differentiation is that of finding the gradient of the curve representing the function under consideration and the differential coefficient or derivative is the slope of the tangent to the curve at a given point.

Consider as an example the function $y = 5x^2$. To differentiate this function from first principles we have

$$y = 5x^2,$$
$$y + \delta y = 5(x + \delta x)^2.$$

By subtraction,

$$\delta y = 5\{(x + \delta x)^2 - x^2\} = 5(2x + \delta x)\delta x.$$

Hence

$$\frac{\delta y}{\delta x} = 5(2x + \delta x) = 10x + 5\delta x,$$

and, since the limiting value as $\delta x \to 0$ of the expression on the right is $10x$,

$$\frac{dy}{dx} = \lim_{\delta x \to 0} \left(\frac{\delta y}{\delta x}\right) = 10x.$$

Thus, if $y = 5x^2$, the differential coefficient of y with respect to x is $10x$. Alternatively we could express the result as

$$\frac{d}{dx}(5x^2) = 10x,$$

or, if $f(x) = 5x^2$, $\qquad\qquad f'(x) = 10x.$

For a general function $y = f(x)$, we should have

$$\frac{dy}{dx} = \lim_{\delta x \to 0} \left\{\frac{f(x + \delta x) - f(x)}{\delta x}\right\}. \qquad (7.1)$$

7.5. The differential coefficient of x^n, n a positive integer

If $y = x^n$, then

$$y + \delta y = (x + \delta x)^n,$$

and, by subtraction,

$$\delta y = (x + \delta x)^n - x^n.$$

Since n is assumed to be a positive integer, $(x + \delta x)^n$ can be expanded in a terminating series by the binomial theorem and we have

$$\delta y = x^n + nx^{n-1}(\delta x) + \frac{n(n-1)}{2!}x^{n-2}(\delta x)^2 + \ldots + (\delta x)^n - x^n$$

$$= nx^{n-1}(\delta x) + \frac{n(n-1)}{2!}x^{n-2}(\delta x)^2 + \ldots + (\delta x)^n.$$

Division by δx gives

$$\frac{\delta y}{\delta x} = nx^{n-1} + \frac{n(n-1)}{2!}x^{n-2}(\delta x) + \ldots + (\delta x)^{n-1}.$$

Since all the terms on the right except the first contain δx raised to a positive index, the limit of this expression as δx approaches zero is nx^{n-1} and we have the result that if

$$y = x^n, \quad \frac{dy}{dx} = nx^{n-1}. \tag{7.2}$$

The same result holds when n is negative or fractional but we shall delay a proof of this until pp. 141, 142.

7.6. The differential coefficients of sin x and cos x

If we take $y = \sin x$, then

$$y + \delta y = \sin(x + \delta x),$$

and, by subtraction,

$$\delta y = \sin(x + \delta x) - \sin x$$
$$= 2 \cos(x + \tfrac{1}{2}\delta x) \sin \tfrac{1}{2}\delta x,$$

by formula (5.30). Division by δx and a slight rearrangement gives

$$\frac{\delta y}{\delta x} = \cos(x + \tfrac{1}{2}\delta x).\left(\frac{\sin\tfrac{1}{2}\delta x}{\tfrac{1}{2}\delta x}\right).$$

Now, from (5.34), as $\delta x \to 0$, $(\sin\tfrac{1}{2}\delta x)/(\tfrac{1}{2}\delta x)$ tends to unity, and $\cos(x + \tfrac{1}{2}\delta x)$ tends to $\cos x$, so that if

$$y = \sin x, \quad \frac{dy}{dx} = \cos x. \tag{7.3}$$

Similarly if $y = \cos x$, $y + \delta y = \cos(x + \delta x)$ and

$$\delta y = \cos(x + \delta x) - \cos x.$$

This can be written in the form

$$\frac{\delta y}{\delta x} = -\sin(x + \tfrac{1}{2}\delta x).\left(\frac{\sin\tfrac{1}{2}\delta x}{\tfrac{1}{2}\delta x}\right),$$

and the limit of $\delta y/\delta x$ as δx tends to zero is now $-\sin x$. Hence if

$$y = \cos x, \quad \frac{dy}{dx} = -\sin x. \tag{7.4}$$

Example 1. *Differentiate from first principles, $x^3 + x^2$.*

Let $y = x^3 + x^2$, then $y + \delta y = (x + \delta x)^3 + (x + \delta x)^2$.

By subtraction,
$$\delta y = (x + \delta x)^3 - x^3 + (x + \delta x)^2 - x^2$$
$$= 3x^2(\delta x) + 3x(\delta x)^2 + (\delta x)^3 + 2x(\delta x) + (\delta x)^2.$$

Hence $\dfrac{\delta y}{\delta x} = 3x^2 + 3x(\delta x) + (\delta x)^2 + 2x + \delta x,$

and the limit of this as δx tends to zero is $3x^2 + 2x$. Hence the derivative of $x^3 + x^2$ is $3x^2 + 2x$.

It should be noted that the differential coefficient of this function is the sum of the separate differential coefficients of x^3 and x^2. This is a special case of a general result that the differential coefficient of the sum (or difference) of two functions is the sum (or difference) of their separate differential coefficients (see § 8.2).

Example 2. *Differentiate $1/x^2$ from first principles.*

Here we let $y = 1/x^2$, so that $y + \delta y = 1/(x + \delta x)^2$.

Hence
$$\delta y = \frac{1}{(x + \delta x)^2} - \frac{1}{x^2}$$
$$= \frac{x^2 - (x + \delta x)^2}{x^2(x + \delta x)^2}$$
$$= \frac{-2x(\delta x) - (\delta x)^2}{x^2(x + \delta x)^2}.$$

Thus $\dfrac{\delta y}{\delta x} = \dfrac{-2x - (\delta x)}{x^2(x + \delta x)^2},$

and the limit as $\delta x \to 0$ is $-2x/x^4$. Thus the derivative of $1/x^2$ is $-2/x^3$. It should be noted that this can be expressed by saying that the derivative of x^{-2} is $-2x^{-3}$. Hence the result that if $y = x^n$, $\dfrac{dy}{dx} = nx^{n-1}$, proved for positive integral n, holds when $n = -2$.

Example 3. *Differentiate $3x^2 + \cos 2x$ from first principles.*

Let $y = 3x^2 + \cos 2x$, so that $y + \delta y = 3(x + \delta x)^2 + \cos 2(x + \delta x)$.

Then, by subtraction,
$$\delta y = 3\{(x + \delta x)^2 - x^2\} + \cos 2(x + \delta x) - \cos 2x$$
$$= 3(2x + \delta x)\delta x - 2\sin(2x + \delta x)\sin \delta x.$$

Thus $\dfrac{\delta y}{\delta x} = 3(2x + \delta x) - 2\sin(2x + \delta x).\left(\dfrac{\sin \delta x}{\delta x}\right),$

and, taking the limit as δx tends to zero, we have
$$\frac{dy}{dx} = 6x - 2\sin 2x.$$

Notice that this result is the sum of the differential coefficients of $3x^2$ and $\cos 2x$ and that the differential coefficient of $3x^2$ is 3 times the differential coefficient of x^2.

EXERCISES 7 (b)

Differentiate from first principles:—

1. $5x^3$.
2. $x^4 - x^2$.
3. $1/(2x^2)$.
4. $\sin 2x$.
5. $\cos 3x$.
6. $x + \sin x$.

7.7. The differential coefficient as a rate measurer

Suppose that a body moves a distance s in time t and that s is a function of t given by $s = f(t)$. We have already seen that if s, s_1 are the distances moved in times t, t_1 respectively, the average speed of the body over the time interval $(t_1 - t)$ is $(s_1 - s)/(t_1 - t)$, or in the incremental notation, $\delta s/\delta t$. The limit of this quantity as δt approaches zero is ds/dt. This is the rate of change of distance with respect to time and is the instantaneous speed of the body at time t.

Similarly if a body is moving in a straight line with velocity v, the rate of change of velocity with respect to time is dv/dt and this is the acceleration of the body at the instant considered. In general, if a variable quantity y is a function of another variable x, the differential coefficient dy/dx can be regarded as giving not only the gradient of the graph of the function but also the rate of increase of y with respect to x. The sign convention adopted is that the function is increasing where its differential coefficient is positive and decreasing where its differential coefficient is negative.

Example 4. *A body moves in a straight line a distance s metres in t seconds and* $s = t^3$. *Find the velocity and acceleration of the body after 3 seconds.*

$$\text{Velocity,} \quad v = \frac{ds}{dt} = 3t^2.$$

$$\text{Acceleration} = \frac{dv}{dt} = 6t.$$

Putting $t = 3$, the velocity and acceleration after 3 seconds are respectively 27 m/s and 18 m/s².

Example 5. *The diameter of an expanding smoke ring at time t is proportional to* t^2. *If the diameter is 0·06 m after 6 seconds, at what rate is it then changing?*

If D is the diameter, $D = kt^2$ where k is a constant. Since $D = 0·06$ when $t = 6$, $0·06 = 36k$ so that $k = 1/600$ and $D = t^2/600$.

The rate of change of $D = \dfrac{dD}{dt} = \dfrac{t}{300}$ m/s.

When $t = 6$, the rate of change of D is therefore 0·02 m/s.

7.8. Approximations

If $y = f(x)$ and if δx, δy are respectively the increment in x and the corresponding increment in y, the limiting value of the ratio $\delta y/\delta x$ when δx approaches zero is, by definition, dy/dx or $f'(x)$. Hence we can write

$$\frac{\delta y}{\delta x} = f'(x) + \sigma,$$

where σ is a quantity which tends to zero as δx tends to zero. This can be written

$$\delta y = \{f'(x) + \sigma\}\delta x,$$

and as δx approaches zero, the second term on the right becomes more and more insignificant compared with the first. Hence we may use the equation

$$\delta y \coloneqq f'(x)\,\delta x \qquad (7.5)$$

as an *approximate* formula to find the effect on the value of a function of a *small* change in the value of the independent variable.

Example 6. *If* $y = x^3$, *find the approximate percentage increase in* y *due to an increase of* 0.1 *per cent. in* x.

Here $\qquad\qquad\qquad\qquad f(x) = x^3, \quad f'(x) = 3x^2$

and $\qquad\qquad\qquad\qquad \delta y \coloneqq 3x^2\,\delta x.$

Hence $\qquad\qquad\qquad \dfrac{\delta y}{y} \coloneqq \dfrac{3x^2\,\delta x}{x^3} = 3\dfrac{\delta x}{x},$

dividing by the equal quantities y and x^3. Now $\delta x/x$ is the ratio of the change in x to x and the percentage change in x is therefore $100\,\delta x/x$. Similarly the percentage change in y is $100\,\delta y/y$ and we can write

$$\text{percentage change in } y \coloneqq 3 \times \text{percentage change in } x.$$

The percentage change in x being 0.1, the approximate change in y is therefore 0.3 per cent.

Example 7. *Find the approximate error made in calculating the area of a triangle in which two of the sides are accurately measured as* 0.18 *m and* 0.25 *m, while the included angle is measured as* $60°$, *but is* $\frac{1}{2}°$ *wrong.*

Taking the given sides as b, c and the included angle as A, the area Δ is given by $\Delta = \frac{1}{2}bc \sin A$. If A is in error by δA, the area will be in error by $\delta\Delta$ where

$$\delta\Delta \coloneqq \tfrac{1}{2}bc\frac{d}{dA}(\sin A)\,\delta A = \tfrac{1}{2}bc\cos A\,\delta A.$$

Now $b = 0.18$, $c = 0.25$, $A = 60° = \pi/3$ rad., $\delta A = \frac{1}{2}° = \pi/360$ rad., so that

$$\delta\Delta \coloneqq \frac{0.18 \times 0.25 \times \cos \pi/3}{2} \times \frac{\pi}{360} = \frac{5\pi}{16} \times 10^{-4}\ \text{m}^2.$$

EXERCISES 7 (c)

1. A body moves in a straight line so that the distance moved s metres is given in terms of the time t seconds by $s = t^3 - t^2$. Find an expression for the velocity of the body at time t and find the times at which the body is at rest.

2. The velocity v m/s at time t seconds of a body moving in a straight line is proportional to t^4. Find the acceleration of the body when $t = 2$ seconds if its velocity is then 16 m/s.

3. After t seconds, the area A m² of an ink stain is given by $A = 10^{-4}(t+t^3)$. Find the rate at which the area is increasing after 2 seconds.

4. If the side of a square can be measured accurately to 0·1 mm, what is the possible error in the area of a square whose side is measured to be 200 mm?

5. Find the approximate percentage change in the square of a quantity when the quantity itself changes by 0·1 per cent. Hence calculate an approximate value for $(10·01)^2$.

6. Find the increase in the area of a circle when its radius changes from 5 m to 5·1 m.

EXERCISES 7 (d)

1. If $f(x) = 3x^2 + \sin x$, find the values of $f(0), f(\pi/2)$ and $f(-\pi)$.

2. If $4x^3y - 2x^4 + 3xy = 0$, express y explicitly in terms of x and find the value of y when $x = 2$.

3. If $f(x) = ax^2 + bx + c$ where a, b, c are constants, find an expression for $f(x + 1)$.

4. $f(x)$ denotes an expression of the third degree in x. If $f(-1) = 6, f(0) = 9$, $f(2) = 19, f(3) = 11$, find the expression for $f(x)$.

5. If $f(x) = \log x$, show that:—

$$\text{(i)} \quad f(ab) = f(a) + f(b),$$
$$\text{(ii)} \quad f(a/b) = f(a) - f(b).$$

6. P is the point (x, y) and Q the point $(x + \delta x, y + \delta y)$ on the graph of $y = \sqrt{x}$. Show that

$$\frac{\delta y}{\delta x} = \frac{1}{\sqrt{(x + \delta x)} + \sqrt{x}},$$

and hence find the gradient of the curve at the point P.

7. Find the slope of the curve $y = ax^2 + bx + c$, where a, b, c are constants, at a point whose abscissa is x. At what point is the tangent to the curve parallel to the x-axis?

8. Calculate the gradient of the curve $y = -x^3 + 4x^2 - 3x$ at each of the points where it crosses the axis of x.   (L.U.)

9. Show that the gradient of the curve $y = -1 + 3x - (x^2/4)$ when $x = 2$ is double that when $x = 4$. Find also the abscissa of the point on the curve at which the gradient is -1.   (O.C.)

10. An expression of the second degree is denoted by $f(x)$. If $f(1) = 7, f(2) = 23$, $f(3) = 17$, find the gradient of the graph of $f(x)$ at $x = 2$.   (O.C.)

11. Find the value of the constant c so that the tangent at the origin of coordinates to the curve $y = x(c - x^2)$ makes an angle of 45° with the x-axis.

12. For what values of x is the tangent to the curve

$$y = \tfrac{2}{3}x^3 + \tfrac{1}{2}x^2$$

equally inclined to the two coordinate axes?

13. Differentiate from first principles:—

(i) $4x^2 + x$,   (ii) $1/x$.

14. Differentiate from first principles:—

(i) $\sin ax$, where a is a constant,
(ii) $\frac{1}{2}\cos 2x$.

15. Obtain from first principles the differential coefficient of $1/(x + 2)$ with respect to x.   (L.U.)

16. If $f(x) = 3x^2 + \sin x$, find $f'(x)$. What is the value of $f'(0)$?

17. An excavator removes V cubic metres of soil in t minutes, where

$$V = 0.42t - 0.00047t^2$$

At what rate is the soil being removed after 20 minutes?   (L.U.)

18. Find the gradients of the curve $y = x^3 + 24x^2 + 192x$ at the points where (i) $x = 2$, (ii) $x = 4$.

19. If the distance s travelled by a particle in time t is given by

$$s = ut + \tfrac{1}{2}at^2,$$

where u and a are constants, show that the velocity at time t is $u + at$ and that the acceleration is constant.

20. The distance s m which a body has travelled in t seconds is given by $s = \frac{1}{2}t^2 - t$. Find when the body is at rest and the acceleration at that time.

21. Assuming that the path traced out by a mortar bomb is given by the equation $y = 4x - (x^2/3.2)$, the x- and y-axes being horizontal and vertical lines through the point of projection, find the angle of projection of the bomb.

22. A particle moves along a straight line in such a manner that its distance x from a fixed point in the line at time t is given by $x = \cos t$. Prove that its velocity is equal to $-\sqrt{(1 - x^2)}$ and that its acceleration is $-x$.

23. A point moves along a straight line and, at the end of t seconds, its distance (s) from a fixed point in the line is given by

$$s = t^3 - 9t^2 + 24t - 18.$$

Show that the velocity vanishes for two values of t and the acceleration for one value of t. Find also the value of the velocity when the acceleration vanishes and the values of the acceleration when the velocity vanishes.
                                                                (O.C.)

24. The distance moved in a straight line by a particle in t seconds is $5t^3$. Show that v, the velocity, and f, the acceleration, are connected by the relation $f^2 = 60v$.   (O.C.)

25. A body moves in a straight line so that the distance travelled s metres in time t seconds is given by $s = 8t^2 + 3t$. Find the approximate space travelled during $\frac{1}{10}$ seconds after 5 seconds of motion and deduce the average velocity in that interval of time.

CHAPTER 8

TECHNICAL PROCESSES IN THE DIFFERENTIAL CALCULUS

8.1. Introduction

In the last chapter, where an attempt was made to introduce the basic ideas of the differential calculus, examples were chosen which involved only very simple functions. The differential coefficients of more complicated functions can be found from first principles in the same way but the labour involved in some cases is considerable. Fortunately certain general theorems can be set up and these, together with the differential coefficients of quite a small number of standard functions, enable the differential coefficients of more complicated functions to be found readily.

It is the object of this chapter to deal with the technical processes involved in finding the differential coefficients of functions which can be considered as sums or differences, products or quotients, etc., of simpler functions. The student should work through a large number of examples on differentiation until he is thoroughly familiar with these processes.

8.2. The differentiation of a sum

(a) Let $y = u + C$, where u is a given function of x and C is a constant. Then, if x increases to $x + \delta x$, u increases to $u + \delta u$ and y increases to $y + \delta y$, so that

$$y + \delta y = u + \delta u + C.$$

By subtraction, $\delta y = \delta u$, and division by δx gives

$$\frac{\delta y}{\delta x} = \frac{\delta u}{\delta x}.$$

In the limit as δx tends to zero, we have

$$\frac{dy}{dx} = \frac{du}{dx}.$$

Thus an additive constant disappears on differentiation.

(b) Let $y = u + v$, where u and v are given functions of x. If x increases to $x + \delta x$, u and v increase to $u + \delta u$ and $v + \delta v$ respectively and y increases to $y + \delta y$. Hence

$$y + \delta y = u + \delta u + v + \delta v,$$

and by subtraction

$$\delta y = \delta u + \delta v.$$

Dividing by δx and proceeding to the limit as δx tends to zero we have

$$\frac{dy}{dx} = \frac{du}{dx} + \frac{dv}{dx},\qquad(8.1)$$

showing that *the differential coefficient of a sum of two functions is the sum of the differential coefficients of the separate functions.* It is clear that the plus sign can be replaced throughout by a minus sign and that *the differential coefficient of the difference of two functions is the difference of the differential coefficients of the separate functions.*

(c) If $y = u + v + w$, where w is a third function of x, we can write this as $y = (u + v) + w$, and application of the preceding result gives

$$\frac{dy}{dx} = \frac{d}{dx}(u + v) + \frac{dw}{dx}$$

$$= \frac{du}{dx} + \frac{dv}{dx} + \frac{dw}{dx}.\qquad(8.2)$$

In this way we can show, step by step, that *the differential coefficient of the sum of any finite number of functions is the sum of the differential coefficients of the separate functions.*

Example 1. *Find the differential coefficient of the function*

$$x^5 - x^4 + x^2 - x + 3.$$

Since the differential coefficient of x^n is nx^{n-1}, application of the above rules gives for the differential coefficient, $5x^4 - 4x^3 + 2x - 1$.

8.3. The differentiation of a product

(a) Let $y = Cu$, where u is a function of x and C is a constant. Then

$$y + \delta y = C(u + \delta u),$$

where δy, δu are the increments in y and u respectively corresponding to an increment δx in x. By subtraction we find $\delta y = C\,\delta u$, and division by δx gives

$$\frac{\delta y}{\delta x} = C\frac{\delta u}{\delta x}.$$

In the limit as δx tends to zero, we have

$$\frac{dy}{dx} = C\frac{du}{dx},\qquad(8.3)$$

and *the differential coefficient of a constant multiplied by a function of x is equal to the constant multiplied by the differential coefficient of the function.*

(b) Let $y = uv$, where u and v are given functions of x. If δy, δu, δv are the increments in y, u and v respectively corresponding to an increment δx in x,

$$y + \delta y = (u + \delta u)(v + \delta v)$$
$$= uv + v\,\delta u + u\,\delta v + \delta u\,\delta v.$$

By subtraction,

$$\delta y = v\,\delta u + u\,\delta v + \delta u\,\delta v.$$

A slight rearrangement and division by δx gives

$$\frac{\delta y}{\delta x} = v\frac{\delta u}{\delta x} + (u + \delta u)\frac{\delta v}{\delta x}.$$

As δx tends to zero, δu tends to zero and $\delta u/\delta x$, $\delta v/\delta x$, $\delta y/\delta x$ tend respectively to $\dfrac{du}{dx}, \dfrac{dv}{dx}$ and $\dfrac{dy}{dx}$. Hence

$$\frac{dy}{dx} = v\frac{du}{dx} + u\frac{dv}{dx}, \qquad (8.4)$$

showing that *the differential coefficient of the product of two functions of x is equal to the second function multiplied by the differential coefficient of the first plus the first function multiplied by the differential coefficient of the second.*

(c) The differential coefficient of $y = uvw$ where w is a third function of x can be found as follows. If we divide the left-hand side of (8.4) by y and the right-hand side by $uv(=y)$ we can express the result in the form

$$\frac{1}{y}\frac{dy}{dx} = \frac{1}{u}\frac{du}{dx} + \frac{1}{v}\frac{dv}{dx}.$$

If now we apply this result to $y = (uv)w$ we have

$$\frac{1}{y}\frac{dy}{dx} = \frac{1}{uv}\frac{d}{dx}(uv) + \frac{1}{w}\frac{dw}{dx}.$$

But, by (8.4),

$$\frac{d}{dx}(uv) = v\frac{du}{dx} + u\frac{dv}{dx},$$

so that

$$\frac{1}{y}\frac{dy}{dx} = \frac{1}{uv}\left(v\frac{du}{dx} + u\frac{dv}{dx}\right) + \frac{1}{w}\frac{dw}{dx}$$
$$= \frac{1}{u}\frac{du}{dx} + \frac{1}{v}\frac{dv}{dx} + \frac{1}{w}\frac{dw}{dx}.$$

Multiplication by y, or uvw, then gives

$$\frac{dy}{dx} = vw\frac{du}{dx} + uw\frac{dv}{dx} + uv\frac{dw}{dx}.$$

For a more general function $y = uvw \ldots$, we can obtain similarly

$$\frac{dy}{dx} = vw \ldots \frac{du}{dx} + uw \ldots \frac{dv}{dx} + uv \ldots \frac{dw}{dx}, \qquad (8.5)$$

so that *the derivative can be obtained by differentiating each function separately multiplying by the remaining functions and adding the results.*

Example 2. *Find the differential coefficients of* (i) $6x^4$ *and* (ii) $\frac{\sin x}{4}$.

(i) The differential coefficient of $6x^4$ is 6 times the differential coefficient of x^4 and so is $6 \times 4x^3$, or $24x^3$.

(ii) The differential coefficient of $\frac{\sin x}{4}$ is one-quarter of the differential coefficient of $\sin x$ and is therefore $\frac{\cos x}{4}$.

Example 3. *Find* $\frac{dy}{dx}$ *if* (i) $y = x^2 \sin x$, (ii) $y = x \sin x \cos x$.

(i) By (8.4),

$$\frac{dy}{dx} = \sin x \frac{d}{dx}(x^2) + x^2 \frac{d}{dx}(\sin x)$$
$$= 2x \sin x + x^2 \cos x.$$

(ii) By (8.5),

$$\frac{dy}{dx} = \sin x \cos x \frac{d}{dx}(x) + x \cos x \frac{d}{dx}(\sin x) + x \sin x \frac{d}{dx}(\cos x)$$
$$= \sin x \cos x + x \cos x \cdot \cos x + x \sin x(-\sin x)$$
$$= \sin x \cos x + x(\cos^2 x - \sin^2 x).$$

EXERCISES 8 (a)

Differentiate the following functions of x:—

1. $4x^3 - \sin x + 2$.
2. $10 \sin x \cos x$.
3. $\sin x - x^3 \cos x$.
4. $x(1 - x)$.
5. $(1 + x^2)(1 - 2x^2)$.

6. $x^3 \cos x$.
7. $(3x + 2)^3$.
8. $x^2 \sin x \cos x$.
9. $\cos^2 x$.
10. $10x^4 - \sin^2 x + x \cos x$.

8.4. The differentiation of a quotient

Suppose that $y = u/v$ where u and v are given functions of x. Then $u = yv$ and application of the rule for differentiating the product yv gives

$$\frac{du}{dx} = v\frac{dy}{dx} + y\frac{dv}{dx}.$$

Solving for $\dfrac{dy}{dx}$ we have

$$\frac{dy}{dx} = \frac{1}{v}\frac{du}{dx} - \frac{y}{v}\frac{dv}{dx}.$$

Since $y = u/v$, $y/v = u/v^2$, and hence,

$$\frac{dy}{dx} = \frac{1}{v}\frac{du}{dx} - \frac{u}{v^2}\frac{dv}{dx},$$

or,

$$\frac{dy}{dx} = \frac{v\dfrac{du}{dx} - u\dfrac{dv}{dx}}{v^2}. \qquad (8.6)$$

Thus *the differential coefficient of the quotient of two functions of x is equal to the denominator times the differential coefficient of the numerator minus the numerator times the differential coefficient of the denominator all divided by the square of the denominator.*

Example 4. Find $\dfrac{dy}{dx}$ when (i) $y = \dfrac{1-x}{1+x}$, (ii) $y = \dfrac{\sin x}{x}$.

(i) Using (8.6),

$$\frac{dy}{dx} = \frac{(1+x)\dfrac{d}{dx}(1-x) - (1-x)\dfrac{d}{dx}(1+x)}{(1+x)^2}$$

$$= \frac{(1+x)(-1) - (1-x)(1)}{(1+x)^2} = \frac{-2}{(1+x)^2}.$$

(ii) Again using (8.6),

$$\frac{dy}{dx} = \frac{x\dfrac{d}{dx}(\sin x) - \sin x\dfrac{d}{dx}(x)}{x^2}$$

$$= \frac{x\cos x - \sin x}{x^2}.$$

8.5. The differential coefficients of tan *x*, cot *x*, cosec *x* and sec *x*

The differential coeffficients of tan x, cot x, cosec x and sec x can be derived from those of sin x and cos x and the rule for differentiating a quotient. Thus

$$\frac{d}{dx}(\tan x) = \frac{d}{dx}\left(\frac{\sin x}{\cos x}\right)$$

$$= \frac{\cos x\dfrac{d}{dx}(\sin x) - \sin x\dfrac{d}{dx}(\cos x)}{\cos^2 x}$$

$$= \frac{\cos x(\cos x) - \sin x(-\sin x)}{\cos^2 x}$$

$$= \frac{\cos^2 x + \sin^2 x}{\cos^2 x}$$

$$= \frac{1}{\cos^2 x} = \sec^2 x. \qquad (8.7)$$

Similarly

$$\frac{d}{dx}(\cot x) = \frac{d}{dx}\left(\frac{\cos x}{\sin x}\right)$$

$$= \frac{\sin x\frac{d}{dx}(\cos x) - \cos x\frac{d}{dx}(\sin x)}{\sin^2 x}$$

$$= \frac{-\sin^2 x - \cos^2 x}{\sin^2 x}$$

$$= -\frac{1}{\sin^2 x} = -\operatorname{cosec}^2 x. \tag{8.8}$$

Also

$$\frac{d}{dx}(\operatorname{cosec} x) = \frac{d}{dx}\left(\frac{1}{\sin x}\right)$$

$$= \frac{\sin x\frac{d}{dx}(1) - 1.\frac{d}{dx}(\sin x)}{\sin^2 x}$$

$$= \frac{\sin x.(0) - (1)\cos x}{\sin^2 x}$$

$$= -\frac{\cos x}{\sin^2 x} = -\operatorname{cosec} x \cot x. \tag{8.9}$$

It is left as an exercise for the student to show in the same way that

$$\frac{d}{dx}(\sec x) = \sec x \tan x.$$

EXERCISES 8 (b)

Differentiate the following functions with respect to x:—

1. $\dfrac{x}{x^2 + 1}$.

2. $\dfrac{2 - x}{1 + 2x}$.

3. $\dfrac{1 - 3x^2}{2 + 4x^2}$.

4. $(3 - 2x^2)^{-2}$.

5. $\dfrac{1 + \sin x}{1 - \sin x}$.

6. $\dfrac{\sin x - \cos x}{\sin x + \cos x}$.

7. $\dfrac{x}{\tan x}$.

8. $\sec^2 x$.

9. $\dfrac{\sin x}{1 + \tan x}$.

10. $\cot^2 x$.

8.6. Differentiation of a function of a function

A function like $y = (x + 3)^2$ is a function of a function for $(x + 3)$ is a function of x and $(x + 3)^2$ is a function of $(x + 3)$. Other examples are $\sin ax$, $\tan(x^3)$, etc. The object of this section is to establish a very simple rule for the rapid determination of the differential coefficients of such functions.

The differential coefficient of a function like our first example can, of course, be found by first squaring out the right-hand side and differentiating term by term. Thus

$$y = (x + 3)^2$$
$$= x^2 + 6x + 9.$$

$$\frac{dy}{dx} = 2x + 6 = 2(x + 3).$$

It should be observed that the result is exactly the same as if we had treated $(x + 3)$ as if it were x and used the standard result for the differential coefficient of x^2. Similarly if $y = (x + a)^3$, where a is a constant,

$$y = (x + a)^3$$
$$= x^3 + 3ax^2 + 3a^2x + a^3.$$

$$\frac{dy}{dx} = 3x^2 + 6ax + 3a^2.$$

$$= 3(x^2 + 2ax + a^2) = 3(x + a)^2,$$

and again the result is the same as if we had treated $(x + a)$ as if it were x and used the standard result for the differential coefficient of x^3.

Now consider $y = (2x + 3)^2$. Working as before

$$y = 4x^2 + 12x + 9,$$

$$\frac{dy}{dx} = 8x + 12 = 4(2x + 3),$$

so that the result is not now $2(2x + 3)$ but twice this. A rough explanation is that whereas $(x + 3)$ changes at the same rate as x, $(2x + 3)$ changes twice as fast. Similarly if $y = (cx + d)^2$, where c and d are constants,

$$y = c^2x^2 + 2cdx + d^2,$$

$$\frac{dy}{dx} = 2c^2x + 2cd = 2c(cx + d)$$

and we observe that the result is the same as if we had treated $(cx + d)$ as if it were x, used the standard result for the differential coefficient of x^2 and then multiplied by c, the differential coefficient of $cx + d$.

This suggests that if y is a function of u where u is a function of x, the formula giving $\dfrac{dy}{dx}$ might well be

$$\frac{dy}{dx} = \frac{dy}{du} \times \frac{du}{dx}. \qquad (8.10)$$

Applied to some of the examples already given this would give:—

(a) $y = (x + 3)^2$, or $y = u^2$ where $u = x + 3$.

$$\frac{dy}{du} = 2u, \quad \frac{du}{dx} = 1,$$

$$\frac{dy}{dx} = \frac{dy}{du} \times \frac{du}{dx} = 2u \times 1 = 2(x + 3).$$

(b) $y = (x + a)^3$, or $y = u^3$ where $u = x + a$.

$$\frac{dy}{du} = 3u^2, \quad \frac{du}{dx} = 1,$$

$$\frac{dy}{dx} = \frac{dy}{du} \times \frac{du}{dx} = 3u^2 \times 1 = 3(x + a)^2.$$

(c) $y = (cx + d)^2$, or $y = u^2$ where $u = cx + d$.

$$\frac{dy}{du} = 2u, \quad \frac{du}{dx} = c,$$

$$\frac{dy}{dx} = \frac{dy}{du} \times \frac{du}{dx} = 2u \times c = 2c(cx + d).$$

A strict proof of the important formula (8.10) is rather beyond the scope of the present book. The following, although it assumes a result not already proved, must suffice. If y is a function of u and u is a function of x, let δu be the increment in u corresponding to an increment δx in x, and let δy be the increment in y corresponding to the increment δu in u. Then, provided $\delta u \neq 0$,

$$\frac{\delta y}{\delta x} = \frac{\delta y}{\delta u} \times \frac{\delta u}{\delta x}$$

and, assuming that the limit of a product is the product of the limits, this gives

$$\frac{dy}{dx} = \frac{dy}{du} \times \frac{du}{dx} \qquad (8.10)$$

since $\dfrac{dy}{dx}, \dfrac{dy}{du}, \dfrac{du}{dx}$ are respectively the limiting values of $\dfrac{\delta y}{\delta x}, \dfrac{\delta y}{\delta u}$ and $\dfrac{\delta u}{\delta x}$ as δx tends to zero.

This formula is a most important one and the student should work through many examples of its use. At first it is probably best to introduce the auxiliary variable u as in the examples below but with practice this soon becomes unnecessary and the results can be written down at once.

Example 5. *Find* $\dfrac{dy}{dx}$ *when* (i) $y = (1 - 3x^2)^5$, (ii) $y = \left(\dfrac{1 + 2x}{1 + x}\right)^2$.

(i) Let $u = 1 - 3x^2$, so that $y = u^5$. Then

$$\frac{du}{dx} = -6x \quad \text{and} \quad \frac{dy}{du} = 5u^4.$$

Hence
$$\frac{dy}{dx} = \frac{dy}{du} \times \frac{du}{dx} = 5u^4 \times (-6x)$$
$$= -30xu^4 = -30x(1 - 3x^2)^4.$$

(ii) Let $u = \dfrac{1 + 2x}{1 + x}$, so that $y = u^2$.

Here $\dfrac{dy}{du} = 2u$, but to find $\dfrac{du}{dx}$ we have to apply the rule (8.6) for differentiating a quotient. This gives

$$\frac{du}{dx} = \frac{(1 + x)\dfrac{d}{dx}(1 + 2x) - (1 + 2x)\dfrac{d}{dx}(1 + x)}{(1 + x)^2}$$

$$= \frac{(1 + x)(2) - (1 + 2x)(1)}{(1 + x)^2} = \frac{1}{(1 + x)^2}.$$

Hence
$$\frac{dy}{dx} = \frac{dy}{du} \times \frac{du}{dx}$$

$$= 2u \times \frac{1}{(1 + x)^2}$$

$$= 2\left(\frac{1 + 2x}{1 + x}\right) \times \frac{1}{(1 + x)^2} = \frac{2(1 + 2x)}{(1 + x)^3}.$$

Example 6. *Find* (i) $\dfrac{d}{dt}\left\{\sin\left(4t - \dfrac{\pi}{5}\right)\right\}$, (ii) $\dfrac{d}{d\theta}\left\{\cos^4\left(2\theta - \dfrac{\pi}{5}\right)\right\}$.

(i) Let $u = 4t - \pi/5$, so that $\dfrac{du}{dt} = 4$. Then

$$\frac{d}{dt}\{\sin(4t - \pi/5)\} = \frac{d}{dt}(\sin u) = \frac{d}{du}(\sin u) \times \frac{du}{dt}$$

$$= \cos u \times 4 = 4\cos(4t - \pi/5).$$

(ii) Let $y = \cos^4(2\theta - \pi/5)$ and let $u = 2\theta - \pi/5$. Then $y = \cos^4 u$. This is still a function of a function and we now let $\cos u = v$.

Then $y = v^4$ and $\dfrac{dy}{dv} = 4v^3$. By (8.10),

$$\frac{dy}{du} = \frac{dy}{dv} \times \frac{dv}{du} = 4v^3 \times (-\sin u),$$

since $v = \cos u$ and so $\dfrac{dv}{du} = -\sin u$.

Also
$$\frac{dy}{d\theta} = \frac{dy}{du} \times \frac{du}{d\theta} = -4v^3 \sin u \times 2,$$

since $u = 2\theta - \pi/5$, so that $\dfrac{du}{d\theta} = 2$.

Replacing the values $u = 2\theta - \pi/5$, $v = \cos u = \cos(2\theta - \pi/5)$, we have

$$\frac{dy}{d\theta} = -8\cos^3\left(2\theta - \frac{\pi}{5}\right)\sin\left(2\theta - \frac{\pi}{5}\right).$$

EXERCISES 8 (c)

Find $\dfrac{dy}{dx}$ when :—

1. $y = (4x - 5)^3$.

2. $y = (x^2 + 3x)^5$.

3. $y = \sin 2(x - \alpha)$, α constant.

4. $y = \tan 2x$.

5. $y = \sec 3x$.

6. $y = \tan^2 (3x + 1)$.

7. $y = \sin^3 (2 - x)$.

8. $y = x^2 (1 - x)^2$.

9. $y = \sin^3 x \sin 3x$.

10. $y = \dfrac{1 + \sin^2 x}{1 - \sin^2 x}$.

11. If $u = \sin^m \theta \cos^n \theta$, find the value of $\dfrac{du}{d\theta}$ if m and n are positive integers.

12. Find the differential coefficient of s with respect to t if $s = \sin^2 (a/t)$ and a is a constant.

13. If m is a positive integer, find the differential coefficients with respect to x of:—

$$\text{(i) } \sin^m x, \quad \text{(ii) } \sin (x^m), \quad \text{(iii) } \sin (\cos x).$$

14. If $y = (\tan x + \sec x)^m$, where m is a positive integer, show that $\dfrac{dy}{dx} = my \sec x$.

15. Show that the differential coefficient with respect to x of

$$\tan x(1 + 2 \sec^2 x) - 3x \sec^2 x$$

is $6 \sec^2 x \tan x(\tan x - x)$. Hence show that, if x is a positive acute angle, this differential coefficient is positive.

8.7. The differential coefficient of x^n when n is negative or fractional

It was shown in the last chapter (§ 7.5) that, if n is a positive integer, the differential coefficient with respect to x of x^n is nx^{n-1}. The rules established in §§ 8.4, 8.6 for differentiating a quotient and a function of a function enable us to show that the same result is true when n is negative or fractional.

(a) Let n be a negative integer and let $n = -m$ so that m is a positive integer. Then if $y = x^n$, we have

$$y = x^{-m} = \frac{1}{x^m}.$$

By the rule for differentiating a quotient, since m is a positive integer and therefore the differential coefficient of x^m is mx^{m-1}, we have

$$\frac{dy}{dx} = \frac{x^m \times 0 - 1 \times mx^{m-1}}{x^{2m}}$$

$$= \frac{-m}{x^{m+1}} = -mx^{-m-1}$$

$$= nx^{n-1},$$

since $n = -m$.

(b) Let $n = p/q$ where p and q are integers. There is no loss of generality in assuming that q is positive. Then, if $y = x^n$, we have

$$y = x^{p/q} = (x^{1/q})^p.$$

Put $x^{1/q} = u$, then $y = u^p$ and $x = u^q$, so that

$$\frac{dy}{du} = pu^{p-1}, \quad \frac{dx}{du} = qu^{q-1}$$

by the result already established.

The rule for differentiating a function of a function can be written in the form

$$\frac{dy}{du} = \frac{dy}{dx} \times \frac{dx}{du},$$

so that

$$pu^{p-1} = \frac{dy}{dx} \times qu^{q-1}.$$

giving

$$\frac{dy}{dx} = \frac{p}{q}u^{p-q}.$$

Since $u = x^{1/q}$, this can be written

$$\frac{dy}{dx} = \frac{p}{q}(x^{1/q})^{p-q}$$

$$= \frac{p}{q}x^{(p/q)-1}$$

$$= nx^{n-1},$$

since $n = p/q$.

Hence $\dfrac{d}{dx}(x^n) = nx^{n-1}$ for all rational values of n.

Example 7. Find $\dfrac{dy}{dx}$ when (i) $y = \left(x - \dfrac{3}{x^2}\right)^2$, (ii) $y = \sqrt{\left(\dfrac{1+x}{1-x}\right)}$.

(i) Let $u = x - 3/x^2 = x - 3x^{-2}$. Then $y = u^2$ and

$$\frac{dy}{du} = 2u, \quad \frac{du}{dx} = 1 - 3(-2)x^{-3} = 1 + 6/x^3.$$

Hence

$$\frac{dy}{dx} = \frac{dy}{du} \times \frac{du}{dx}$$

$$= 2u(1 + 6/x^3)$$

$$= 2(x - 3/x^2)(1 + 6/x^3).$$

(ii) Let $u = (1 + x)/(1 - x)$. Then $y = \sqrt{u} = u^{1/2}$ and

$$\frac{dy}{du} = \tfrac{1}{2}u^{-1/2} = \frac{1}{2\sqrt{u}}$$

and, by the rule for the differential coefficient of a quotient,

$$\frac{du}{dx} = \frac{1 - x + (1 + x)}{(1 - x)^2} = \frac{2}{(1 - x)^2}.$$

Hence
$$\frac{dy}{dx} = \frac{dy}{du} \times \frac{du}{dx}$$

$$= \frac{1}{2\sqrt{u}} \cdot \frac{2}{(1-x)^2}$$

$$= \sqrt{\left(\frac{1-x}{1+x}\right)} \cdot \frac{1}{(1-x)^2}$$

$$= \frac{1}{(1+x)^{1/2}(1-x)^{3/2}}.$$

EXERCISES 8 (d)

Differentiate with respect to x:—

1. $\left(x^4 - \dfrac{1}{x^2}\right)^3.$

2. $(2 - 5x^3)^{-2}.$

3. $\sqrt{(1+x)}.$

4. $1/\sqrt{(1+x)}.$

5. $(1-x)\sqrt{(1+x^2)}.$

6. $\sin(\sqrt{x}).$

7. If $y = \sqrt{(1 + \sin x)}$, show that $\dfrac{dy}{dx} = \frac{1}{2}\sqrt{(1 - \sin x)}.$

8. If $y = \sqrt{\left(\dfrac{1 + \sin x}{1 - \sin x}\right)}$, show that $\dfrac{dy}{dx} = \dfrac{1}{1 - \sin x}.$

8.8. Differentiation of inverse functions

If in Fig. 48, PT is the tangent at the point P to the curve repre-

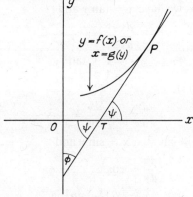

FIG. 48

senting the function $y = f(x)$, and if PT makes an angle ψ with the x-axis,

$$\tan \psi = \frac{dy}{dx}.$$

If the equation $y = f(x)$ is written in the form $x = g(y)$, the curve of Fig. 48 also represents this function. If PT makes an angle ϕ with the y-axis, we have

$$\tan \phi = \frac{dx}{dy}.$$

But, from the figure, $\phi + \psi = 90°$, so that $\phi = 90° - \psi$, and

$$\tan \phi = \tan (90° - \psi) = \cot \psi = \frac{1}{\tan \psi},$$

so that $$\frac{dx}{dy} = 1 \Big/ \frac{dy}{dx}. \tag{8.11}$$

An analytical proof of formula (8.11) would run as follows. If $y = f(x)$ and x is the inverse function given by $x = g(y)$, we have

$$y = f[g(y)].$$

Differentiating this with respect to y as a function of a function

$$1 = f'[g(y)]g'(y),$$

or, since $x = g(y)$,

$$1 = f'(x).g'(y).$$

Hence $g'(y) = 1/f'(x)$, and this can be written

$$\frac{dx}{dy} = 1 \Big/ \frac{dy}{dx}.$$

8.9. The differential coefficients of $\tan^{-1} x$ and $\sin^{-1} x$

If $y = \tan^{-1} x$, we have $x = \tan y$ and

$$\frac{dx}{dy} = \sec^2 y$$
$$= 1 + \tan^2 y$$
$$= 1 + x^2.$$

Hence, using (8.11),

$$\frac{dy}{dx} = 1 \Big/ \frac{dx}{dy} = \frac{1}{1 + x^2}. \tag{8.12}$$

Similarly, if $y = \sin^{-1} x$, $x = \sin y$ and

$$\frac{dx}{dy} = \cos y$$
$$= \sqrt{(1 - \sin^2 y)}$$
$$= \sqrt{(1 - x^2)},$$

giving $$\frac{dy}{dx} = 1 \Big/ \frac{dx}{dy} = \frac{1}{\sqrt{(1 - x^2)}}. \tag{8.13}$$

The differential coefficients of the other inverse trigonometrical functions can be found in the same way and are left as exercises for the student.

Example 8. *If* $y = sec^{-1} x$, *show that* $\dfrac{dy}{dx} = \dfrac{1}{x\sqrt{(x^2 - 1)}}$.

If $y = sec^{-1} x$, $x = sec\ y$, so that

$$\frac{dx}{dy} = sec\ y\ tan\ y = sec\ y\sqrt{(sec^2 y - 1)} = x\sqrt{(x^2 - 1)}.$$

Hence $\quad \dfrac{dy}{dx} = 1 \Big/ \dfrac{dx}{dy} = \dfrac{1}{x\sqrt{(x^2 - 1)}}$.

Example 9. *If* $y = cos^{-1}\left(\dfrac{1 - x^2}{1 + x^2}\right)$, *show that* $\dfrac{dy}{dx} = \dfrac{2}{1 + x^2}$.

Let $u = \dfrac{1 - x^2}{1 + x^2}$, so that $y = cos^{-1} u$, or $u = cos\ y$.

$$\frac{du}{dy} = -sin\ y = -\sqrt{(1 - cos^2 y)} = -\sqrt{(1 - u^2)}$$

$$= -\sqrt{\left\{1 - \left(\frac{1 - x^2}{1 + x^2}\right)^2\right\}} = -\frac{2x}{1 + x^2}.$$

Also, $\qquad \dfrac{du}{dx} = \dfrac{d}{dx}\left(\dfrac{1 - x^2}{1 + x^2}\right)$

$$= \frac{(1 + x^2)(-2x) - (1 - x^2)(2x)}{(1 + x^2)^2}$$

$$= \frac{-4x}{(1 + x^2)^2}$$

But $\qquad \dfrac{dy}{dx} = \dfrac{dy}{du} \times \dfrac{du}{dx} = \left(1\Big/\dfrac{du}{dy}\right) \times \dfrac{du}{dx}$

$$= -\frac{(1 + x^2)}{2x} \times \frac{-4x}{(1 + x^2)^2}$$

$$= \frac{2}{1 + x^2}.$$

EXERCISES 8 (e)

1. Show that $\dfrac{d}{dx}(cot^{-1} x) = -\dfrac{1}{1 + x^2}$.

2. If a is a constant, show that $\dfrac{d}{dx}\left(cos^{-1}\dfrac{x}{a}\right) = -\dfrac{1}{\sqrt{(a^2 - x^2)}}$.

3. Show that $\dfrac{d}{dx}(cosec^{-1} x) = -\dfrac{1}{x\sqrt{(x^2 - 1)}}$.

4. If $y = tan^{-1}\left(\dfrac{2x}{1 - x^2}\right)$, show that $\dfrac{dy}{dx} = \dfrac{2}{1 + x^2}$.

5. Differentiate $x\ sin^{-1} x$ with respect to x.

6. If $y = (sin^{-1} x)/\sqrt{(1 - x^2)}$, show that $(1 - x^2)\dfrac{dy}{dx} - xy = 1$.

8.10. Differentiation of implicit functions

So far we have established rules for finding the differential coefficients of *explicit* functions only. When the dependent variable y is not given explicitly in terms of the independent variable x it is not necessary, nor indeed is it usually possible, to solve for y in terms of x.

Suppose, for example, that y is given *implicitly* in terms of x by the equation

$$x^2 + y^2 = 2x. \tag{8.14}$$

Since y^2 is a function of y and y is a function of x, the rule for differentiating a function of a function (8.10) gives

$$\frac{d}{dx}(y^2) = \frac{d}{dy}(y^2) \times \frac{dy}{dx} = 2y\frac{dy}{dx}.$$

Hence differentiating each term of equation (8.14) with respect to x we have

$$2x + 2y\frac{dy}{dx} = 2,$$

giving
$$\frac{dy}{dx} = \frac{1-x}{y}.$$

In this example we can first solve equation (8.14) for y to give

$$y = \sqrt{(2x - x^2)},$$

and then find dy/dx from the equations $y = \sqrt{u}$ where $u = 2x - x^2$. This procedure would give

$$\frac{dy}{du} = \frac{1}{2\sqrt{u}}, \quad \frac{du}{dx} = 2 - 2x,$$

and then,
$$\frac{dy}{dx} = \frac{dy}{du} \times \frac{du}{dx} = \frac{1}{2\sqrt{u}}(2 - 2x)$$

$$= \frac{1-x}{\sqrt{u}} = \frac{1-x}{y},$$

as before, but this method is more laborious.

As an example of a case where it is not possible to express y explicitly in terms of x, consider the equation

$$x + y + \cos y = 4. \tag{8.15}$$

The differential coefficients of y and $\cos y$ with respect to x are respectively $\frac{dy}{dx}$ and $-\sin y\frac{dy}{dx}$, so term by term differentiation of (8.15) gives

$$1 + \frac{dy}{dx} - \sin y\frac{dy}{dx} = 0,$$

leading to
$$\frac{dy}{dx} = \frac{1}{\sin y - 1}.$$

Example 10. *Find dy/dx when (i) $\sqrt{x} + \sqrt{y} = 2$, (ii) $x^m y^n = const.$*

(i) $\sqrt{x} + \sqrt{y} = 2$, so that

$$\frac{1}{2\sqrt{x}} + \frac{1}{2\sqrt{y}}\frac{dy}{dx} = 0,$$

giving

$$\frac{dy}{dx} = -\sqrt{\left(\frac{y}{x}\right)}.$$

(ii) $x^m y^n = const.$, and the rule for the differentiation of a product gives

$$mx^{m-1}y^n + x^m . ny^{n-1}\frac{dy}{dx} = 0,$$

so that

$$\frac{dy}{dx} = -\frac{mx^{m-1}y^n}{nx^m y^{n-1}} = -\frac{my}{nx}.$$

8.11. List of standard forms

The differential coefficients of certain standard functions and the rules for differentiation established in this chapter are most important. They are here collected for easy reference. In all cases C denotes a constant and u, v are functions of x.

$$\frac{d}{dx}(u + v) = \frac{du}{dx} + \frac{dv}{dx}.$$

$$\frac{d}{dx}(Cu) = C\frac{du}{dx}.$$

$$\frac{d}{dx}(uv) = v\frac{du}{dx} + u\frac{dv}{dx}.$$

$$\frac{d}{dx}\left(\frac{u}{v}\right) = \frac{v\frac{du}{dx} - u\frac{dv}{dx}}{v^2}.$$

$$\frac{dy}{dx} = \frac{dy}{du} \times \frac{du}{dx}.$$

$$\frac{dx}{dy} = 1\bigg/\frac{dy}{dx}.$$

$$\frac{d(C)}{dx} = 0. \qquad\qquad \frac{d}{dx}(\sin x) = \cos x.$$

$$\frac{d}{dx}(x^n) = nx^{n-1}. \qquad\qquad \frac{d}{dx}(\cos x) = -\sin x.$$

$$\frac{d}{dx}(\tan x) = \sec^2 x. \qquad\qquad \frac{d}{dx}(\cot x) = -\operatorname{cosec}^2 x.$$

$$\frac{d}{dx}(\operatorname{cosec} x) = -\operatorname{cosec} x \cot x. \qquad \frac{d}{dx}(\sec x) = \sec x \tan x.$$

$$\frac{d}{dx}(\sin^{-1} x) = \frac{1}{\sqrt{(1 - x^2)}}. \qquad \frac{d}{dx}(\tan^{-1} x) = \frac{1}{1 + x^2}.$$

8.12. Higher derivatives

If y be a function of x, the differential coefficient will itself be a function of x. The result of differentiating dy/dx with respect to x is called the *second differential coefficient of y with respect to x* or the *second derivative*. Proceeding further, the differential coefficient of the second derivative is called the third *differential coefficient* or *third derivative* and so on.

A conventional notation for the first, second, third, ... and nth differential coefficients of y with respect to x is

$$\frac{dy}{dx}, \frac{d^2y}{dx^2}, \frac{d^3y}{dx^3}, \cdots, \frac{d^ny}{dx^n}.$$

If y is a function of x given by $y = f(x)$, the notation

$$f'(x), f''(x), f'''(x), \ldots, f^{(n)}(x),$$

is also sometimes used for the first, second, third, ..., nth derivatives.

There are but few cases in which the general expression for the nth derivative of a function can be found. Here we shall be concerned only with the first few differential coefficients and shall not attempt to discuss the general derivative.

Example 11. *Find* $\dfrac{d^2y}{dx^2}$ *and* $\dfrac{d^3y}{dx^3}$ *when* (i) $y = x^{10}$, (ii) $y = \cos 2x$.

(i) $\dfrac{dy}{dx} = 10x^9$, $\dfrac{d^2y}{dx^2} = 90x^8$, $\dfrac{d^3y}{dx^3} = 720x^7$.

(ii) $\dfrac{dy}{dx} = -2\sin 2x$, $\dfrac{d^2y}{dx^2} = -4\cos 2x$, $\dfrac{d^3y}{dx^3} = 8\sin 2x$.

Example 12. *Find* $\dfrac{d^2y}{dx^2}$ *when* (i) $y = x^2(1-x)^2$, (ii) $y = x \sin x$.

(i) $y = x^2(1-x)^2 = x^2 - 2x^3 + x^4$,

$$\frac{dy}{dx} = 2x - 6x^2 + 4x^3,$$

$$\frac{d^2y}{dx^2} = 2 - 12x + 12x^2.$$

(ii) $y = x \sin x$.

$$\frac{dy}{dx} = \sin x + x \cos x,$$

$$\frac{d^2y}{dx^2} = \cos x + \cos x - x \sin x = 2\cos x - x \sin x.$$

EXERCISES 8 (f)

Find dy/dx when:—

1. $x^3y^2 - x = 0$.
2. $y^2 - \sin 2x = 4$.

3. $x^2 + xy + y^2 = a^2$, (a constant).

4. Find $dr/d\theta$ when $r^2 \cos \theta = \text{const.}$

5. If $y = \tan x$, show that

$$\frac{d^2 y}{dx^2} = 2 \tan x + 2 \tan^3 x.$$

6. Show that

$$\frac{d^3(UV)}{dx^3} = \frac{d^3 U}{dx^3} V + 3\frac{d^2 U}{dx^2}\frac{dV}{dx} + 3\frac{dU}{dx}\frac{d^2 V}{dx^2} + U\frac{d^3 V}{dx^3},$$

where U and V are functions of x. (Q.E.)

7. Evaluate $\dfrac{d^2}{dx^2}\{(1 + 4x + x^2) \sin x\}$. (Q.E.)

8. If $y = (\cos x)/x$, prove that

$$\frac{d^2 y}{dx^2} + \frac{2}{x}\frac{dy}{dx} + y = 0.$$ (L.U.)

EXERCISES 8 (g)

1. Find (i) $\dfrac{d}{dx}(\tfrac{2}{3}x^3 - \tfrac{3}{2}x^2 + 5)$, (ii) $\dfrac{d}{dt}(3t^{3/2} - 2t^{1/2} + 6t)$,

(iii) $\dfrac{d}{d\theta}(\theta^{-2} - 4\theta^{-1/2})$.

2. Find dy/dx when (i) $y = \cos x + x \sin x$, (ii) $y = (3x - 1)(x - 3)$. (O.C.)

3. Find $d\theta/dt$ when (i) $\theta = \sin t \sin 3t$, (ii) $\theta = t^2 \sin^{-1} t$. (L.U.)

4. Differentiate with respect to x:—

(i) $(2 - x^2) \cos x + 2x \sin x$, (ii) $(1 - 1/x) \tan x$.

5. Find (i) $\dfrac{d}{dx}\{x^2(1 - 3x)^3\}$, (ii) $\dfrac{d}{d\theta}\{\theta^2 \cos^2 2\theta\}$.

6. Find dy/dx when

(i) $y = \dfrac{x^3 + x}{x + 1}$, (ii) $y = \dfrac{x^{3/2}}{1 + x^{1/2}}$, (iii) $y = \dfrac{1 - \sqrt{x}}{1 + \sqrt{x}}$.

7. Differentiate with respect to x:—

(i) $\dfrac{x - 1}{x^2 - 4}$, (ii) $\dfrac{(x - 2)(x - 4)}{x - 3}$, (iii) $\dfrac{\cos x}{1 + x}$.

8. Find dy/dx when:—

(i) $y = \dfrac{x}{\sin x}$, (ii) $y = \dfrac{\tan x}{x}$, (iii) $y = \dfrac{\sin x}{2 + 3 \cos x}$.

9. Differentiate the following functions with respect to x and simplify the resulting functions as far as possible.

(i) $x^n \tan nx$, (ii) $\dfrac{\cos (x^2 + 1)}{\sin (x^2 - 1)}$. (Q.E.)

10. If y is a function of x, find the differential coefficients with respect to x of:—

$$\text{(i) } xy^2, \quad \text{(ii) } (3y+2)^2, \quad \text{(iii) } x/y, \quad \text{(iv) } y/x.$$

11. Differentiate with respect to x:—

$$\text{(i) } (x^2-x)^6, \quad \text{(ii) } \cos(2-3x), \quad \text{(iii) } x^2 \sin 2x. \quad \text{(L.U.)}$$

12. Find

$$\text{(i) } \frac{d}{dt}\left\{\frac{t}{\sqrt{(1+t^2)}}\right\}, \quad \text{(ii) } \frac{d}{d\theta}\left\{\sin^{-1}\left(\frac{1}{1+\theta}\right)\right\},$$

$$\text{(iii) } \frac{d}{dx}\left\{\frac{2x^2}{\sqrt{(x^2-1)}}\right\}.$$

13. (i) If $y = \tan x + \frac{1}{3}\tan^3 x$, show that $\frac{dy}{dx} = \sec^4 x$.

(ii) If $y = \sin^{-1}(\cos x)$, show that $\frac{dy}{dx} = -1$.

14. Find the value of:—

$$\text{(i) } \frac{d}{dx}\left\{\sin^{-1}\left(\tan\frac{x}{2}\right)\right\}, \quad \text{(ii) } \frac{d}{dx}\left\{\tan^{-1}\left(\sin\frac{x}{2}\right)\right\}.$$

15. Find $d\theta/dt$ when:—

$$\text{(i) } \theta = \cos^{-1}(1-2t^2), \quad \text{(ii) } \theta = \sin^{-1}(2t^3-1).$$

16. Differentiate with respect to x:—

$$\text{(i) } \cos(1/x), \quad \text{(ii) } \tan(x^2), \quad \text{(iii) } \sin^{-1}\left(\frac{1-x^2}{1+x^2}\right). \quad \text{(Q.E.)}$$

17. Show that $x = (g \sin 6\pi t)/4(6g - 36\pi^2)$ satisfies the relation

$$\frac{d^2x}{dt^2} + 6gx = \frac{g}{4}\sin 6\pi t. \quad \text{(Q.E.)}$$

18. If α and β are constants, show that the derivative with respect to x of each of the functions

$$2\sin^{-1}\sqrt{\left(\frac{x-\beta}{\alpha-\beta}\right)} \quad \text{and} \quad 2\tan^{-1}\sqrt{\left(\frac{x-\beta}{\alpha-x}\right)},$$

is $\{(\alpha-x)(x-\beta)\}^{-1/2}$.

19. If $y = \sin^{-1}(3x - 4x^3)$, show that $\sqrt{(1-x^2)}\frac{dy}{dx} = 3$.

20. If $u = \theta^2 + (\sin^{-1}\theta)^2 - 2\theta\sqrt{(1-\theta^2)}\sin^{-1}\theta$, show that

$$\sqrt{(1-\theta^2)}\frac{du}{d\theta} = 4\theta^2\sin^{-1}\theta.$$

21. Find dy/dx when $y^3 - 3yx^2 + 2x^3 = 0$.

22. If $y^2 - 2y\sqrt{(1+x^2)} + x^2 = 0$, show that

$$\frac{dy}{dx} = \frac{x}{\sqrt{(1+x^2)}}.$$

23. Find dy/dx when y is given by:—

    (i) $y^2 + x^2 = 4x + 1$, (ii) $4y^2 + xy - 3x^2 = 0$.

24. Find dy/dx when y is given by:—

    (i) $x^2 \sin y - y \cos x = 0$, (ii) $x \cos y - y^2 \sin x = 0$.

25. Find the slope of the tangent to the curve

$$xy^3 - 2x^2y^2 + x^4 - 1 = 0$$

    at the point where $x = 1$, $y = 2$.

26. Find d^2y/dx^2 when:—

    (i) $y = x^3 \sin x$, (ii) $y = x \tan^{-1} x$, (iii) $y = \dfrac{x^2}{1 + x}$.

27. If $y = \tan^2 x$, prove that

$$\frac{d^2y}{dx^2} = 2(1 + y)(1 + 3y).$$
        (O.C.)

28. If $y = \dfrac{\sin x}{x^2}$, find $\dfrac{dy}{dx}$ and $\dfrac{d^2y}{dx^2}$, and prove that

$$x^2\frac{d^2y}{dx^2} + 4x\frac{dy}{dx} + (x^2 + 2)y = 0.$$
        (L.U.)

29. If $y = (\tan^{-1} x)^2$, prove that

$$\frac{d}{dx}\left\{(1 + x^2)\frac{dy}{dx}\right\} = \frac{2}{1 + x^2}.$$

30. Show that if $u = \tan^{-1} \theta$, then

$$(1 + \theta^2)\frac{d^2u}{d\theta^2} + 2\theta\frac{du}{d\theta} = 0.$$

CHAPTER 9

SOME APPLICATIONS OF THE DIFFERENTIAL CALCULUS

9.1. Introduction

We have seen in Chapter 7 that the derivative is a measure of the slope of the tangent to the curve representing a function. The process of differentiation has therefore the geometrical application of finding the slope of a tangent to a curve but we shall delay giving examples of this application until we discuss the methods of coordinate geometry (Chapters 16, 17).

Other important applications already discussed are the use of the derivative as a rate measurer and in finding velocities and accelerations in dynamical problems. More eleaborate examples can be given now that the technical processes of differentiation have been studied. Further uses of the differential calculus occur in finding maximum and minimum values, and in curve tracing. These applications are discussed in the paragraphs which follow.

9.2. Some examples of the derivative as a rate measurer

We give below two examples in which the methods of the last chapter can be used in solving problems on rates of change.

Example 1. *The volume of a solid cube increases uniformly at k^3 cubic metres per second. Find an expression for the rate of increase of its surface-area when the area of a face is b^2 square metres.*

If x is the length of an edge of the cube at time t, the volume V is given by $V = x^3$. Differentiating with respect to t, we have

$$\frac{dV}{dt} = 3x^2\frac{dx}{dt}.$$

But dV/dt is the rate of increase of volume and this is k^3, so that

$$k^3 = 3x^2\frac{dx}{dt},$$

giving

$$\frac{dx}{dt} = \frac{k^3}{3x^2}.$$

The surface-area S of the cube is $6x^2$, so that the rate of increase of surface-area at time t is

$$\frac{dS}{dt} = 12x\frac{dx}{dt}.$$

Substituting for dx/dt we find

$$\frac{dS}{dt} = 12x \times \frac{k^3}{3x^2} = \frac{4k^3}{x}.$$

When the area of a face is b^2, $x = b$ and the required rate of increase of surface-area at this instant is obtained by writing $x = b$ in the expression for dS/dt, giving $4k^3/b$.

Example 2. *A hollow circular cone with vertical angle 90° and height 0·36 m is inverted and filled with water. This water begins to leak away through a small hole in the vertex. If the level of the water begins to sink at the rate of 0·01 metres in 120 seconds, and the water continues to leak away at the same rate, at what rate is the level sinking when the water is 0·24 m from the top? (The volume of a cone is one-third the area of the base times the height.)*

Let (Fig. 49) the height of the water at time t be h, and let the volume of the water then be V. Then, $V = \frac{1}{3}\pi r^2 h$, where r is the radius of the water surface.

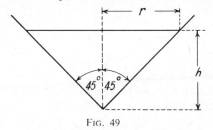

FIG. 49

Since the semi-vertical angle of the cone is 45°, it is clear that $r = h$, so that $r = h$, so that

$$V = \tfrac{1}{3}\pi h^3.$$

By the rule for differentiating a function of a function,

$$\frac{dV}{dt} = \frac{dV}{dh} \times \frac{dh}{dt}$$

$$= \pi h^2 \frac{dh}{dt}. \tag{9.1}$$

Since the level of the water is *decreasing* at the rate of $(0·01/120)$ m/s when the cone is full, i.e. when $h = 0·36$ m,

$$\frac{dV}{dt} = \pi \times (0·36)^2 \times \frac{0·01}{120} \text{ m}^3/\text{s}$$

(a negative sign meaning a negative rate of increase or a rate of decrease). dV/dt remains constant at this value. When the water is 0·24 m from the top, $h = 0·36 - 0·24 = 0·12$ m, and substitution in (9.1) gives

$$\pi \times (0 36)^2 \times \frac{0·01}{120} = \pi \times (0·12)^2 \frac{dh}{dt},$$

so that

$$\frac{dh}{dt} = -\frac{0·01}{120} \times \left(\frac{0·36}{0·12}\right)^2 = -7·5 \times 10^{-4},$$

showing that the water is sinking at 0·75 mm/s at this instant.

9.3. Some dynamical applications

We have already seen (§ 7.7) that if a body, moving in a straight line, has travelled a distance s and acquired a velocity v in time t, then

$$v = \frac{ds}{dt},\qquad (9.2)$$

and its acceleration (a) at time t is given by

$$a = \frac{dv}{dt}.\qquad (9.3)$$

Alternative expressions for the acceleration can be found as follows. Firstly, combining (9.2) and (9.3),

$$a = \frac{d}{dt}\left(\frac{ds}{dt}\right) = \frac{d^2 s}{dt^2}.\qquad (9.4)$$

Secondly, since v is a function of s and s is a function of t,

$$a = \frac{dv}{dt} = \frac{dv}{ds} \times \frac{ds}{dt} = \frac{dv}{ds} \times v = v\frac{dv}{ds}.\qquad (9.5)$$

Thus the acceleration may be expressed in any one of the three equivalent forms

$$\frac{dv}{dt},\quad \frac{d^2 s}{dt^2},\quad v\frac{dv}{ds}.$$

In mechanical applications, differential coefficients with respect to the *time* are often denoted by dots placed above the dependent variable. Thus ds/dt, $d^2 s/dt^2$, dv/dt are denoted by $\dot{s}$, $\ddot{s}$ and $\dot{v}$ respectively. In this notation, equations (9.2), (9.3) and (9.4) would be written

$$v = \dot{s},\quad a = \dot{v},\quad a = \ddot{s}.$$

Example 3. *The distance s moved in a straight line by a particle in time t is given by $s = at^2 + bt + c$, where a, b and c are constants. If v is the velocity of the particle at time t, show that $4a(s - c) = v^2 - b^2$.* (L.U.)

In the notation just given,

$$v = \dot{s} = 2at + b,$$

so that

$$4a(s - c) = 4a(at^2 + bt + c - c) = 4a(at^2 + bt)$$
$$= 4a^2 t^2 + 4abt = (2at + b)^2 - b^2$$
$$= v^2 - b^2.$$

Example 4. *If the velocity of a body varies inversely as the square root of the distance, prove that the acceleration varies as the fourth power of the velocity.*

Denoting the distance travelled by s, the velocity and acceleration by v and a we have,

$$v = \frac{k}{\sqrt{s}}, \text{ where } k \text{ is a constant.}$$

$$a = v\frac{dv}{ds} = \frac{k}{\sqrt{s}}\cdot\left(\frac{-k}{2s^{3/2}}\right)$$

$$= -\frac{k^2}{2s^2} = -\frac{k^2}{2}\cdot\frac{v^4}{k^4} = -\frac{v^4}{2k^2},$$

since $\sqrt{s} = k/v$, and we have established the result required.

EXERCISES 9 (a)

1. A conical vessel has a vertical angle of $60°$. If liquid is poured in at a rate of 2×10^{-5} m³/s, find the rate at which the level is rising when the depth of liquid in the vessel is 0·05 m.

2. The radius of a sphere is r, the area of its surface is $4\pi r^2$, and its volume is $\frac{4}{3}\pi r^3$; if, when the radius of the sphere is 21 m, it is increasing at the rate of 0·01 m/s, find the rates at which the surface and volume are increasing at the same time.

3. The inner and outer radii of a cylindrical tube of constant length change in such a way that the volume of the material forming the tube remains constant. Find the rate of increase of the outer radius at the instant when the radii are 0·03 m and 0·05 m and the rate of increase of the inner radius is 0·5 mm/s.

4. The displacement x at time t of a moving particle is given by
$$x = a \sin 2t + b \cos 2t,$$
where a and b are constants. If v is the speed at time t, prove that $v = 2\sqrt{(a^2 + b^2 - x^2)}$.   (L.U.)

5. If the velocity of a body varies as the square of the distance travelled, show that the acceleration of the body varies as the cube of the distance.

6. s is the distance moved and v the velocity acquired by a body moving in a straight line at time t. If (i) $v = u + ft$, (ii) $v^2 = u^2 + 2fs$, where u and f are constants, show that in each case the acceleration of the body is f.

9.4. Maximum and minimum

Suppose the graph of $y = f(x)$ is as shown in Fig. 50. Points such as A, B, C are called *turning points*. As x increases, the values taken by y increase until the point A is reached, decrease from A to B, increase again from B to C and then decrease. At A, B and C, y is neither increasing nor decreasing.

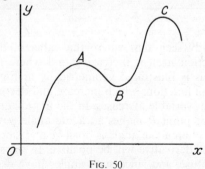

FIG. 50

Sometimes the points *A*, *B* and *C* are referred to as points of *maximum* or *minimum* values, maxima at *A* and *C* and minimum at *B*. It should be noted that a maximum or minimum value is the greatest or least value *in the neighbourhood*, but it need not be the absolutely greatest or least value. For example, there are points on the left of the curve of Fig. 50 for which the values of *y* are less than the "minimum value" at *B* and there are points on the right where the values are greater than the maximum value at *A*. Again in Fig. 51, which shows the graph of $y = 2x^2 - x^4$, there are maximum values

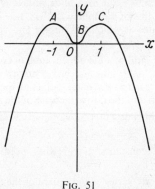

FIG. 51

where $x = \pm 1$ and these are also greatest values, but the minimum value at $x = 0$ is not a least value, for there are points on the curve with smaller values of *y* than that at *O*.

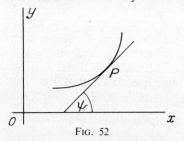

FIG. 52   FIG. 53

We have already seen that where the value of the derivative is positive, the function itself is increasing and where it is negative, it is decreasing. This is illustrated geometrically in Figs. 52, 53 which show the graphs of functions which increase and decrease respectively as the independent variable increases. In the first diagram, the tangent at a representative point *P* makes an acute angle ψ with the x-axis; since the tangent of an acute angle is positive and since, by definition, $dy/dx = \tan \psi$, the derivative will be positive. In the second diagram, the angle ψ is obtuse, and since such angles have negative tangents, the derivative will be negative. At points like A, B or C in Figs. 50 or 51, the tangents to the curves will be *parallel* to the x-axis. At such points, the tangent will make a zero angle with the x-axis and we shall have

$$\frac{dy}{dx} = 0.$$

Returning to Fig. 50, shown again in Fig. 54, the signs of the derivative have now been marked in. Immediately to the left of the point A, the function is increasing and its derivative is positive. At point A,

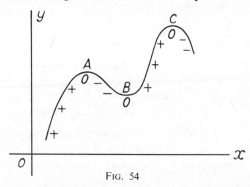

FIG. 54

the function is neither increasing nor decreasing and the derivative is zero. Immediately to the right of point A, the function is decreasing and so has a negative derivative, and so on. We can formulate the following rules for determining the position of turning points and distinguishing between maximum and minimum values:—

(*a*) At a turning point, $dy/dx = 0$.

(*b*) At a point giving a *maximum* value, dy/dx changes from *positive* to *negative* as x takes values just less and just greater respectively than the value at the turning point.

(*c*) At a point giving a *minimum* value, dy/dx changes from *negative* to *positive* as x takes values just less and just greater respectively than the value at the turning point.

Example 5. *Find the turning point on the curve* $y = x^2 - 2x$ *and determine whether it is a point of maximum or minimum* y.

Here
$$\frac{dy}{dx} = 2x - 2,$$

and dy/dx vanishes when $2x - 2 = 0$, i.e., when $x = 1$. Hence the point $x = 1$, $y = (1)^2 - 2(1) = -1$ is a turning point on the curve. For $x = 0.9$ (a value a little less than the value $x = 1$ at the turning point),

$$\frac{dy}{dx} = 2 \times 0.9 - 2 = -0.2.$$

For $x = 1.1$ (a value a little greater than $x = 1$),

$$\frac{dy}{dx} = 2 \times 1.1 - 2 = +0.2.$$

The derivative therefore changes from negative to positive, so the point $x = 1$, $y = -1$ gives a point of minimum y. The graph of $y = x^2 - 2x$ is shown in Fig. 55.

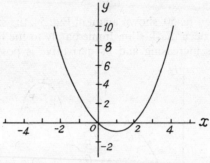

FIG. 55

An alternative method of discrimination between maximum and minimum values can be obtained as follows. If we plot on the same diagram the graphs of $y = f(x)$ and $y = f'(x)$ (the derived function),

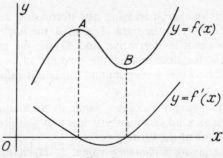

FIG. 56

we shall obtain a diagram such as that shown in Fig. 56. For points to the left of the maximum value at A, $f(x)$ is increasing and $f'(x)$ is positive, for points between A and the minimum value at B, $f(x)$ is decreasing and $f'(x)$ is negative, and for points to the right of B, $f(x)$ is again increasing and $f'(x)$ is positive. The derived function $f'(x)$ is zero for values of x corresponding to the points A and B. Considering the graph $y = f'(x)$, we see that for a value of x corresponding to the point A, $f'(x)$ is decreasing and therefore has a negative derivative. Thus at point A, $f''(x)$ (or d^2y/dx^2) is negative. Similarly at point B, $f'(x)$ is increasing and its derivative $f''(x)$ is positive. Hence at turning points giving maximum values

$$\frac{d^2y}{dx^2} < 0,$$

and at turning points giving minimum values

$$\frac{d^2y}{dx^2} > 0.$$

Applied to Example 5, $y = x^2 - 2x$, $dy/dx = 2x - 2$ and there is a turning point where $2x - 2 = 0$, i.e., where $x = 1$. For this curve

$$\frac{d^2y}{dx^2} = 2,$$

and, this being positive, the turning point is one giving a minimum value to y. This method is further illustrated in the following example.

Example 6. *Find the maximum and minimum ordinates of the curve $y = x^2(x + 1)$.*

$$y = x^2(x + 1) = x^3 + x^2.$$

$$\frac{dy}{dx} = 3x^2 + 2x, \quad \frac{d^2y}{dx^2} = 6x + 2.$$

dy/dx vanishes when $3x^2 + 2x = 0$, i.e., when $x = 0$ and when $x = -2/3$. When $x = 0$, $d^2y/dx^2 = 2$; this being positive, $x = 0$ gives a minimum value to y, the minimum ordinate being $y = 0$.

When $x = -2/3$, $d^2y/dx^2 = 6(-2/3) + 2 = -2$; this being negative, $x = -2/3$ gives a maximum ordinate of

$$(-2/3)^2(-2/3 + 1), \quad \text{or} \quad 4/27.$$

EXERCISES 9 (b)

1. Find the values of x for which the expression $(x - 2)(x - 3)^2$ has maximum and minimum values and discriminate between them.

2. If $y = (x - 2)(x + 1)^2$, find the maximum and minimum values of y, stating which is which.

3. Find the maximum and minimum values of the expression

$$\frac{3x}{(x - 1)(x - 4)}.$$

4. Find the maximum and minimum values of the function

$$(x - 1)(x - 2)/x$$

and illustrate your result by drawing the graph of the function between $x = -3$ and $x = 3$.

5. Find the maximum and minimum values of the function $2 \sin t + \cos 2t$ and discriminate between them.

6. Show that the function $x^3 - 6x^2 + 18x + 5$ increases with x for all values of x. Find the value of the function when the rate of increase is least.

9.5. Applications to practical problems

Many practical problems can be solved by the method of the last section. It sometimes happens that the quantity whose maximum or minimum value is sought appears at first to be a function of more than one variable. In such cases it is often possible, by means of geometrical or other relations between the variables, to eliminate all but one of these variables. Once the quantity has been expressed in terms of a single variable, the procedure is identical to that given in the previous section. We differentiate with respect to the single variable remaining and the values which make the derivative vanish include those giving maximum and minimum values to the quantity under discussion. In many cases it is unnecessary to examine the changes in sign of the derivative (or the sign of the second derivative, if that method is used) to discriminate between maximum and minimum values, for it is often possible to see at once on physical grounds whether the solution leads to a maximum or minimum. Some illustrative examples follow.

Example 7. *Find the height of the right circular cylinder of greatest volume which can be cut from a sphere of radius a.* (L.U.)

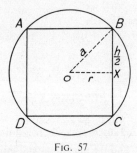

FIG. 57

In Fig. 57, O is the centre of the sphere and $ABCD$ is a plane section of the cylinder through its axis. If X is the mid-point of the generator BC of the cylinder and if the radius and height of the cylinder are respectively r and h, the right-angled triangle BOX gives

$$r^2 + \tfrac{1}{4}h^2 = a^2. \qquad (9.6)$$

The volume V of the cylinder* is given by

$$V = \pi r^2 h.$$

Since, from (9.6),

$$r^2 = a^2 - \tfrac{1}{4}h^2,$$
$$V = \pi h(a^2 - \tfrac{1}{4}h^2) = \pi(a^2 h - \tfrac{1}{4}h^3),$$

the volume is now expressed in terms of the single variable h. V is a maximum

---

* See § 20.5.

or minimum when $dV/dh = 0$, i.e., when

$$\pi(a^2 - \tfrac{3}{4}h^2) = 0$$

or when $h = 2a/\sqrt{3}$.

This value of h does in fact give a *maximum* value for the volume of the cylinder since

$$\frac{d^2V}{dh^2} = -\frac{3}{2}\pi h = -\sqrt{3}\pi a,$$

which is a negative quantity.

Example 8. *A despatch rider is in open country at a distance of 6 kilometres from the nearest point P of a straight road. He wishes to proceed as quickly as possible to a point Q on the road 20 kilometres from P. If his maximum speed, across country, is 40 kilometres per hour and, along the road, 50 kilometres per hour, find at what distance from P he should strike the road.* (L.U.)

In Fig. 58, the rider starts from a point A and strikes the road at a point B, x kilometres along the road from P. Then $AP = 6$, $PB = x$, $PQ = 20$, $BQ = PQ - PB = 20 - x$ kilometres.

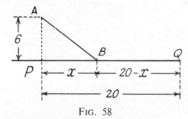

FIG. 58

Since the angle APB is a right angle,

$$AB = \sqrt{(36 + x^2)} \text{ kilometres.}$$

Along AB, the rider's speed is 40 km/h and along BQ it is 50 km/h, so that the times taken to traverse AB and BQ are respectively

$$\frac{\sqrt{(36 + x^2)}}{40} \quad \text{and} \quad \frac{20 - x}{50} \text{ hours.}$$

The total time T for the journey is therefore given by

$$T = \frac{\sqrt{(36 + x^2)}}{40} + \frac{20 - x}{50},$$

and for the journey to be accomplished as quickly as possible, this must be a minimum, or $dT/dx = 0$. Now

$$\frac{dT}{dx} = \frac{1}{40} \times \frac{2x}{2\sqrt{(36 + x^2)}} - \frac{1}{50}.$$

This vanishes when $\sqrt{(36 + x^2)} = 5x/4$, i.e., when $36 + x^2 = 25x^2/16$ or when $9x^2/16 = 36$, giving $3x/4 = 6$ or $x = 8$ kilometres.

Example 9. *A straight line AB has its ends on two fixed perpendicular lines OX, OY and passes through a fixed point C whose distances from the fixed lines are a, b. Find the position of AB which makes the triangle AOB of minimum area and calculate that minimum area.* (L.U.)

Draw CD, CE perpendicular respectively to OX and OY. Let AB make an angle θ with OX. Then

$$AD = a \cot \theta, \quad BE = b \tan \theta.$$

The area of the triangle AOB is the sum of the areas of the rectangle $ODCE$ and the two triangles DAC, ECB.

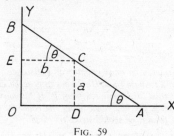

FIG. 59

Hence if Δ is the area of the triangle AOB,

$$\Delta = ab + \tfrac{1}{2}a \cot \theta \cdot a + \tfrac{1}{2}b \cdot b \tan \theta$$
$$= ab + \tfrac{1}{2}a^2 \cot \theta + \tfrac{1}{2}b^2 \tan \theta. \qquad (9.7)$$

This is a minimum when $d\Delta/d\theta = 0$, i.e., when

$$-\tfrac{1}{2}a^2 \operatorname{cosec}^2 \theta + \tfrac{1}{2}b^2 \sec^2 \theta = 0,$$

or when $\tan \theta = a/b$.

The value of the minimum area is obtained by substituting the value of θ given above in (9.7), giving

$$\Delta_{\min} = ab + \tfrac{1}{2}a^2(b/a) + \tfrac{1}{2}b^2(a/b)$$
$$= 2ab.$$

EXERCISES 9 (c)

1. An open cylindrical vessel is to be constructed from a given amount of uniform thin material. Show that it contains the greatest possible volume when its height is equal to the radius of its base.   (O.C.)

2. A piece of wire, which forms the circumference of a circle of 0·12 m radius, is cut and bent so as to form two new circles. Find the radius of each circle in order that the sum of the areas of the two circles shall be as small as possible.   (O.C.)

3. Square pieces are cut out of a square sheet of metal as shown in the figure, and the remainder is folded about the dotted lines so as to form an open

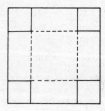

FIG. 60

box. If the length of the edge of the square sheet is 0·18 m, find the maximum volume of the box.

4. A cylindrical tin canister without a lid is made of sheet metal. If S is the area of the sheet used, without waste, V the volume of the canister and r the radius of the cross-section, prove that
$$2V = Sr - \pi r^3.$$
If S is given, prove that the volume of the canister is greatest when the ratio of the height to the diameter is $1:2$.   (L.U.)

5. $ABCD$ is a square ploughed field of side 132 metres, with a path along its perimeter. A man can walk at 8 km/h along the path, but only at 5 km/h across the field. He starts from A along AB, leaves AB at a point P, and walks straight from P to C. Find the distance of P from A, if the time taken is the least possible.   (L.U.)

6. An isosceles triangle of vertical angle 2θ is inscribed in a circle of radius a. Show that the area of the triangle is $4a^2 \sin\theta \cos^3\theta$ and hence that the area is a maximum when the triangle is equilateral.   (L.U.)

7. The sum of the perimeters of two rectangles is 1·98 m. The ratio of length to breadth is $3:2$ for one rectangle and $4:3$ for the other. Find the minimum value for the sum of their areas.

8. A piece of wire 0·1 m long is cut into two parts one of which is bent into a circle, and the other into a square. If the sum of the areas of the circle and of the square is to be a minimum, find the radius of the circle.   (Q.E.)

9.6. Points of inflexion

Consider the function $y = (x - 1)^3(12x^2 - 9x - 43)$. By the rule for differentiating a product,

$$\frac{dy}{dx} = 3(x - 1)^2(12x^2 - 9x - 43) + (x - 1)^3(24x - 9)$$

$$= (3(x - 1)^2\{12x^2 - 9x - 43 + (x - 1)(8x - 3)\}$$
$$= 3(x - 1)^2\{20x^2 - 20x - 40\}$$
$$= 60(x - 1)^2(x + 1)(x - 2).$$

Hence dy/dx vanishes when $x = -1$, $x = 1$ and $x = 2$. When $x < -1$, dy/dx is positive, when x lies between -1 and 1, dy/dx is negative, when x lies between 1 and 2, dy/dx is negative and when $x > 2$, dy/dx is positive. Since dy/dx changes sign from positive to negative as x increases through $x = -1$, $x = -1$ gives a maximum value and since dy/dx changes from negative to positive as x passes through $x = 2$, $x = 2$ gives a minimum value to y. Although dy/dx vanishes when $x = 1$, dy/dx does not change sign as x passes through this value and, although the tangent to the curve is parallel to the x-axis at $x = 1$, this point is neither a maximum nor a minimum. Such a point is called a *point of inflexion*: a rough graph of the function is shown in Fig. 61 and the tangents at and adjacent to the critical points A, B and C are shown in the subsidiary diagrams.

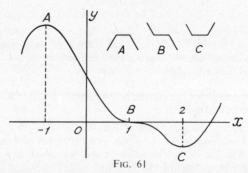

FIG. 61

As we pass through the point of inflexion B, the derivative changes from negative, through zero, to negative again. A graph of the derivative would therefore show a maximum at B and another method of finding points of inflexion would be to seek maximum (or minimum) values of the derivative. Thus, at a point of inflexion at which the tangent to the curve is parallel to the x-axis both

$$\frac{dy}{dx} = 0 \quad \text{and} \quad \frac{d^2y}{dx^2} = 0,$$

and the second derivative changes sign as we pass through the critical point.

At a point of inflexion, the curve "crosses its tangent" and such points can, of course, occur when the tangent is not parallel to the x-axis. At such points the second derivative can be shown to vanish but the first derivative is, of course, not itself zero. It is beyond the scope of the present book to consider these points in further detail and we simply state the rule that *the second derivative changes sign as we pass through and vanishes at a point of inflexion.*

To sum up the results of the last three sections we have:—

$$\frac{dy}{dx} = 0, \quad \frac{d^2y}{dx^2} \text{ positive; minimum value for } y.$$

$$\frac{dy}{dx} = 0, \quad \frac{d^2y}{dx^2} \text{ negative; maximum value for } y.$$

$$\qquad \frac{d^2y}{dx^2} = 0, \quad \frac{d^2y}{dx^2} \text{ changing sign, point of inflexion.}$$

Example 10. *Find the turning points and point of inflexion on the curve*
$$y = x^5 - 5x^4 + 5x^3 - 1.$$

Here,

$$\frac{dy}{dx} = 5x^4 - 20x^3 + 15x^2 = 5x^2(x^2 - 4x + 3)$$

$$= 5x^2(x - 1)(x - 3),$$

$$\frac{d^2y}{dx^2} = 20x^3 - 60x^2 + 30x = 10x(2x^2 - 6x + 3).$$

Hence the first derivative vanishes when $x = 0$, 1 and 3. When $x = 0$, the second derivative vanishes, when $x = 1$, the second derivative is -10, and when $x = 3$ it is 90. Hence $x = 1$ gives a maximum, $x = 3$ gives a minimum. Since the second derivative is negative for small negative values of x, positive for small positive values of x and zero at $x = 0$, there is a point of inflexion for this value of x. Since the first derivative also vanishes when $x = 0$, the tangent to the curve is parallel to the x-axis at the point of inflexion.

9.7. Curve sketching

It is often useful to be able to make a rough sketch of a curve without going to the labour of actually plotting a large number of points on it. The following procedure, either wholly or in part, should enable a good idea of the shape of a curve to be obtained.

  (i) Determine if the curve is symmetrical about either or both axes of coordinates. Symmetry about the x-axis occurs if the equation contains only even powers of y and about the y-axis if the equation contains only even powers of x.

 (ii) Determine if there is symmetry about the origin. Such symmetry occurs when a change in the sign of x causes a change in the sign of y without altering its numerical value.

(iii) Seek values of x which make y^2 negative and therefore y imaginary. No real points occur on the curve for such values of x.

 (iv) Find where the curve crosses the axes of coordinates. The curve cuts the x-axis at points for which $y = 0$ and it cuts the y-axis where $x = 0$. It passes through the origin if $y = 0$ when $x = 0$.

  (v) Find values of x (if any) which make y very large and values of y (if any) which make x very large.

 (vi) If the curve passes through the origin, its behaviour in this neighbourhood can sometimes be decided by studying the value of the ratio y/x. If this ratio is small the curve keeps close to the x-axis near the origin, if y/x is nearly unity, the direction of the curve bisects the angle between the axes, while if y/x is large the curve keeps near the y-axis. An alternative and better method is to study the value of the derivative dy/dx near the origin. Since this quantity measures the slope of the tangent to the curve, a small value means that the curve lies near the x-axis, a large value that it lies near the y-axis, while a value near unity means that the tangent at the origin approximately bisects the angle between the axes.

(vii) Find turning points and points of inflexion (if any) by the methods of this chapter.

Some illustrative examples follow.

Example 11. *Sketch the curve* $y = x^4 - 6x^2 + 8x + 10$.

Since odd values of x and y occur, there is no symmetry about the coordinate axes. A change in sign of x alters the value of y so that there is no symmetry about the origin. Points exist on the curve for all values of x. The curve crosses the y-axis where $y = 10$. It is not convenient to find quickly the points at which the curve crosses the x-axis for this requires the solution of a quartic equation. The curve does not pass through the origin. When x is large, the dominant term is x^4 and this is positive whatever the sign of x. Hence y is large and positive when x is large and positive or large and negative.

$$\frac{dy}{dx} = 4x^3 - 12x + 8$$

$$= 4(x - 1)^2(x + 2).$$

$$\frac{d^2y}{dx^2} = 12x^2 - 12.$$

Hence turning points or points of inflexion occur when $x = 1$ and $x = -2$.

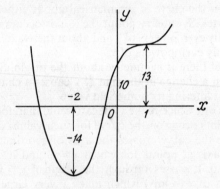

FIG. 62

When $x = 1$, the second derivative vanishes and changes sign, so $x = 1$ gives a point of inflexion where the tangent is parallel to the x-axis. When $x = 1$, the value of y is easily found to be 13. When $x = -2$, the second derivative is positive, so that $x = -2$ gives a minimum value to y, the value of y at this point being -14. A rough sketch of the curve is shown in Fig. 62.

Example 12. *Sketch the curve $y^2 = x^3$.*

Since only even powers of y occur, the curve is symmetrical about the x-axis. It is not symmetrical about the y-axis for an odd power of x occurs in the

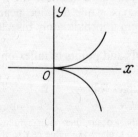

FIG. 63

equation. When x is negative, y^2 is negative and there are no real points of the curve for such values of x. The curve passes through the origin and since its equation can be written in the form $(y/x)^2 = x$, the ratio y/x is small when x is small. The curve therefore lies close to the x-axis near the origin. As x becomes large so does y. A sketch is given in Fig. 63.

EXERCISES 9 (d)

1. Find the values of x at the points of inflexion (if any) of the curve $y = 3x^4 - 4x^3 + 1$.

2. For what values of x are there points of inflexion on the curve $y = x + \sin x$ at which the tangent is parallel to the x-axis?

3. Discuss the nature of the points on the curve
$$y = 3x^4 - 8x^3 - 24x^2 + 96x$$
at which the tangent to the curve is parallel to the x-axis.

4. Give a rough sketch of the curve $a^2y = 4x^2(3a - 4x)$, where a is constant. (L.U.)

5. Sketch the graph of the curve $y^2 = x(5 - x)^2$. (L.U.)

6. Find (i) the slopes of the tangents to the curve whose equation is $y^2 = x^2(1 - x^2)$ at the points where $x = 0$ and $x = 1$, and (ii) the values of x at the turning points of the curve. Sketch the curve. (O.C.)

EXERCISES 9 (e)

1. A vessel is constructed so that the volume of water contained in it is
$$\frac{\pi}{192}(x^3 + 24x^2 + 192x),$$
when the depth is x. What is the rate of increase of volume per unit increase of x when (i) $x = 2$, (ii) $x = 4$? How many times faster does the surface rise when $x = 2$ than when $x = 4$, if water is poured in at a constant rate? (O.C.)

2. If the volume of a cone remains constant while the radius of its base is increasing at the rate of 1 per cent. per second, find the percentage rate per second at which its height is diminishing. (O.C.)

3. A trough 3 m long has its cross-section in the form of an isosceles triangle. The depth of the trough is 0·2 m and it is 0·25 m wide at the top. If water runs into it at the steady rate of 6×10^{-4} m³/s at what rate is the surface rising when the depth of water is 0·1 m?

4. A body moves in a straight line so that its distance s m from a fixed point O at time t seconds is given by
$$s = (t - 2)^2(2t - 7).$$
Find when the body passes through O and the velocity and acceleration each time it passes. Find also the minimum value of the velocity.

5. The velocity v m/s of a particle which has travelled a distance s metres from a fixed point is given by $v^2 = 16s$. Find the acceleration of the particle.

6. Find the values of x and y at the turning point of the curve
    $$ay = bx^2 + cx,$$
    where a, b and c are positive constants. Is the turning point a maximum or minimum?

7. Find the values of x at the turning points of the curve
    $$y = \frac{x^2 - x + 1}{x^2 + x + 1}.$$
    State which is a maximum and which a minimum.

8. Show that maximum and minimum values of $\cos^3 x \sin x$ occur when $\sin^2 x = 1/4$.

9. Show that the minimum value of $a \sec \theta - b \tan \theta$ is $\sqrt{(a^2 - b^2)}$.

10. Determine the value of x for which
    $$\sqrt{\left(x^{\frac{2}{n}} - x^{\frac{n+1}{n}} \right)}$$
    is a maximum; n is a constant.   (Q.E.)

11. A prism of square section contains 64 cubic metres of clay, the side of the square being x metres. Express the length of the prism in terms of x and find the total area of its faces.
    Show that the total area is a minimum when the prism is a cube.

12. A piece of wire of length l is cut into two portions of lengths x and $(l - x)$. Each portion is then cut into twelve equal parts which are soldered together so as to form the edges of a cube. Find an expression for the sum of the volumes of the two cubes so formed. What is the least value of the sum of the volumes?   (O.C.)

13. Find the height of the right circular cone of maximum volume, the sum of the height and radius of the base being 0·12 m.

14. Post Office regulations restrict parcels to a maximum length of 1·07 m and a maximum girth of 1·83 m. Find the maximum permissible volume of a rectangular parcel.
    Find also the length of the longest thin rod which can be packed inside a parcel of maximum permissible volume.

15. The point A is 7 kilometres due north of a point B. One man starts from A and walks due east at the rate of 3 km/h. Simultaneously a second man starts from B and walks due north at the rate of 4 km/h. Find the rate at which the distance between them is increasing when they are 15 kilometres apart. Find also the minimum distance between them.

16. A water tank with an open top and square horizontal cross-section is to contain 1 cubic metre. Find the cost of lining the tank with sheet lead at £0·37½ per square metre when that cost is the least possible.

17. In a triangle ABC, the angles B and C are equal. Prove that the maximum value of $\cos A + \cos B$ is 9/8.

18. Find the dimensions of the rectangle of greatest area which can be inscribed in a circle of radius r.

19. Given that the stiffness of a beam of rectangular cross-section varies directly as its breadth and as the cube of its depth, find the breadth of the stiffest beam of rectangular cross-section which can be cut from a cylindrical log of diameter 0·6 m. (Q.E.)

20. Find the quantity which, when added to the square of its reciprocal, gives a minimum sum.

21. Find the maximum and minimum value of y when
$$y = x^3 - 4x^2 - 3x + 1.$$
Find also the value of x at the point of inflexion. (Q.E.)

22. Find the abscissa of the point of inflexion on the curve
$$y = ax^3 + bx^2 + cx + d,$$
where a, b, c and d are constants.

23. Sketch the curve $(a - x)y^2 = (a + x)x^2$, where a is constant.

24. If a is a constant, trace the curve $x^2y = 4a^2(2a - y)$.

25. Sketch the graph of $y = x/(x^2 + 1)$, finding the maximum and minimum values of y.

   Prove that the graph lies entirely within the region bounded by the lines $y = \pm \frac{1}{2}$. (L.U.)

CHAPTER 10

INTRODUCTION TO THE IDEAS OF THE INTEGRAL CALCULUS

10.1. The nature of the fundamental problem of the integral calculus

The preceding chapters on the differential calculus have been concerned with the rate of variation of various *known* functions. The integral calculus is concerned with the *inverse* problem—*if the rate of variation of a function is known, what is the function itself?* In symbols, we have to find a function y of x when the derivative dy/dx is known, i.e., we have to find y from the equation

$$\frac{dy}{dx} = \phi(x), \tag{10.1}$$

where $\phi(x)$ is a known function of x.

As an example, suppose we know that the velocity at time t of a particle moving in a straight line is $(u + at)$ where u and a are constants, and we wish to find an expression for the distance s travelled by the particle in this time. Since the velocity is expressed by ds/dt, we have to find s from the equation

$$\frac{ds}{dt} = u + at. \tag{10.2}$$

This entails finding a function of t whose derivative with respect to t is $(u + at)$. An inspired guess will lead to the result $s = ut + \frac{1}{2}at^2$ for the derivative of this expression is $u + at$. This, however, is not the only solution to our problem. The function $s = ut + \frac{1}{2}at^2 + C$, where C is any constant whatever, has the same derivative and the general solution of equation (10.2) is

$$s = ut + \frac{1}{2}at^2 + C. \tag{10.3}$$

The conventional way of writing the solution of equation (10.1) is

$$y = \int \phi(x)\,dx, \tag{10.4}$$

and y is called *the indefinite integral of $\phi(x)$ with respect to x.* The origin of this notation will be explained later (§ 10.5); at present we shall merely regard it as a means of expressing y when the derivative of y with respect to x is $\phi(x)$. In this notation the solution of equation (10.2) would be written

$$s = \int (u + at)dt,$$

and this indefinite integral we have seen in (10.3) to be $ut + \frac{1}{2}at^2 + C$, where C is an arbitrary constant.

There is a distinction between *direct* and *inverse* operations in mathematics. Differentiation is a direct operation and can be performed according to definite rules to give an unambiguous result. An inverse operation is of the nature of a question. The operation of integration, i.e., of finding the indefinite integral, asks what function when differentiated will produce an assigned result? We have seen in the example given in (10.2) and (10.3) that if there is one answer there are an infinite number owing to the presence of the arbitrary constant C. To discover under what circumstances there is an answer is beyond the scope of the present book. We shall simply state here that there is an answer, i.e., the indefinite integral exists, for a large class of functions and in this and the subsequent chapter we shall discuss methods for finding it.

10.2. Standard forms

There are no infallible rules by which the indefinite integral $\int \phi(x)\, dx$ of any given function $\phi(x)$ can be found. Integration being an inverse operation, we can only be guided by the results of the direct operation of differentiation. Moreover, although the indefinite integral exists for a large class of functions, it may not be capable of expression in terms of functions normally employed in mathematics. An example of a comparatively simple function where this is the case is

$$\int \frac{\sin x}{x} dx,$$

and such instances can be extended indefinitely.

A first list of integrals is easily obtained from the list of standard forms for the derivative given on page 147. On inversion, each of these will give an indefinite integral. The student should become thoroughly familiar with this list which is fundamental.

$$\frac{d}{dx}(x^n) = nx^{n-1}, \qquad \int x^n\, dx = \frac{x^{n+1}}{n+1} + C,$$

$$\text{(except when } n = -1).$$

$$\frac{d}{dx}(\sin x) = \cos x, \qquad \int \cos x\, dx = \sin x + C.$$

$$\frac{d}{dx}(\cos x) = -\sin x, \qquad \int \sin x\, dx = -\cos x + C.$$

$$\frac{d}{dx}(\tan x) = \sec^2 x, \qquad \int \sec^2 x\, dx = \tan x + C.$$

$$\frac{d}{dx}(\cot x) = -\operatorname{cosec}^2 x, \qquad \int \operatorname{cosec}^2 x \, dx = -\cot x + C.$$

$$\frac{d}{dx}(\sin^{-1} x) = \frac{1}{\sqrt{(1 - x^2)}}, \qquad \int \frac{dx}{\sqrt{(1 - x^2)}} = \sin^{-1} x + C.$$

$$\frac{d}{dx}(\tan^{-1} x) = \frac{1}{1 + x^2}, \qquad \int \frac{dx}{1 + x^2} = \tan^{-1} x + C.$$

In each case C denotes an arbitrary constant.

By differentiating with respect to x the results given on the right of the above list by use of the standard derivatives on the left it will be seen that the result in each case is the function (*the integrand*) included within the sign $\int \dots dx$. Thus, since

$$\frac{d}{dx}\left(\frac{x^{n+1}}{n + 1} + C\right) = \frac{(n + 1)x^n}{n + 1} = x^n,$$

then
$$\int x^n \, dx = \frac{x^{n+1}}{n + 1} + C.$$

And since
$$\frac{d}{dx}(\tan x + C) = \sec^2 x,$$

then
$$\int \sec^2 x \, dx = \tan x + C,$$

and so on. It should be noted that the result given for $\int x^n \, dx$ is invalid for $n = -1$: the integral $\int x^{-1} \, dx$ will be discussed later (§ 13.4). It should also be noted that

$$\int \frac{dx}{\sqrt{(1 - x^2)}} \quad \text{and} \quad \int \frac{dx}{1 + x^2}$$

are conventional ways of writing integrals which should strictly be written

$$\int \frac{1}{\sqrt{(1 - x^2)}} dx \quad \text{and} \quad \int \frac{1}{1 + x^2} dx.$$

Since the differential coefficient of the sum (or difference) of two functions is the sum (or difference) of the differential coefficients of the separate functions, it follows conversely that *the indefinite integral of the sum (or difference) of two functions is the sum (or difference) of the indefinite integrals of the separate functions.* Thus

$$\int \{\phi(x) \pm \chi(x)\} dx = \int \phi(x)dx \pm \int \chi(x)dx, \tag{10.5}$$

and this result can be generalised to cover any finite number of functions.

Again, since the differential coefficient of a constant multiplied by a function is equal to the constant multiplied by the differential coefficient of the function, it follows conversely that *the indefinite integral of a constant multiplied by a function is equal to the constant multiplied by the indefinite integral of the function.* In symbols, if a is a constant,

$$\int a\,\phi(x)dx = a \int \phi(x)dx. \tag{10.6}$$

The standard integrals given in this section and the rules expressed symbolically in (10.5) and (10.6) enable the integrals of quite a large number of functions to be written down. Some examples follow.

Example 1. *Evaluate* $\int \left(x^2 + 2 + \dfrac{1}{x^2}\right)dx$.

$$\int \left(x^2 + 2 + \frac{1}{x^2}\right)dx = \int x^2\,dx + 2\int dx + \int x^{-2}\,dx$$

$$= \frac{x^3}{3} + 2x - x^{-1} + C.$$

It should be noted that the integral is first expressed as the sum of three separate integrals and that $\int 2dx = 2\int dx = 2\int x^0\,dx = 2x$. Also that the three arbitrary constants from the three separate integrals can be combined into a single arbitrary constant C.

Example 2. *Integrate* $(2x - 1)^3$ *with respect to* x.

Since $(2x - 1)^3 = 8x^3 - 12x^2 + 6x - 1$, we can write

$$\int (2x - 1)^3\,dx = 8\int x^3\,dx - 12\int x^2\,dx + 6\int x\,dx - \int dx$$

$$= 8\left(\frac{x^4}{4}\right) - 12\left(\frac{x^3}{3}\right) + 6\left(\frac{x^2}{2}\right) - x + C$$

$$= 2x^4 - 4x^3 + 3x^2 - x + C.$$

Example 3. *Evaluate* $\int (2\theta + \sin\theta)d\theta$.

$$\int (2\theta + \sin\theta)d\theta = 2\int \theta\,d\theta + \int \sin\theta\,d\theta.$$

$$= 2.\frac{\theta^2}{2} - \cos\theta + C$$

$$= \theta^2 - \cos\theta + C.$$

Example 4. *Find* $\int \left(\dfrac{t^4 + 1}{t^2}\right)dt$.

Since $(t^4 + 1)/t^2$ can be written $t^2 + 1/t^2$, we have

$$\int\left(\frac{t^4 + 1}{t^2}\right)dt = \int(t^2 + t^{-2})dt = \int t^2\, dt + \int t^{-2}\, dt$$

$$= \frac{t^3}{3} - t^{-1} + C = \frac{t^3}{3} - \frac{1}{t} + C.$$

EXERCISES 10 (a)

Integrate the following functions with respect to x:—

1. $x^{4/3}$.

5. $\dfrac{x^2 + 1}{x^2}$.   (L.U.)

2. $3/x^2$.

6. $\sin x + \cos x$.

3. $(1 + x)^2$.

7. $2\sec^2 x + \dfrac{1}{1 + x^2}$.

4. $\left(x + \dfrac{1}{x}\right)^2$.

8. $x^5 + 4x^3 - 2x^2 + x - 3$.

Evaluate the following indefinite integrals:—

9. $\int(3t^2 - t + 7)dt$.

11. $\int\dfrac{ax^{-2} + bx^{-1} + c}{x^{-3}}dx$.

10. $\int\left(\dfrac{1}{\theta^3} + \dfrac{1}{\theta^2} - 2\right)d\theta$.

12. $\int(2\theta + \cos\theta)d\theta$.

13. By using the relation $\sec^2\theta = 1 + \tan^2\theta$, find the value of $\int\tan^2\theta\, d\theta$. In a similar manner show that

$$\int\cot^2\theta\, d\theta = C - \theta - \cot\theta.$$

14. Use the relation $\cos x = 2\cos^2\frac{1}{2}x - 1$ to evaluate $\int\cos^2\frac{1}{2}x\, dx$.

15. If $(1 + x^2)\dfrac{dy}{dx} = 1$, find the general value of y.

10.3. Some geometrical and dynamical applications

The problem of finding a function which has a known differential coefficient has many geometrical and dynamical applications. The indefinite integral gives a general solution to this problem but often a particular solution is required which satisfies some geometrical or physical condition obtaining in the specific problem under discussion. Such a condition enables the particular solution to be selected from the general solution by fixing the value of the arbitrary constant in the indefinite integral.

Some illustrative examples are given below.

Example 5. *At a point on a curve the product of the slope of the curve and the square of the abscissa of the point is* 2. *If the curve passes through the point* $x = 1$, $y = -1$, *find its equation.* (L.U.)

The slope of the curve at the point whose coordinates are x and y is dy/dx so that

$$x^2\frac{dy}{dx} = 2.$$

Hence
$$\frac{dy}{dx} = \frac{2}{x^2},$$

and
$$y = \int \frac{2}{x^2}dx = -\frac{2}{x} + C,$$

where C is an arbitrary constant. The equation $y = -2/x + C$ gives, for different values of C, a family of curves in each of which the product of the slope and the square of the abscissa at a point is 2. The particular curve which passes through the point $x = 1$, $y = -1$ is obtained by selecting from this family that curve which passes through the point in question. Since $y = -1$ when $x = 1$

$$-1 = -2/1 + C,$$

so that $C = 1$ and the required equation is $y = 1 - 2/x$.

Example 6. *If* $\frac{dy}{dx} = ax + 2$, *where a is constant, express y as a function of x, given that* $\frac{d^2y}{dx^2} = 6$ *and that* $y = 4$ *when* $x = 0$. (L.U.)

Differentiating the given expression for dy/dx we have

$$\frac{d^2y}{dx^2} = a,$$

and hence $a = 6$. Therefore

$$\frac{dy}{dx} = 6x + 2,$$

and
$$y = \int (6x + 2)dx = 3x^2 + 2x + C.$$

The constant C is found from the condition that $y = 4$ when $x = 0$, so that $C = 4$ and $y = 3x^2 + 2x + 4$.

Example 7. *A particle starts from rest with acceleration* $(30 - 6t)$ *m/s² at time t. When and where will it come to rest again?*

Since acceleration is rate of change of velocity, if v is the velocity at time t,

$$\frac{dv}{dt} = 30 - 6t.$$

Hence
$$v = \int (30 - 6t)dt = 30t - 3t^2 + C,$$

where C is an arbitrary constant.

Since the particle starts from rest, $v = 0$ when $t = 0$, so that $C = 0$. Thus $v = 30t - 3t^2 = 3t(10 - t)$. The body is at rest when $v = 0$ and this occurs when $t = 0$ (the beginning of the motion) and again when $t = 10$ seconds. If s is the distance travelled in t seconds,

$$\frac{ds}{dt} = v = 30t - 3t^2.$$

Hence $\qquad s = \int (30t - 3t^2)dt = 15t^2 - t^3 + C',$

where C' is another arbitrary constant. Here s is the distance travelled from the starting point so that $s = 0$ when $t = 0$ and therefore $C' = 0$. Thus $s = 15t^2 - t^3 = t^2(15 - t)$ and when $t = 10$, the time when the particle is again at rest, the distance travelled will be $10^2(15 - 10)$ or 500 m.

EXERCISES 10 (b)

1. The gradient of a curve at any point is given by $\frac{dy}{dx} = 2x - 1$. If the curve passes through the point $x = 1$, $y = 1$, find the equation of the curve. (L.U.)

2. Find the equation of the curve whose gradient is $1 - 2x^2$ and which passes through the point $x = 0$, $y = 1$. (L.U.)

3. A curve passes through the origin of coordinates and its gradient is $2x - x^2/2$ at the point whose abscissa is x. Find the ordinate of the curve when $x = 2$. (O.C.)

4. A particle is moving along a straight line with acceleration $(2 + 3t)$ m/s^2 at time t seconds. At zero time its distance from the origin is 5 m; at time $t = 1$ its velocity is 10 m/s. Where is it at time $t = 1$?

5. A particle moves on a straight line OA and at time t it is distant x from O, x being taken positive when the point is on the same side of O as A. Write down expressions for the velocity v and acceleration f of the particle at time t.
   Find the distance x at time t if $f = 48t - 24$, given also that $v = 6$ and $x = -1$ when $t = 0$.
   Show that the particle is stationary at O when $t = 1/2$. (O.C.)

6. At time t the velocity of a particle moving in a straight line is increasing at the rate $(4t + 3/t^3)$. When $t = 1$, the velocity is 10 and at that time the particle is at distance 4 from the origin. Where is the particle 2 seconds later and what is its velocity then?

10.4. Calculation of an area as a limit of a sum

As a preliminary to a second interpretation of integration we give below an example of the calculation of an area as the limit of a sum.

If the graph of $y = 1 + x$ be plotted, the graph is seen to be a straight line and a diagram as shown in Fig. 64 results. Consider the trapezium $AOBC$ included between the graph, the x-axis and ordinates OA, BC at $x = 0$ and $x = 10$ respectively. The base of the trapezium is 10 units, $OA = 1$, $BC = 11$, the mean height of the trapezium is 6 and its area A is therefore 60 units of area.

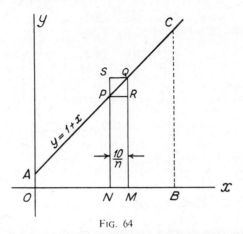

FIG. 64

The area A might also be calculated as follows. Divide the area into n strips of equal width $10/n$ by lines parallel to the y-axis. Suppose the rth strip is $PNMQ$. Since for the first strip PN lies along OA, for the second strip PN is at distance $10/n$ from the y-axis, for the third strip PN is distant $2(10/n)$ from the y-axis and so on, the distance of PN from the y-axis for the rth strip will be $(r-1)(10/n)$. Similarly the distance of QM from the y-axis for the rth strip will be $r(10/n)$. Thus for the strip shown in the figure, the x-coordinate of P is $(r-1)(10/n)$, and since P lies on the graph of $y = 1 + x$, the ordinate PN will be given by

$$PN = 1 + \frac{10(r-1)}{n} = 1 - \frac{10}{n} + \frac{10r}{n}. \tag{10.7}$$

Similarly the length QM is given by

$$QM = 1 + \frac{10r}{n}. \tag{10.8}$$

By drawing PR perpendicular to QM and QS perpendicular to PN, it can be inferred that the area of the trapezium is greater than the area of n rectangles like $PNMR$ and that it is less than the area of n rectangles like $SNMQ$. Since the base of the rectangles is $10/n$,

$$\text{area rectangle } PNMR = \frac{10}{n}\left(1 - \frac{10}{n} + \frac{10r}{n}\right) = \frac{10}{n} - \frac{100}{n^2} + \frac{100r}{n^2},$$

$$\text{area rectangle } SNMQ = \frac{10}{n}\left(1 + \frac{10r}{n}\right) = \frac{10}{n} + \frac{100r}{n^2}.$$

The sum of the n rectangles of which $PNMR$ is typical is therefore

$$\left(\frac{10}{n} - \frac{100}{n^2} + \frac{100}{n^2}\right) + \left(\frac{10}{n} - \frac{100}{n^2} + \frac{200}{n^2}\right)$$

$$+ \left(\frac{10}{n} - \frac{100}{n^2} + \frac{300}{n^2}\right) + \ldots + \left(\frac{10}{n} - \frac{100}{n^2} + \frac{100n}{n^2}\right)$$

$$= \left(\frac{10}{n} - \frac{100}{n^2}\right)(1 + 1 + 1 + \ldots \text{to } n \text{ terms})$$

$$+ \frac{100}{n^2}(1 + 2 + 3 + \ldots + n)$$

$$= \left(\frac{10}{n} - \frac{100}{n^2}\right)n + \frac{100}{n^2} \cdot \frac{n(1 + n)}{2}$$

$$= 10 - \frac{100}{n} + \frac{50(1 + n)}{n} = 60 - \frac{50}{n}.$$

Similarly the sum of the n rectangles of which $SNMQ$ is typical can be found to be $60 + 50/n$, and we have

$$60 - \frac{50}{n} < A < 60 + \frac{50}{n}. \tag{10.9}$$

By taking $n = 10$, i.e., by dividing OB into 10 equal parts, we have
$$55 < A < 65,$$
while if we take 100 strips ($n = 100$),
$$59{\cdot}5 < A < 60{\cdot}5.$$

By using 1000 strips, we should find that A lies between $59{\cdot}95$ and $60{\cdot}05$ and equation (10·9) shows that A lies between $60 - \epsilon$ and $60 + \epsilon$ where ϵ can be made as small as we please by taking n sufficiently large.

It should be noted that as the number n of rectangles such as $PNMR$ and $SNMQ$ increases, their widths ($10/n$) decrease and the area can be estimated with greater precision. The area is in fact the limit to which the sum of the areas of the rectangles approaches as their number increases and their width decreases.

10.5. The integral as a sum

The procedure outlined in § 10.4 could be used to find the area below a curve like $y = 1 + x^2$ but the summations involved would be more complicated. We apply below a similar procedure to the more general curve $y = \phi(x)$ and are led to the idea of an integral being regarded as the limit of a sum.

To simplify matters we consider the function $y = \phi(x)$ as one which is positive and increasing as x increases from $x = a$ to $x = b$. In Fig. 65, $OU = a$, $OV = b$ and $a < b$. We wish to investigate the area $AUVB$.

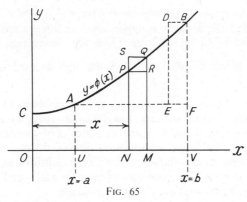

Fig. 65

$PNMQ$ is one of the n strips into which the area is divided in the manner of § 10.4 and we take $ON = x$, $PN = y$, $NM = \delta x$. It is assumed that the area $CONP$ bounded by the axes, the ordinate PN and the curve $y = \phi(x)$ is some function $A(x)$ of x. When x increases to $x + \delta x$, the area $CONP$ increases to the area $COMQ$ and this we take to be $A(x) + \delta A(x)$. The area $PNMQ$ is, by subtraction, $\delta A(x)$. The area of the rectangle $PNMR$ is $y\,\delta x$, and since $QM = y + \delta y$, the area of the rectangle $SNMQ$ is $(y + \delta y)\delta x$. It can be inferred from the figure that

$$\text{area } PNMR < \text{area } PNMQ < \text{area } SNMQ,$$

or, $$y\,\delta x < \delta A(x) < (y + \delta y)\delta x. \qquad (10.10)$$

Employing the symbol Σ to denote the summation of n such strips, we have

$$\Sigma\, y\,\delta x < \text{area } AUVB < \Sigma\,(y + \delta y)\delta x. \qquad (10.11)$$

Suppose we draw AF parallel to the x-axis and construct the rectangle $DEFB$ of Fig. 65 to be of width δx. The difference between the two sums $\Sigma\, y\,\delta x$ and $\Sigma\,(y + \delta y)\delta x$ (or the difference between the sums of rectangles like $SNMQ$ and $PNMR$) can be seen, by sliding rectangles like $SPRQ$ parallel to the x-axis until QR lies along BV, to be equal to the area of the rectangle $DEFB$. This area is $BF.\delta x$, and since $\delta x = UV/n$, the difference $\Sigma\,(\delta y\,\delta x)$ between the two sums is $(BF.UV)/n$ and this can be made arbitrarily small by taking n large enough. Thus $\Sigma\,(\delta y\,\delta x)$ tends to zero with δx and if $\Sigma y\,\delta x$ tends to a limit then $\Sigma\,(y + \delta y)\delta x$ tends to the same limit. In this case the area $AUVB$ lies between sums which have the same limit, and it follows that

$$\text{area } AUVB = \lim_{\delta x \to 0} \Sigma\, y\,\delta x.$$

This limit is denoted by $\displaystyle\int_a^b y\,dx$ *and is called the definite integral of* y

with respect to x taken over the range from x = a to x = b. The letters *a* and *b*, called *the lower and upper limits of integration*, indicate the range in *x* from *UA* to *VB* over which the summation is made. The symbol $\int$ is a specialised form of *S*, the sign of summation used in earlier times.

The definite integral so defined is independent of the idea of differentiation. Except in very simple cases such as that given in § 10.4, it is not practicable to use it as a means of determining an area because of the awkward summations involved. We can, however, connect this definition of the integral with the definition of the indefinite integral (the inverse of the derivative) given in § 10.1 and a practicable method of calculation of area is then available.

To do so, we return to the inequalities of (10.10) which, after division by δx, give

$$y < \frac{\delta A(x)}{\delta x} < y + \delta y.$$

Assuming* that δy tends to zero with δx, this shows that

$$y = \frac{dA}{dx}. \tag{10.12}$$

y is therefore the derivative of *A(x)* and hence *A(x)* is the indefinite integral of *y*. Now *A(x)* measures the area *CONP*, and from Fig. 65,

$$\text{area } AUVB = \text{area } COVB - \text{area } COUA = A(b) - A(a).$$

Hence
$$\int_a^b y\, dx = \lim_{\delta x \to 0} \Sigma\, y\, \delta x = \text{area } AUVB$$
$$= A(b) - A(a), \tag{10.13}$$

where *A(x)* denotes the indefinite integral of *y* with respect to *x* as defined in § 10.1.

The argument given above is for a curve in which *y* is positive and increasing with *x*. When *y* decreases as *x* increases, the inequalities are reversed but it is still true that $\delta A/\delta x$ lies between *y* and $y + \delta y$ and (10.13) still applies. If *y* increases while *x* increases from *a* to *k* and decreases while *x* increases from *k* to *b*, the integral or area can be found in two parts and these can then be summed. It is also assumed above that *y* is positive; when *y* is negative $\Sigma\, y\, \delta x$ and $\int_a^b y\, dx$ are negative (see § 12.2).

---

* For the functions used in this book δy always tends to zero with δx for the values of *x* under discussion. The student should be warned, however, that this is not always the case and equation (10.12) does not then remain valid.

10.6. The example of § 10.4 solved by integration

The area bounded by $y = 1 + x$, the x-axis and ordinates at $x = 0$, $x = 10$ is, by (10.13),

$$A(10) - A(0),$$

where $A(x) =$ indefinite integral of $(1 + x)$ with respect to x, $= x + \frac{1}{2}x^2 + C$. $A(10)$ is therefore $10 + \frac{1}{2}(10^2) + C = 60 + C$ and $A(0)$ is simply C, so that in Fig. 64,

$$\text{area } AOBC = A(10) - A(0) = 60 + C - C$$
$$= 60 \text{ units.}$$

This example shows that in evaluating an area or a definite integral, the arbitrary constant of the indefinite integral may be omitted. If the notation $\left[A(x)\right]_a^b$ is used to denote the difference $A(b) - A(a)$, a convenient way of setting out the working of this example would be

$$\text{area } AOBC = \int_0^{10} (1 + x)dx$$
$$= \left[x + \frac{1}{2}x^2\right]_0^{10}$$
$$= 10 + \frac{1}{2}10^2 - 0 - 0 = 60 \text{ units.}$$

The connection between the limit of a sum and integration has therefore enabled lengthy summation processes to be avoided. In the present example, this may not be a matter of great importance for the area of a trapezium can be otherwise calculated. Nevertheless when we pass to areas bounded by curves, formulae for the area are not generally known, summation processes similar to those of § 10.4 become very complicated and integration provides a simple means of effecting the calculations.

10.7. Illustrative examples of definite integrals and calculation of area

In this section a few examples are given illustrating a method for setting out the calculations involved in evaluating a definite integral. Some applications of the definite integral to the calculation of area are also given.

Example 8. *Evaluate* $\displaystyle\int_a^{2a} \left(\frac{a^3}{x^2} + \frac{x^2}{a}\right)dx.$ (L.U.)

If I denotes the definite integral,

$$I = \int_a^{2a} (a^3x^{-2} + x^2/a)dx$$
$$= \left[-a^3x^{-1} + \frac{x^3}{3a}\right]_a^{2a}$$

$$= -\frac{a^3}{2a} + \frac{(2a)^3}{3a} + \frac{a^3}{a} - \frac{a^3}{3a} = \frac{17a^2}{6}.$$

Example 9. *Evaluate* $\int_1^2 \left(\frac{x^3 - 1}{x^2}\right) dx.$ (L.U.)

Writing the integrand $\frac{x^3 - 1}{x^2}$ in the form $x - \frac{1}{x^2}$ and denoting the integral by I,

$$I = \int_1^2 (x - x^{-2}) dx$$

$$= \left[\frac{x^2}{2} + \frac{1}{x}\right]_1^2$$

$$= \frac{2^2}{2} + \frac{1}{2} - \frac{1^2}{2} - 1 = 1.$$

Example 10. *Evaluate* (i) $\int_0^{\frac{1}{2}\pi} \sin x \, dx,$ (ii) $\int_0^1 \frac{dx}{1 + x^2}.$

(i) $\int_0^{\frac{1}{2}\pi} \sin x \, dx = \left[-\cos x\right]_0^{\frac{1}{2}\pi} = -\cos \frac{1}{2}\pi + \cos 0 = 1.$

(ii) $\int_0^1 \frac{dx}{1 + x^2} = \left[\tan^{-1} x\right]_0^1 = \tan^{-1}(1) - \tan^{-1}(0) = \pi/4.$

Example 11. *Calculate the area between the x-axis and the curve* $y = x(x - 4)$.
(L.U.)

A rough sketch of the curve shows that the area required (shown shaded in Fig. 66) is

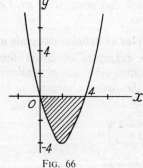

FIG. 66

$$\int_0^4 y \, dx = \int_0^4 x(x - 4) dx$$

$$= \int_0^4 (x^2 - 4x) dx$$

$$= \left[\frac{x^3}{3} - 2x^2\right]_0^4 = \frac{4^3}{3} - 2(4)^2$$

$$= -\frac{32}{3} \text{ units,}$$

(the negative sign occurs because the area lies below the x-axis, see also § 12.2).

Example 12. *The gradient of a curve at the point whose abscissa is x is $1 + \frac{1}{2}x^2$ and the curve passes through the point $x = 1$, $y = 0$. Find the equation of the curve and the area bounded by it, the x-axis and the ordinates $x = 1$, $x = 3$.*
(L.U.)

We are given $\dfrac{dy}{dx} = 1 + \frac{1}{2}x^2$. Hence

$$y = \int (1 + \tfrac{1}{2}x^2)dx = x + \frac{x^3}{6} + C,$$

where C is a constant. Since $y = 0$ when $x = 1$, $0 = 1 + \dfrac{1}{6} + C$, giving

$C = -\dfrac{7}{6}$ and the equation of the curve is $y = \dfrac{x^3}{6} + x - \dfrac{7}{6}$.

The area required $= \displaystyle\int_1^3 \left(\frac{x^3}{6} + x - \frac{7}{6}\right)dx$

$$= \left[\frac{x^4}{24} + \frac{x^2}{2} - \frac{7}{6}x\right]_1^3 = \frac{81}{24} + \frac{9}{2} - \frac{21}{6} - \frac{1}{24} - \frac{1}{2} + \frac{7}{6} = 5.$$

EXERCISES 10 (c)

Evaluate the following definite integrals:—

1. $\displaystyle\int_0^{10} (1 + x^2)dx.$

2. $\displaystyle\int_1^2 x^2\left(1 - \frac{2}{x^4}\right)dx.$ (L.U.)

3. $\displaystyle\int_{-1}^2 x(1 + x^3)dx.$ (L.U.)

4. $\displaystyle\int_0^4 x(x - 1)(x - 2)dx.$

5. $\displaystyle\int_0^1 \frac{d\theta}{\sqrt{(1 - \theta^2)}}.$

6. $\displaystyle\int_0^{\frac{1}{4}\pi} \sec^2 \theta \, d\theta.$

7. $\displaystyle\int_0^{\frac{1}{4}\pi} \tan^2 \theta \, d\theta.$

8. $\displaystyle\int_0^{\pi/3} (\sin \tfrac{1}{2}t + \cos \tfrac{1}{2}t)^2 \, dt.$ (L.U.)

9. Find the area of the space bounded by the curve $y = 1 + 10x - 2x^2$, the x-axis and the ordinates for which $x = 1$ and $x = 5$.   (O.C.)

10. Find the area included by that part of the curve
$$y = x^3 - 6x^2 + 11x - 6,$$
which lies between $x = 1$ and $x = 2$.   (O.C.)

11. Find the area enclosed by the x-axis and that part of the curve $y = 5x - 6 - x^2$ for which y is positive.   (O.C.)

12. The gradient of a curve is $6x - 3x^2$. If the curve passes through the origin find its equation and show that it cuts the x-axis again where $x = 3$. Find also the area bounded by the curve and the x-axis. (L.U.)

13. An ordinate is drawn to the curve $y = x(1 - x^2)$ through the point $x = 1 + p$ where p is positive. Find p so that the area between the x-axis and the curve for x between 1 and $(1 + p)$ may equal in absolute magnitude the area between the x-axis and the curve for x between 0 and 1. (L.U.)

14. The curve $y = 11x - 24 - x^2$ cuts the x-axis at points A, B and PN is the greatest positive ordinate. Show that $2PN \cdot AB$ equals three times the area of that portion of the curve which lies in the first quadrant. (L.U.)

10.8. Volumes of figures of revolution

Another simple application of the definite integral occurs in the calculation of the volume formed by the rotation of a plane curve about an axis. Fig. 67 shows part of the curve $y = \phi(x)$ (again for simplicity shown as positive and increasing with x). That part of the curve between ordinates at $x = a$ and $x = b$ is rotated about the x-axis and we wish to find the volume of the solid so formed.

As in § 10.5, $PNMQ$ is one of the n strips into which the area $AUVB$ is divided and $ON = x$, $PN = y$, $NM = \delta x$. The volume

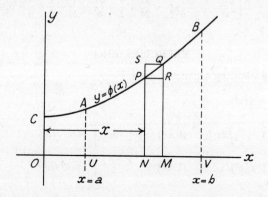

FIG. 67

formed by the rotation about the x-axis of the strip $PNMQ$ will be greater than the volume formed by the rotation of the rectangle $PNMR$ and less than that formed by the rotation of the rectangle $SNMQ$. The body formed by the rotation of the rectangle $PNMR$ will be a right circular cylinder* of radius $PN = y$ and length $NM = \delta x$ and therefore of volume $\pi y^2 \, \delta x$. Similarly the volume

---

* See § 20.5.

formed by the rotation of the rectangle $SNMQ$ will be $\pi(y + \delta y)^2 \, \delta x$, for the radius of this cylinder is $QM = y + \delta y$. If we take the volume formed by the rotation about the x-axis of the area $CONP$ to be $V(x)$, that formed by the rotation of $COMQ$ will be $V(x) + \delta V(x)$ and, by subtraction, the volume formed by the rotation of $PNMQ$ will be $\delta V(x)$. Hence

$$\pi y^2 \, \delta x < \delta V(x) < \pi(y + \delta y)^2 \, \delta x.$$

The summation of n such strips shows that $\Sigma \, \pi y^2 \, \delta x <$ volume (V) formed by rotation of area $AUVB < \Sigma \, \pi(y + \delta y)^2 \, \delta x$, and an argument similar to that used in § 10.5 leads to

$$V = \lim_{\delta x \to 0} \Sigma \, \pi y^2 \, \delta x$$

$$= \int_a^b \pi y^2 \, dx = \pi \int_a^b y^2 \, dx. \qquad (10.14)$$

Again the reasoning has been given for a curve in which y increases with x. Arguments similar to those given at the end of § 10.5 will enable this restriction to be removed.

Example 13. *The portion of the curve $xy = 8$ from $x = 2$ to $x = 2$ to $x = 4$ is rotated about the x-axis: find the volume generated.* (L.U.)

Here $y = 8/x$ and the required volume is given by

$$V = \pi \int_2^4 y^2 \, dx = \pi \int_2^4 (8/x)^2 \, dx$$

$$= 64\pi \int_2^4 \frac{dx}{x^2} = 64\pi \left[-\frac{1}{x} \right]_2^4$$

$$= 64\pi \{ -\tfrac{1}{4} + \tfrac{1}{2} \} = 16\pi \text{ units.}$$

Example 14. *Find the volume of a right circular cone of height h and semi-vertical angle α.*

The graph of $y = \tan \alpha$ is (Fig. 68) a straight line through the origin O making

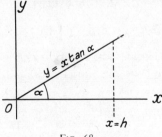

FIG. 68

an angle α with the x-axis. The rotation of this line about the x-axis will generate a right circular cone of semi-vertical angle α. The volume of such a cone of height h is, by (10.14), given by

$$V = \pi \int_0^h y^2 \, dx = \pi \int_0^h (x \tan \alpha)^2 \, dx$$

$$= \pi \tan^2 \alpha \int_0^h x^2 \, dx = \pi \tan^2 \alpha \left[\frac{x^3}{3} \right]_0^h$$

$$= \tfrac{1}{3} \pi h^3 \tan^2 \alpha.$$

EXERCISES 10 (d)

1. Find the volume generated when the area enclosed by the x-axis and the curve $y = 3x^2 - x^3$ is rotated about the x-axis.   (L.U.)

2. Find the volume generated when the area enclosed between the curve $y = x^3/16$, the x-axis and the ordinate at $x = 2$ revolves about the x-axis.

3. Draw a graph of the curve $y^2 = x(5 - x)^2$. If the loop of the curve is rotated about the x-axis, find the volume generated.   (L.U.)

4. Show by an argument similar to that given in § 10.8, that the volume generated when a portion of a curve between $y = c$ and $y = d$ is rotated about the y-axis is $\pi \int_c^d x^2 \, dy$.

   The curve $y^2 = 4x$, contained between the axes and the point where $x = 1, y = 2$ is revolved about (a) the x-axis, (b) the y-axis. Find the difference between the volumes of the solids so formed.   (L.U.)

5. The part of the curve $x^2 y = x^4 + 3$ between the ordinates $x = 1$ and $x = 2$ is rotated about the x-axis. Calculate the volume generated.   (L.U.)

6. Find the volume of the solid formed by the rotation of the curve $y = \cos \tfrac{1}{2} x$ from $x = 0$ to $x = \pi$ about the x-axis.

EXERCISES 10 (e)

1. Evaluate (i) $\int 3x(2x^2 + 5x - 4) dx$, (ii) $\int (1 - x)^2 \sqrt{x} \, dx$.

2. Find the indefinite integrals with respect to x of:—

   (i) $(3x - 4)^2$, (ii) $(3 + x)(x - 1)$.

3. If a is a constant show that

$$\int (a^2 - t^2)^3 \sqrt{t} \, dt = 2t^{3/2} \left(\frac{a^6}{3} - \frac{3a^4 t^2}{7} + \frac{3a^2 y^4}{11} - \frac{t^6}{15} \right) + C.$$

4. By using the addition formula for $\sin (x + \alpha)$, show that

$$\int \sin (x + \alpha) dx = -\cos (x + \alpha) + C,$$

where α is a constant. Show, in a similar way, that

$$\int \cos(x + \alpha)dx = \sin(x + \alpha) + C.$$

5. Use the relation $\cos \theta = 1 - 2\sin^2 \tfrac{1}{2}\theta$ to evaluate $\int \sin^2 \tfrac{1}{2}\theta \, d\theta$.

6. A curve passes through the point where $x = 0$ and $y = 1$ and its gradient at any point is $\tfrac{3}{2} + x - \tfrac{1}{2}x^2$. Find the equation of the curve and the area enclosed by the curve, the x-axis and the ordinates $x = 1$, $x = 3$.   (L.U.)

7. If $\theta = 2$ when $t = \pi/3$ and if $d\theta/dt = \sin t$, express θ in terms of t.

8. The gradient of a curve which passes through the point $x = 3$, $y = 1$ is given by $dy/dx = x^2 - 4x + 3$. Find the equation of the curve and the area enclosed by the curve, the maximum and minimum ordinates and the x-axis.   (L.U.)

9. The velocity (v) of a particle is given by

$$\frac{d(v^2)}{dx} = -18x,$$

and $v = 4$ m/s when $x = 1$ metre. Find the velocity when $x = 0$ and the greatest value reached by x during the motion.

10. Integrate the equation $EI\dfrac{d^2y}{dx^2} = M - Wx$, where E, I, M and W are constants.

    Given that both dy/dx and $y = 0$, when $x = 0$, determine the value of dy/dx and y at the point $x = 1$.   (Q.E.)

11. Evaluate the definite integrals:—

    (i) $\displaystyle\int_1^2 \left(x + \frac{1}{x}\right)^2 dx$, (ii) $\displaystyle\int_1^2 \left(x^2 + \frac{1}{x^2}\right)^3 dx$.   (L.U.)

12. Evaluate

    (i) $\displaystyle\int_0^1 (1 - \sqrt{x})^2 \, dx$, (ii) $\displaystyle\int_{\pi/2}^{\pi} (\cos x + \sin x)dx$.   (L.U.)

13. Find the value of:—

    (i) $\displaystyle\int_0^{1/2} \frac{dx}{\sqrt{(1 - x^2)}}$, (ii) $\displaystyle\int_a^b (x + \cos x)dx$.

14. If $y = x^3$, evaluate $\displaystyle\int_0^4 y \, dx + \int_0^{64} x \, dy$.

15. For what values of a does $\displaystyle\int_1^a (x + \tfrac{1}{2})dx = 2\int_0^{\pi/4} \sec^2 x \, dx$?

16. The equation of a curve is of the form $y = ax^2 + bx + c$. It meets the x-axis where $x = -1$ and $x = 3$: also $y = 12$ when $x = 1$. Find the equation of the curve and the area between it and the x-axis.   (L.U.)

17. Find the area in the first quadrant bounded by the curve $a^2y = 4x^2(3a - 4x)$ and the x-axis.   (L.U.)

18. For the curve $y = 2x^3 - 3x^2 - 12x + 25$, find the area enclosed beween the curve, the x-axis and the ordinates at the maximum and minimum points. (Q.E.)

19. Find the area between the curve $y^2 = 4x^3$, the x-axis and the ordinate at $x = 4$. If the ordinate at $x = c$ halves this area, find the value of c.

20. From the point where $x = 6$, $y = 27$ on the curve $8y = x^3$, perpendiculars are drawn to the axes of coordinates. Show that the rectangle formed by these perpendiculars with the axes is divided by the curve into portions the areas of which are in the ratio $1:3$. (O.C.)

21. A vessel for holding flowers is of the shape formed by the revolution of the curve $10y^2 = 4x$ about the y-axis. Its height is 0.2 m. Find its volume.

22. Liquid in a cylindrical can of radius r which is rotating rapidly about its axis is bounded by the surface formed by rotating the curve $y^2 = 4ax$ about the x-axis (the x-axis of the curve coincides with the axis of the cylinder and the origin is at the centre of the base of the can). The liquid just clears the centre of the base of the can and just does not overflow. Find an expression for the height of the can and also the volume of the liquid in terms of r and a.

23. Show that the volume generated by the revolution about the x-axis of the area included between that axis and the curve $cy = (x - a)(x - b)$ is

$$\frac{\pi}{30} \cdot \frac{(a - b)^5}{c^2}.$$

24. Find the volume generated when the loop of the curve $3x^2 = y(3 - y)^2$ revolves about the y-axis.

25. The part of the curve $y = \sqrt{(20x - x^2)}$ between $x = 0$ and $x = 10$ is a quadrant of a circle. The arc of the curve from the origin to a point P whose abscissa is a is revolved about the x-axis. If the volume of the bowl thus obtained is one-half of the bowl obtained by revolving the whole quadrant, show that $30a^2 - a^3 = 1000$. (N.U.)

CHAPTER 11

SOME METHODS OF INTEGRATION

11.1. Introduction

In the introduction to the basic ideas of the integral calculus in the last chapter, examples were chosen which involved only a few simple integrals: these were obtained as the inverses of the differential coefficients given in the standard list on p. 147. The scope of the integral calculus will, of course, be extended when other integrals can be determined and it is the object of the present chapter to indicate methods which help in this direction.

The student will gradually realise that although he can usually expect to be able to find the derivative of any function, there are many functions which he cannot integrate. In certain cases this may arise from a lack of experience or ingenuity but in others it will result from the fact that many functions cannot be integrated in terms of functions which are familiar to the student at this stage of his knowledge of mathematics.

A systematic study of methods of integration is most important and the student should not regard integration as a purely tentative process. Nevertheless it does require a certain maturity of judgment to decide whether to apply a standard method or to rely on some special device in a particular case and the student will be well advised to work through many examples in his early work. The exercises given at the ends of the various sections of this chapter will usually be on the methods given in the preceding sections: the set at the end of the chapter is not arranged in any definite order and the student will have to decide for himself what standard method or special device he is going to adopt.

The methods of integration given in this chapter are by no means all that are available. There are methods which depend on a knowledge of $\int x^{-1}\,dx$ which are delayed until this integral is discussed in Chapter 13, and there are many other methods which are beyond the scope of this book.

11.2. Generalisation of the list of standard integrals

Suppose the differential coefficient of $f(x)$ with respect to x is $f'(x)$, then the definition of the indefinite integral gives

$$\int f'(x)dx = f(x) + C, \tag{11.1}$$

where C is an arbitrary constant. The rule for the differentiation of a function of a function (§ 8.6) shows that, if a and b are constants, the differential coefficient with respect to x of $f(ax + b)$ is $af'(ax + b)$. Hence

$$\int f'(ax + b)dx = \frac{1}{a}f(ax + b) + C. \tag{11.2}$$

A comparison of (11.1) and (11.2) enables the integral of a function of $(ax + b)$ to be written down when the integral of the *same* function of x is known. We can in fact say that *if the integral of a function of x is known, the integral of the same function of $(ax + b)$ is of the same form but it is divided by a.* For the special case of $a = 1$, we have the rule that *the addition of a constant to x makes no difference in the form of the integral.*

As an example of the first of these important rules,

$$\int (2x + 3)^3 \, dx = \tfrac{1}{8}(2x + 3)^4 + C,$$

for the integral of x^3 is $\tfrac{1}{4}x^4$ and so the integral of $(2x + 3)^3$ will be $\tfrac{1}{4}(2x + 3)^4$ divided by 2, the coefficient of x in $(2x + 3)$. Other examples are

$$\int (x + 1)^2 \, dx = \tfrac{1}{3}(x + 1)^3 + C,$$

$$\int (1 - x)^3 \, dx = -\tfrac{1}{4}(1 - x)^4 + C,$$

$$\int \frac{dx}{(2x + 3)^2} = -\frac{1}{2(2x + 3)} + C.$$

These rules enable the following revised and more general list of integrals to be drawn up and it is this list that the student should try to memorise. In the list a and b are constants and C is the arbitrary constant of integration.

$$\int (ax + b)^n \, dx = \frac{(ax + b)^{n+1}}{(n + 1)a} + C, \text{ except when } n = -1.$$

$$\int \cos ax \, dx = \frac{1}{a}\sin ax + C, \qquad \int \sin ax \, dx = -\frac{1}{a}\cos ax + C,$$

$$\int \sec^2 ax \, dx = \frac{1}{a}\tan ax + C, \qquad \int \mathrm{cosec}^2 ax \, dx = -\frac{1}{a}\cot ax + C,$$

$$\int \frac{dx}{\sqrt{(a^2 - x^2)}} = \sin^{-1}\frac{x}{a} + C, \qquad \int \frac{dx}{a^2 + x^2} = \frac{1}{a}\tan^{-1}\frac{x}{a} + C.$$

The above integrals all follow directly from those given in § 10.2 except the last two which are derived as follows. Since

$$\sqrt{(a^2 - x^2)} = a\sqrt{\{1 - (x/a)^2\}},$$

$$\int \frac{dx}{\sqrt{(a^2 - x^2)}} = \frac{1}{a} \int \frac{dx}{\sqrt{\{1 - (x/a)^2\}}}$$

$$= \frac{1}{a} \cdot \frac{1}{(1/a)} \sin^{-1} \frac{x}{a} + C = \sin^{-1} \frac{x}{a} + C,$$

and

$$\int \frac{dx}{a^2 + x^2} = \frac{1}{a^2} \int \frac{dx}{1 + (x/a)^2}$$

$$= \frac{1}{a^2} \cdot \frac{1}{(1/a)} \tan^{-1} \frac{x}{a} + C = \frac{1}{a} \tan^{-1} \frac{x}{a} + C.$$

It should be carefully noted that the rules given in this section apply only when x is replaced by $(ax + b)$, i.e., by an expression linear (of the first degree) in x. They do not apply to integrals like $\int (x^2 + 4)^2 \, dx$ or $\int \sin 2x^3 \, dx$ where x is replaced by expressions of degrees other than the first in x. It is also important to be careful not to omit the dividing factor a, a common mistake when the process of integration is first begun. It is perhaps as well in the early stages to check the correctness of an integration by differentiating the expression obtained. This should, of course, give back the function which was to be integrated (the integrand).

One other point may be noticed here. Earlier in this section we found that, apart from the arbitrary constant,

$$\int (x + 1)^2 \, dx = \tfrac{1}{3}(x + 1)^3;$$

if we find this integral by first squaring $(x + 1)$, we have

$$\int (x + 1)^2 \, dx = \int (x^2 + 2x + 1) dx$$

$$= \tfrac{1}{3}x^3 + x^2 + x.$$

This can be written $\tfrac{1}{3}(x^3 + 3x^2 + 3x + 1) - \tfrac{1}{3}$, or $\tfrac{1}{3}(x + 1)^3 - \tfrac{1}{3}$, and the two results differ by $1/3$. This apparent discrepancy can be removed by including the quantity $-1/3$ in the arbitrary constant which so far we have omitted. The arbitrary constants in the expressions obtained by the two methods will differ by $1/3$ but all the terms containing x will be identical. Expressions for indefinite integrals obtained by different methods often differ at first sight: provided no error has been made it will be found, on examination, that all terms containing the variable are identical and the results differ only by a constant.

Example 1. *Evaluate* $\int_0^1 \left\{ \frac{1}{\sqrt{(x + 3)}} + \sqrt{(x + 3)} \right\} dx.$ (L.U.)

The given integral

$$= \int_0^1 \{(x + 3)^{-1/2} + (x + 3)^{1/2}\}dx$$

$$= \left[2(x + 3)^{1/2} + \tfrac{2}{3}(x + 3)^{3/2} \right]_0^1$$

$$= 2\sqrt{4} + \tfrac{2}{3}\sqrt{(4)^3} - 2\sqrt{3} - \tfrac{2}{3}\sqrt{(3)^3}$$

$$= 4 + \tfrac{16}{3} - 2\sqrt{3} - 2\sqrt{3} = \tfrac{28}{3} - 4\sqrt{3}.$$

Example 2. *Evaluate* $\displaystyle\int_0^{\pi/2} (\cos x + 2 \cos 2x)dx.$ (L.U.)

The integral

$$= \left[\sin x + \frac{2}{2} \sin 2x \right]_0^{\pi/2}$$

$$= \sin \frac{\pi}{2} + \sin \pi - \sin (0) - \sin (0) = 1.$$

Example 3. *Find the value of* $\displaystyle\int_2^3 \frac{dx}{x^2 - 4x + 13}.$ (L.U.)

Since $x^2 - 4x + 13 = x^2 - 4x + 4 + 9 = (x - 2)^2 + 3^2$, we have

$$\int_2^3 \frac{dx}{x^2 - 4x + 13} = \int_2^3 \frac{dx}{3^2 + (x - 2)^2}$$

$$= \left[\tfrac{1}{3} \tan^{-1}\left(\frac{x - 2}{3}\right) \right]_2^3$$

$$= \tfrac{1}{3}\{\tan^{-1}(\tfrac{1}{3}) - \tan^{-1}(0)\} = \tfrac{1}{3} \tan^{-1}\tfrac{1}{3}.$$

The method used above of expressing $x^2 - 4x + 13$ as the sum of the square of a linear expression in x and a constant should be noticed. Integrals involving quadratic denominators without factors can often be made in this way to depend on the integral $\displaystyle\int \frac{dx}{a^2 + x^2}$. When the denominator can be expressed as the product of linear factors, separation into partial fraction can be used to make the integral depend on $\displaystyle\int \frac{dx}{x + a}$, but a discussion of this case must wait until Chapter 13. In a similar way some integrals can be made to depend on $\displaystyle\int \frac{dx}{\sqrt{(a^2 - x^2)}}.$

EXERCISES 11 (a)

Integrate the following expressions with respect to x:—

1. $(2 - x)^3.$

2. $(1 - x)^{10}.$

3. $(2x - 1)^{-3}.$

4. $\dfrac{1}{\sqrt{(5x - 7)}}.$

Evaluate the following definite integrals:—

5. $\displaystyle\int_0^{\pi/6} \sin\left(\frac{\pi}{3} - 2\theta\right) d\theta.$ 6. $\displaystyle\int_0^{\pi/4} \cos 2t\, dt.$

7. Integrate with respect to x:—
   (i) $\cos 2x - \sin 2x,$ (ii) $\sec^2 (2x - 1).$

8. Evaluate $\displaystyle\int_0^{2/3} \frac{du}{4 + 9u^2}.$ (L.U.)

Integrate the following expressions with respect to x:—

9. $\dfrac{1}{\sqrt{(4 - x^2)}}.$ 11. $\dfrac{1}{\sqrt{(5 + 4x - x^2)}}.$

10. $\dfrac{1}{x^2 + x + 1}.$ 12. $\dfrac{1}{3x^2 - 4x + 7}.$

11.3. The integration of products of sines and cosines

The product of two sines, two cosines or a sine and a cosine may be integrated by first expressing the product as a sum by means of the trignometrical identities (§ 5.5),

$$\left.\begin{array}{l} 2 \sin A \sin B = \cos (A - B) - \cos (A + B), \\ 2 \cos A \cos B = \cos (A - B) + \cos (A + B), \\ 2 \sin A \cos B = \sin (A - B) + \sin (A + B). \end{array}\right\} \quad (11.3)$$

Thus,

$$\int \sin mx \sin nx\, dx = \frac{1}{2} \int \{\cos (m - n)x - \cos (m + n)x\}dx$$

$$= \frac{\sin (m - n)x}{2(m - n)} - \frac{\sin (m + n)x}{2(m + n)} + C,$$

$$\int \cos mx \cos nx\, dx = \frac{1}{2} \int \{\cos (m - n)x + \cos (m + n)x\}dx$$

$$= \frac{\sin (m - n)x}{2(m - n)} + \frac{\sin (m + n)x}{2(m + n)} + C,$$

$$\int \sin mx \cos nx\, dx = \frac{1}{2} \int \{\sin (m - n)x + \sin (m + n)x\}dx$$

$$= -\frac{\cos (m - n)x}{2(m - n)} - \frac{\cos (m + n)x}{2(m + n)} + C.$$

It is not recommended that these last three integrals should be memorised—it is better to remember the method and apply it to specific examples.

Example 4. *Integrate cos 3x cos 2x with respect to x.*

$$\int \cos 3x \cos 2x \, dx = \frac{1}{2} \int (\cos x + \cos 5x) dx$$

$$= \frac{1}{2} \sin x + \frac{1}{10} \sin 5x + C.$$

Example 5. *Evaluate* $\int_0^{\pi/2} \sin 2x \cos x \, dx.$ (L.U.)

$$\int_0^{\pi/2} \sin 2x \cos x \, dx = \frac{1}{2} \int_0^{\pi/2} (\sin x + \sin 3x) dx$$

$$= \frac{1}{2} \left[-\cos x - \frac{1}{3} \cos 3x \right]_0^{\pi/2} = \frac{2}{3}.$$

The integrals of $\sin^2 mx$, $\cos^2 mx$ can be found from the double-angle formulae

$$\sin^2 mx = \tfrac{1}{2}(1 - \cos 2mx),$$
$$\cos^2 mx = \tfrac{1}{2}(1 + \cos 2mx),$$

which can be deduced by writing $A = B = mx$ in the first two of equations (11.3) and which have been given also in equations (5.11) and (5.12). Again it is best to apply these identities to the particular example under discussion rather than to attempt to remember the general integrals.

Example 6. *Evaluate* $\int_0^{\pi/4} \cos^2 x \, dx.$ (L.U.)

$$\int_0^{\pi/4} \cos^2 x \, dx = \frac{1}{2} \int_0^{\pi/4} (1 + \cos 2x) dx$$

$$= \frac{1}{2} \left[x + \frac{1}{2} \sin 2x \right]_0^{\pi/4}$$

$$= \frac{1}{2} \left(\frac{\pi}{4} + \frac{1}{2} \right) = \frac{\pi}{8} + \frac{1}{4}.$$

Example 7. *Evaluate* $\int_0^{\pi/2} \sin^2 2x \, dx.$ (L.U.)

$$\int_0^{\pi/2} \sin^2 2x \, dx = \frac{1}{2} \int_0^{\pi/2} (1 - \cos 4x) dx$$

$$= \frac{1}{2} \left[x - \frac{1}{4} \sin 4x \right]_0^{\pi/2} = \frac{\pi}{4}.$$

EXERCISES 11 (b)

Evaluate the following integrals:—

1. $\int_0^{\pi/2} 2 \sin^2 x \, dx.$ 2. $\int_0^a \cos^2 (x - \alpha) dx.$

3. $\displaystyle\int_0^{\pi/2} (2\cos^2\theta + 3\sin^2\theta)d\theta.$

7. $\displaystyle\int_0^{\pi/2} \sin x \cos x \, dx.$

4. $\displaystyle\int_0^{\pi/2} (2\cos^2\theta + 3\sin 3\theta)d\theta.$

8. $\displaystyle\int_{\pi/8}^{3\pi/8} \sin 3\theta \cos\theta \, d\theta.$

5. $\displaystyle\int_0^{\pi/3} 2\sin 3x \cos x \, dx.$

9. $\displaystyle\int_0^{\pi/2} (\sin 3x + \sin x \cos x)dx.$

6. $\displaystyle\int_0^{\pi/4} \sin x \cos x \, dx.$

10. $\displaystyle\int_0^{2\pi} \sin m\theta \sin n\theta \, d\theta$, where m and n are integers and $m \neq n$.

11.4. Integration by change of variable

A very powerful method of integration is that of changing the independent variable, or, as it is sometimes called, integration by substitution. It will be seen from the derivation of equation (11.7) below that this method is the converse of the method of differentiating a function of a function (§ 8.6).

Suppose we wish to change the variable from x to t in the indefinite integral

$$I = \int \phi(x)dx. \tag{11.4}$$

By the definition of the integral

$$\frac{dI}{dx} = \phi(x), \tag{11.5}$$

and, if x is a given function of t, the rule for the differentiation of a function of a function (8.10) gives

$$\frac{dI}{dt} = \frac{dI}{dx} \times \frac{dx}{dt}.$$

Substitution of dI/dx from (11.5) leads to

$$\frac{dI}{dt} = \phi(x).\frac{dx}{dt},$$

and hence, if we again use the definition of an integral,

$$I = \int \phi(x).\frac{dx}{dt}dt. \tag{11.6}$$

Equations (11.4), (11.6) together give the important result

$$\int \phi(x)dx = \int \phi(x)\frac{dx}{dt}dt. \tag{11.7}$$

The following are important cases:—

(i) Suppose we wish to find $\displaystyle\int \phi(ax + b)dx$ where a and b are constants.

Write $ax + b = t$ so that $a(dx/dt) = 1$ and $dx/dt = 1/a$. Then (11.7) gives

$$\int \phi(ax + b)dx = \int \phi(t).\frac{1}{a}dt$$

$$= \frac{1}{a}\int \phi(t)dt,$$

a symbolic expression of the important rule of § 11.2.

(ii) Integrals of the form $\int \phi(x^2)x\,dx$ can often be found by using the substitution $x^2 = t$. In this case, $2x(dx/dt) = 1$, giving $dx/dt = 1/(2x)$, and from (11.7)

$$\int \phi(x^2)x\,dx = \int \phi(t).x.\frac{1}{2x}dt$$

$$= \frac{1}{2}\int \phi(t)dt,$$

so that $\int \phi(x^2)x\,dx$ can be determined when $\int \phi(t)dt$ is known.

The presence of the "extra x" in $\int \phi(x^2)x\,dx$ should be noted. It is this term which enables the integral to be reduced to the simpler form $\frac{1}{2}\int \phi(t)dt$. If the "extra x" were absent, the corresponding result would be

$$\int \phi(x^2)dx = \frac{1}{2}\int \phi(t)t^{-1/2}\,dt,$$

and this is not usually a specially helpful transformation.

Example 8. *Integrate* $x\sqrt{(1 + x^2)}$ *with respect to x.*
By writing $x^2 = t$, we find as above that

$$\int x\sqrt{(1 + x^2)}dx = \frac{1}{2}\int \sqrt{(1 + t)}dt$$

$$= \frac{1}{2}.\frac{(1 + t)^{3/2}}{3/2} + C$$

$$= \frac{1}{3}(1 + x^2)^{3/2} + C,$$

when we replace t by x^2 after the integration.

Example 9. *Find* $\int \dfrac{x\,dx}{1 + x^4}$.
Using the same substitution $x^2 = t$,

$$\int \frac{x\,dx}{1 + x^4} = \frac{1}{2}\int \frac{dt}{1 + t^2} = \frac{1}{2}\tan^{-1}t + C$$

$$= \frac{1}{2}\tan^{-1}x^2 + C.$$

For other types of integral, the choice of a successful substitution is a matter of some judgment. It is only possible here to give a few hints and examples.

(a) If the integrand contains $(a^2 - x^2)$, the substitutions $x = a \sin t$ or $x = \cos t$ are often helpful.

For example let us put $x = a \sin t$ to evaluate the integral $I = \int \dfrac{dx}{\sqrt{(a^2 - x^2)}}$. If $x = a \sin t$, $dx/dt = a \cos t$, and (11.7) gives

$$I = \int \frac{1}{\sqrt{(a^2 - a^2 \sin^2 t)}} . a \cos t \, dt = \int 1 . dt = t + C,$$

for $\sqrt{(a^2 - a^2 \sin^2 t)} = a \cos t$. Since $x = a \sin t$, $t = \sin^{-1}(x/a)$ and we therefore have

$$\int \frac{dx}{\sqrt{(a^2 - x^2)}} = \sin^{-1}\frac{x}{a} + C, \tag{11.8}$$

as given in the list of § 11.2.

If we had determined this integral by means of the substitution $x = a \cos t$, we should have had $dx/dt = -a \sin t$, and by (11.7)

$$I = \int \frac{1}{\sqrt{(a^2 - a^2 \cos^2 t)}}(-a \sin t)dt = \int (-1)dt = -t + C'.$$

Replacing t by its value $\cos^{-1}(x/a)$, the final result would have been

$$\int \frac{dx}{\sqrt{(a^2 - x^2)}} = -\cos^{-1}\frac{x}{a} + C'. \tag{11.9}$$

The apparent discrepancy between the two results (11.8) and (11.9) is explained by the relation $\cos^{-1}x + \sin^{-1}x = \frac{1}{2}\pi$ obtained in Example 12 of Chapter 5. Using this, the value of the integral given in (11.9) can be written $\sin^{-1}(x/a) - \frac{1}{2}\pi + C'$ and agreement with (11.8) can be secured by writing $C = C' - \frac{1}{2}\pi$. This is another example of indefinite integrals obtained by different methods differing by a constant quantity.

(b) When $(a^2 + x^2)$ occurs in the integrand, the substitution $x = a \tan t$ is indicated. With this substitution $dx/dt = a \sec^2 t$ and $a^2 + a^2 \tan^2 t = a^2 \sec^2 t$ so that, using it to evaluate $\int \dfrac{dx}{a^2 + x^2}$ for example,

$$\int \frac{dx}{a^2 + x^2} = \int \frac{1}{a^2 \sec^2 t} . a \sec^2 t \, dt = \frac{1}{a} \int dt$$

$$= \frac{t}{a} + C = \frac{1}{a}\tan^{-1}\frac{x}{a} + C,$$

for $t = \tan^{-1}(x/a)$.

(c) Products of the type $\sin^m x \cos^n x$ where m and n are positive integers and one at least of them is odd can be integrated by the following devices. If n is odd, $n - 1$ will be even and $\cos^{n-1} x$ is expressed in terms of $\sin x$ by means of $\cos^2 x = 1 - \sin^2 x$: then we use $\sin x = t$. If m is odd, $\sin^{m-1} x$ can be expressed in terms of $\cos x$ by means of $\sin^2 x = 1 - \cos^2 x$ and then the substitution $\cos x = t$ used. Some examples follow.

Example 10. *Find* $\int \sin^3 x \cos x \, dx$

Here the powers of the sine and cosine are both odd, so either of the above methods can be used. Choosing the first (it is preferable here), write $\sin x = t$, so that $\cos x (dx/dt) = 1$ and $dx/dt = \sec x$. Using (11.7) we therefore have

$$\int \sin^3 x \cos x \, dx = \int t^3 . \cos x . \sec x \, dt = \int t^3 \, dt$$

$$= \tfrac{1}{4}t^4 + C = \tfrac{1}{4} \sin^4 x + C.$$

Example 11. *Integrate* $\cos^5 x$ *with respect to* x.

Here it is essential to use the first method for $\cos^5 x$ is a particular case of the product $\sin^m x \cos^n x$ with $m = 0$ (even) and $n = 5$ (odd). We write $\sin x = t$, and therefore, as in Example 10, $dx/dt = \sec x$. $\cos^5 x$ is written as $\cos^4 x . \cos x$, or $(1 - \sin^2 x)^2 . \cos x$, so that

$$\int \cos^5 x \, dx = \int (1 - \sin^2 x)^2 . \cos x \, dx$$

$$= \int (1 - t^2)^2 . \cos x . \sec x \, dt$$

$$= \int (1 - t^2)^2 \, dt$$

$$= \int (1 - 2t^2 + t^4) dt$$

$$= t - \tfrac{2}{3}t^3 + \tfrac{1}{5}t^5 + C$$

$$= \sin x - \tfrac{2}{3} \sin^3 x + \tfrac{1}{5} \sin^5 x + C.$$

Example 12. *Find* $\int \sin^3 x \cos^2 x \, dx$.   (Q.E.)

Here the sine is raised to an odd power so we set $\cos x = t$. This gives $-\sin x (dx/dt) = 1$ and $dx/dt = -\operatorname{cosec} x$. $\sin^3 x$ is   written   as $(1 - \cos^2 x) \sin x$ and hence

$$\int \sin^3 x \cos^2 x \, dx = \int (1 - \cos^2 x) \sin x \cos^2 x \, dx$$

$$= \int (1 - t^2) \sin x . t^2 . (-\operatorname{cosec} x) dt$$

$$= - \int (t^2 - t^4) dt = -\tfrac{1}{3}t^3 + \tfrac{1}{5}t^5 + C.$$

$$= - \tfrac{1}{3} \cos^3 x + \tfrac{1}{5} \cos^5 x + C.$$

(*d*) There are many other substitutions which, with (11.7), are effective in particular cases. For example, even positive integral powers of sec *x* may be integrated by writing $t = \tan x$ and those of cosec *x* by using $t = \cot x$. Examples will be found among the exercises. As previously stated, the choice of a correct substitution is not always evident and often depends on judgment and experience: in some of the exercises hints are given.

11.5. Definite integrals by change of variable

In the evaluation of a definite integral when the integration is performed by changing the variable, two methods are available. In the first, the indefinite integral is found and expressed in the original variable and then the limits are inserted: in the second (and usually preferable) method, everything, including the limits, is expressed in terms of the new variable. Both methods are used in the first example below: the second method only is shown in the remaining examples.

Example 13. *Evaluate* $\displaystyle\int_0^{\pi/2} \sin^3 x \, dx$.

Method 1. First find $\displaystyle\int \sin^3 x \, dx$, by writing

$$\cos x = t, \quad \sin^3 x = \sin x(1 - \cos^2 x).$$

Then $-\sin x(dx/dt) = 1$, and $dx/dt = -\operatorname{cosec} x$. Therefore

$$\int \sin^3 x \, dx = \int \sin x(1 - \cos^2 x)(-\operatorname{cosec} x)dt = -\int (1 - t^2)dt$$

$$= -t + \tfrac{1}{3}t^3 = -\cos x + \tfrac{1}{3}\cos^3 x.$$

Hence

$$\int_0^{\pi/2} \sin^3 x \, dx = \left[-\cos x + \tfrac{1}{3}\cos^3 x \right]_0^{\pi/2} = 1 - \frac{1}{3} = \frac{2}{3}.$$

Method 2. Using the substitution $\cos x = t$, the indefinite integral can as before be reduced to $-\displaystyle\int (1 - t^2)dt$. At the lower limit of the given definite integral $x = 0$, so that since $\cos x = t$, the lower limit of the integral in t will be given by $t = \cos 0 = 1$. Similarly, since $x = \pi/2$ at the upper limit of the integral in x, the upper limit of the transformed integral will be $t = \cos(\pi/2) = 0$. Hence

$$\int_0^{\pi/2} \sin^3 x \, dx = -\int_1^0 (1 - t^2)dt = -\left[t - \tfrac{1}{3}t^3 \right]_1^0 = 1 - \frac{1}{3} = \frac{2}{3},$$

as before.

Example 14. *Evaluate* $\displaystyle\int_0^1 \frac{x \, dx}{\sqrt{(1 - x^2)}}$ *by means of the substitution* $x = \sin t$.

If $x = \sin t$, $dx/dt = \cos t$. When $x = 0$, $\sin t = 0$ and $t = 0$, when $x = 1$, $\sin t = 1$ and $t = \pi/2$, so that the lower and upper limits of the integral in t are respectively 0 and $\pi/2$. Hence

$$\int_0^1 \frac{x\,dx}{\sqrt{(1 - x^2)}} = \int_0^{\pi/2} \frac{\sin t}{\sqrt{(1 - \sin^2 t)}} \cdot \cos t\,dt.$$

$$= \int_0^{\pi/2} \sin t\,dt$$

$$= \left[-\cos t\right]_0^{\pi/2} = 1.$$

Example 15. *Evaluate* $\displaystyle\int_0^{\pi/4} \sec^4 x\,dx$.

Put $t = \tan x$ so that $\sec^2 x(dx/dt) = 1$ and $dx/dt = \cos^2 x$. When $x = 0$, $t = \tan 0 = 0$ and when $x = \pi/4$, $t = \tan(\pi/4) = 1$. Hence

$$\int_0^{\pi/4} \sec^4 x\,dx = \int_0^{\pi/4} \sec^2 x \sec^2 x\,dx$$

$$= \int_0^{\pi/4} (1 + \tan^2 x)\sec^2 x\,dx$$

$$= \int_0^1 (1 + t^2)\sec^2 x \cdot \cos^2 x\,dt$$

$$= \int_0^1 (1 + t^2)dt$$

$$= \left[t + \tfrac{1}{3}t^3\right]_0^1 = \frac{4}{3}.$$

EXERCISES 11 (c)

Integrate the following functions with respect to x:—

1. $x \cos(x^2 - 1)$.

2. $\dfrac{x^2}{(x^3 + 8)^4}$ (Hint, put $x^3 + 8 = t$).

3. $\dfrac{1}{\sqrt{x}} \cos \sqrt{x}$ (Hint, put $\sqrt{x} = t$).

4. $\sin x \cos^4 x$.

5. $\sin^3 2x$.

6. $\dfrac{x - 1}{\sqrt{(1 - 9x^2)}}$ (Hint, put $3x = \sin t$).

7. $\dfrac{x^3}{1 + x^8}$ (Hint, put $x^4 = t$).

8. $\sqrt{(4 - x^2)}$.

Evaluate the following definite integrals:—

9. $\displaystyle\int_0^1 x\sqrt{(4 + x^2)}dx$.   (L.U.)

10. $\displaystyle\int_0^{\pi/3} \sin^2 x \cos x \, dx.$　　　　　　　　　　　(O.C.)

11. $\displaystyle\int_0^1 \frac{4x \, dx}{(2 - x^2)^{3/2}}.$

12. $\displaystyle\int_0^a x\sqrt{(a^2 - x^2)}dx.$

13. $\displaystyle\int_0^{\pi/2} \frac{\cos\theta \, d\theta}{1 + \sin^2\theta}$ (Hint, put $\sin\theta = t$).

14. $\displaystyle\int_0^a \sqrt{(a^2 - x^2)}dx.$

15. $\displaystyle\int_{-\pi/4}^{\pi/4} \tan^2\theta \sec^2\theta \, d\theta$ (Hint, put $\tan\theta = t$).

16. $\displaystyle\int_0^{\pi/2} \frac{\cos x \, dx}{\sqrt{(\sin x)}}.$

11.6. Integration by parts

A useful method of integration, known as "integration by parts", results from the inverse of the formula (8.4) for the differential coefficient of the product uv of two functions u and v of x. This is

$$\frac{d}{dx}(uv) = v\frac{du}{dx} + u\frac{dv}{dx},$$

which, with the fundamental definition of the integral, gives

$$uv = \int\left(v\frac{du}{dx} + u\frac{dv}{dx}\right)dx.$$

This can be written

$$uv = \int v\frac{du}{dx}dx + \int u\frac{dv}{dx}dx,$$

which can be transposed into

$$\int u\frac{dv}{dx}dx = uv - \int v\frac{du}{dx}dx. \qquad (11.10)$$

The integral on the right-hand side of (11.10) is often easier to evaluate than the one on the left. Some judgment must be used in the choice of u and v when employing this method which is particularly useful when one of the functions is an inverse trigonometrical function. In this case this function should be taken as the "u" in the integral on the left of (11.10) for then du/dx in the integral on the right becomes a fairly simple algebraic function. This and some other devices useful in this method of integration are illustrated in Examples 16–19 below.

Example 16. *Integrate* $x \tan^{-1} x$ *with respect to x.*

Take $u = \tan^{-1} x$, $\dfrac{dv}{dx} = x$, so that $\dfrac{du}{dx} = \dfrac{1}{1 + x^2}$ and $v = \frac{1}{2}x^2$.

Then (11.10) gives

$$\int x \tan^{-1} x \, dx = \frac{1}{2}x^2 \tan^{-1} x - \int \frac{1}{2}x^2 \left(\frac{1}{1 + x^2}\right) dx. \qquad (11.11)$$

The integral on the right, $\dfrac{1}{2}\displaystyle\int \dfrac{x^2}{1 + x^2} dx$, can be written

$$\frac{1}{2}\int \left(1 - \frac{1}{1 + x^2}\right) dx,$$

and this is $\frac{1}{2}(x - \tan^{-1} x)$. Replacing this in (11.11) we have

$$\int x \tan^{-1} x \, dx = \frac{1}{2}x^2 \tan^{-1} x - \frac{1}{2}(x - \tan^{-1} x) + C$$

$$= \frac{1}{2}(x^2 + 1) \tan^{-1} x - \frac{1}{2}x + C.$$

Example 17. *Find* $\displaystyle\int x \cos x \, dx$.

If we take u as $\cos x$ and dv/dx as x, $du/dx = -\sin x$ and $v = \frac{1}{2}x^2$, so that the integral on the right of (11.10) is $\displaystyle\int \frac{1}{2}x^2 \sin x \, dx$ and this is more complicated than the original integral.

If, however, we take $u = x$, $dv/dx = \cos x$, $du/dx = 1$, $v = \sin x$, and (11.10) gives

$$\int x \cos x \, dx = x \sin x - \int \sin x \, dx$$

$$= x \sin x + \cos x + C.$$

Example 18. *Find* $\displaystyle\int \sin^2 x \, dx$ *by the method of integration by parts.*

Although integration by parts is not the best method for finding $\displaystyle\int \sin^2 x \, dx$, the working below illustrates a device which is sometimes useful.
Take $u = \sin x$, $dv/dx = \sin x$, so that $du/dx = \cos x$, $v = -\cos x$.
Then (11.10) gives

$$\int \sin^2 x \, dx = -\sin x \cos x - \int (-\cos x) \cos x \, dx$$

$$= -\sin x \cos x + \int \cos^2 x \, dx.$$

Since $\cos^2 x = 1 - \sin^2 x$, this can be written

$$\int \sin^2 x \, dx = -\sin x \cos x + \int (1 - \sin^2 x) dx$$

$$= -\frac{1}{2}\sin 2x + \int dx - \int \sin^2 x \, dx,$$

giving, when the last term on the right is transposed to the left,

$$2 \int \sin^2 x \, dx = -\tfrac{1}{2} \sin 2x + \int dx$$

$$= -\tfrac{1}{2} \sin 2x + x + C.$$

Thus
$$\int \sin^2 x \, dx = \frac{x}{2} - \frac{1}{4} \sin 2x + C',$$

where $C' (= \tfrac{1}{2}C)$ is an arbitrary constant.

Example 19. *Find* $\int x^2 \sin x \, dx$.   (Q.E.)

Take $u = x^2$, $dv/dx = \sin x$, so that $du/dx = 2x$, $v = -\cos x$. Then by (11.10),

$$\int x^2 \sin x \, dx = -x^2 \cos x - \int (-\cos x).\, 2x \, dx$$

$$= -x^2 \cos x + 2 \int x \cos x \, dx.$$

Thus, one application of the rule of integration by parts has connected the integral of $x^2 \sin x$ with that of $x \cos x$. As in Example 17, a second application of the rule gives

$$\int x \cos x \, dx = x \sin x - \int \sin x \, dx$$

and hence

$$\int x^2 \sin x \, dx = -x^2 \cos x + 2x \sin x - 2 \int \sin x \, dx.$$

Two applications of the rule have therefore related $\int x^2 \sin x \, dx$ to $\int \sin x \, dx$, a known integral, and finally

$$\int x^2 \sin x \, dx = -x^2 \cos x + 2x \sin x + 2 \cos x + C.$$

In a similar way, if n is a positive integer, $\int x^n \sin x \, dx$ can be related to $\int x^{n-2} \sin x \, dx$ and this latter integral can similarly be related to $\int x^{n-4} \sin x \, dx$ and so on. We shall finally be left with either $\int x \sin x \, dx$ or $\int \sin x \, dx$ to find in order to determine completely the original integral. Both these integrals can in fact be found and integration performed in this way is known as integration by successive reduction. It is a method of considerable importance but it is beyond the scope of this book to pursue it except in very simple special cases such as the example considered above.

In using the method of integration by parts to evaluate a definite integral, the limits of integration can be inserted as the working proceeds. Thus if we require $\int_0^{\pi/2} x \cos x \, dx$ and perform the integration as in Example 17 above, the working could be set out

$$\int_0^{\pi/2} x \cos x \, dx = \left[x \sin x \right]_0^{\pi/2} - \int_0^{\pi/2} \sin x \, dx$$

$$= \frac{\pi}{2} - \left[-\cos x \right]_0^{\pi/2}$$

$$= \frac{\pi}{2} - 1.$$

EXERCISES 11 (d)

Find the following by the method of integration by parts:—

1. $\int x \sin x \, dx.$

2. $\int \cos^2 x \, dx.$

3. $\int \sin x \cos x \, dx.$

4. $\int \sin^{-1} x \, dx.$

5. $\int \dfrac{x \sin^{-1} x}{\sqrt{(1 - x^2)}} dx.$

6. $\int_0^{\pi} x^2 \cos x \, dx.$ (Q.E.)

7. If $U_n = \int x^n \cos x \, dx$ and $V_n = \int x^n \sin x \, dx$, show that

$$U_n = x^n \sin x - nV_{n-1},$$
$$V_n = -x^n \cos x + nU_{n-1}.$$

8. Evaluate

$$\text{(i)} \int_0^{\pi/2} x^2 \cos \tfrac{1}{2}x \, dx, \quad \text{(ii)} \int_0^1 \cos^{-1} x \, dx.$$

11.7. Approximate methods of integration

If the value of the definite integral $\int_a^b f(x)dx$ is required and we are unable to find a function whose derivative is $f(x)$, methods of approximation can be used. There are several methods available.

Since $\int_a^b f(x)dx$ measures the area bounded by the curve $y = f(x)$, the x-axis and ordinates at $x = a, x = b$, probably the simplest method is to draw the curve on squared paper and to estimate the area by counting the squares enclosed. This is often laborious and, unless a very large-scale drawing is used, not very accurate.

Another method is to divide the area into a number of strips and replace each of these by a trapezium. In Fig. 69, the area $PABQ$ is to be found and the area has been split into strips by dividing AB into six equal parts and erecting ordinates $y_1, y_2, \ldots, y_5$ at each point of subdivision. y_0 and y_6 are used to denote the ordinates at $x = a$ and

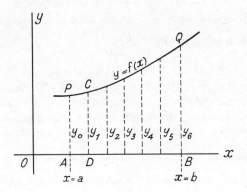

FIG. 69

$x = b$. If C and D are respectively the points of intersection of the ordinate y_1 with the curve and the x-axis, it is clear from the enlarged diagram shown in Fig. 70 that the area of the trapezium $PADC$ is slightly greater than the area bounded by the arc PC of the curve and

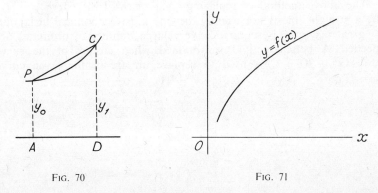

FIG. 70   FIG. 71

the lines PA, AD, DC, and that the area of the trapezium could be used as an approximation to the area below the arc PC of the curve. Now the mean height of the trapezium is $\frac{1}{2}(y_0 + y_1)$ and if we denote AD by h, its area is $\frac{1}{2}h(y_0 + y_1)$. Treating the other strips of Fig. 69 in the same way and summing we see that the area $PABQ$ is approximately equal to

$$\tfrac{1}{2}h(y_0 + y_1) + \tfrac{1}{2}h(y_1 + y_2) + \tfrac{1}{2}h(y_2 + y_3) + \ldots + \tfrac{1}{2}h(y_5 + y_6)$$

or,

$$h\{\tfrac{1}{2}(y_0 + y_6) + y_1 + y_2 + y_3 + y_4 + y_5\},$$

and since $AB = b - a$ and we have used six strips, $h = \frac{1}{6}(b - a)$.

In the same way, using n strips, we should find that the area $PABQ$ or $\int_a^b y\,dx$ is approximately equal to

$$h\{\tfrac{1}{2}(y_0 + y_n) + y_1 + y_2 + \ldots + y_{n-2} + y_{n-1}\}, \qquad (11.12)$$

where now $h = (b - a)/n$. The formula (11.12) can be expressed by saying that *if the range of integration $(b - a)$ is divided into n equal parts each of width $h = (b - a)/n$, called the interval, and ordinates are erected at $x = a$, $x = b$ and at each point of sub-division, then*

$$\int_a^b y\,dx \eqsim \textit{interval} \times \textit{(half the sum of the first and last ordinates}$$

$$+ \textit{the sum of the remaining ordinates}).$$

This is known as the *trapezoidal rule* for approximate integration. It is clear that the approximation improves as the number of strips is increased (or as the interval is decreased). It will also be clear that the rule overestimates the integral when $y = f(x)$ is a curve of the shape shown in Fig. 69. In other cases, for example when the curve $y = f(x)$ is of the shape shown in Fig. 71, it will underestimate the integral.

The above method of replacing each piece of the curve $y = f(x)$ by a straight line is not usually the one most economical in labour. If great accuracy is required a very large number of ordinates are needed. A better method is obtained when an arc of the curve $y = Ax^2 + Bx + C$ is used to replace an arc of the given curve $y = f(x)$.

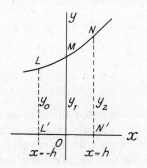

FIG. 72

Suppose in Fig. 72 that LMN is an arc of the curve $y = f(x)$ intersecting the y-axis at M and that LL', NN' are ordinates at $x = -h$, $x = h$. Let $LL' = y_0$, $MO = y_1$, $NN' = y_2$. Since the equation $y = Ax^2 + Bx + C$ contains three constants the curve represented by this equation can be made to pass through three given points. If we take these points to be L, M and N and if h is not too large, it is to be expected that the area below this curve will be a good approximation to the area below the curve *LMN* shown in the diagram. Hence

$$\int_{-h}^{h} f(x)dx \doteqdot \int_{-h}^{h} (Ax^2 + Bx + C)dx$$

$$= \left[\tfrac{1}{3}Ax^3 + \tfrac{1}{2}Bx^2 + Cx \right]_{-h}^{h}$$

$$= \tfrac{2}{3}Ah^3 + 2Ch. \tag{11.13}$$

If the curve $y = Ax^2 + Bx + C$ passes through *L*, *M* and *N*, since $y = y_0$ when $x = -h$, $y = y_1$ when $x = 0$ and $y = y_2$ when $x = h$,

$$y_0 = Ah^2 - Bh + C,$$
$$y_1 = C,$$
$$y_2 = Ah^2 + Bh + C.$$

The addition of the first and last of these relations gives

$$y_0 + y_2 = 2Ah^2 + 2C,$$

and since from the middle relation, $C = y_1$,

$$2Ah^2 = y_0 + y_2 - 2y_1.$$

Substituting for Ah^2 and C in (11.13) we have

$$\int_{-h}^{h} f(x)dx \doteqdot \frac{h}{3}(y_0 + y_2 - 2y_1) + 2hy_1$$

$$= \frac{h}{3}(y_0 + 4y_1 + y_2). \tag{11.14}$$

This formula gives an approximation to the area of a strip of width $2h$ when an arc of the curve $y = f(x)$ is replaced by an arc of $y = Ax^2 + Bx + C$ which passes through three points on the given curve whose successive abscissae differ by h. If this formula is applied to Fig. 69, the addition of three such strips gives

$$\int_{a}^{b} y\,dx = \text{area } PABQ$$

$$\doteqdot \frac{h}{3}(y_0 + 4y_1 + y_2) + \frac{h}{3}(y_2 + 4y_3 + y_4) + \frac{h}{3}(y_4 + 4y_5 + y_6)$$

$$= \frac{h}{3}\{y_0 + y_6 + 4(y_1 + y_3 + y_5) + 2(y_2 + y_4)\}.$$

In a similar way, if we divide the area into an even number $2n$ of strips, we find that $\displaystyle\int_{a}^{b} y\,dx$ is approximately equal to

$$\frac{h}{3}\{y_0 + y_{2n} + 4(y_1 + y_3 + y_5 + \ldots + y_{2n-1})$$

$$+ 2(y_2 + y_4 + \ldots + y_{2n-2})\}, \tag{11.15}$$

where $h = (b - a)/2n$. This formula, which is known as *Simpson's rule*, can be expressed thus:—*if the range of integration $(b - a)$ is divided into an even number $2n$ of equal parts each of width $h = (b - a)/2n$, called the interval, and ordinates are erected at $x = a$, $x = b$ and at each point of sub-division, then*

$$\int_a^b y \, dx \doteqdot \tfrac{1}{3} \times \text{interval} \times (\text{the sum of first and last ordinates}$$
$$+ \text{ four times the sum of all the odd ordinates}$$
$$+ \text{ twice the sum of the remaining even ordinates}).$$

Both the trapezoidal rule (11.12) and Simpson's rule (11.15) can be used to evaluate definite integrals when the integrand is given by a numerical table or by a mathematical formula. If the integrand is given as a formula, a table giving values of the function (the ordinates y) at equal intervals of x has first to be calculated. Two examples are given below. In the first, the integrand is only given numerically; the second, in which the integrand is $4/(1 + x^2)$, has been chosen so that the approximate results can be compared with the known exact answer.

Example 20. *A function y of x is given by the following table:—*

x	0	0·1	0·2	0·3	0·4	0·5	0·6
y	0·0000	0·0499	0·0995	0·1483	0·1960	0·2423	0·2867

Find $\displaystyle\int_0^{0\cdot6} y \, dx$ by both the trapezoidal and Simpson's rules.

The working can be arranged as follows:—

x	First and last ordinates	Odd ordinates	Remaining even ordinates
0·0	0·0000		
0·1		0·0499	
0·2			0·0995
0·3		0·1483	
0·4			0·1960
0·5		0·2423	
0·6	0·2867		
Sums	0·2867	0·4405	0·2955

The interval $h = 0\cdot1$, and the sum of the ordinates omitting the first and last $= 0\cdot4405 + 0\cdot2955 = 0\cdot7360$.

Hence the trapezoidal rule (11.12) gives

$$\int_0^{0\cdot6} y \, dx \doteqdot 0\cdot1(\tfrac{1}{2} \times 0\cdot2867 + 0\cdot7360) = 0\cdot08794.$$

Simpson's rule (11.15) gives

$$\int_0^{0\cdot6} y\,dx \fallingdotseq \frac{0\cdot1}{3}(0\cdot2867 + 4 \times 0\cdot4405 + 2 \times 0\cdot2955) = 0\cdot08799.$$

Thus to four places of decimals, the values of the definite integral by the two methods are respectively 0·0879 and 0·0880. If the graph of y against x be drawn it will be found that the curve is of the shape shown in Fig. 71 and therefore the result 0·0879 obtained by the trapezoidal rule must be expected to be a little low.

Example 21. *Use the trapezoidal and Simpson's rules with five ordinates to find approximate values for* $\displaystyle\int_0^1 \frac{4\,dx}{1 + x^2}$. *Compare your results with the exact value of the integral.*

x	$1 + x^2$	Ordinates $y = 4/(1 + x^2)$		
		First and last ordinates	Odd ordinates	Remaining even ordinates
0·00	1·0000	4·0000		
0·25	1·0625		3·7647	
0·50	1·2500			3·2000
0·75	1·5625		2·5600	
1·00	2·0000	2·0000		
Sums		6·0000	6·3247	3·2000

Here the interval $h = 0\cdot25$, and the sum of the ordinates omitting the first and last $= 6\cdot3247 + 3\cdot2000 = 9\cdot5247$.

Hence, by the trapezoidal rule

$$\int_0^1 \frac{4\,dx}{1 + x^2} \fallingdotseq 0\cdot25(\tfrac{1}{2} \times 6\cdot0000 + 9\cdot5247) = 3\cdot1312.$$

By Simpson's rule,

$$\int_0^1 \frac{4\,dx}{1 + x^2} \fallingdotseq \frac{0\cdot25}{3}(6\cdot0000 + 4 \times 6\cdot3247 + 2 \times 3\cdot2000) = 3\cdot1416.$$

The exact value of the integral is

$$4\int_0^1 \frac{dx}{1 + x^2} = 4\left[\tan^{-1} x\right]_0^1 = 4(\pi/4) = \pi = 3\cdot14159\ldots,$$

so that the result obtained by Simpson's rule is correct to the fourth place of decimals. To obtain similar accuracy with the trapezoidal rule more ordinates, and therefore more labour, would be necessary.

EXERCISES 11 (e)

1. Use the trapezoidal and Simpson's rules to evaluate $\displaystyle\int_9^{10} y\,dx$ when y is given in terms of x by the following table:—

x	9·0	9·25	9·5	9·75	10·0
y	0·1111	0·1081	0·1053	0·1026	0·1000

2. Use eleven ordinates and Simpson's rule to find an approximate value of $\int_1^2 \dfrac{dx}{x}$.

3. Equidistant ordinates of a curve are 1·0, 0·6667, 0·5, 0·4 and 0·333. Estimate the area bounded by the curve, the x-axis and the extreme ordinates which are at $x = 0$ and $x = 2$.

4. The ordinates of a curve $y = f(x)$ are given by the table

x	0	1	2	3	4	5	6
y	0	2	2·5	2·3	2	1·7	1·5

Use Simpson's rule to estimate the volume generated when the area bounded by this curve and the ordinates at $x = 0$, $x = 6$ revolves about the x-axis.

5. Use Simpson's rule and eleven ordinates to obtain an approximate value of the definite integral $\int_0^{1/2} \dfrac{dx}{\sqrt{(1 - x^2)}}$. Compare your result with the exact value of the integral.

EXERCISES 11 (f)

Find the following indefinite integrals:—

1. $\int 2x(x^2 - 3)^2 \, dx$. (L.U.)

2. $\int \tan x \sec^2 x \, dx$. (L.U.)

3. $\int \dfrac{x \, dx}{\sqrt{(4 - x^2)}}$.

4. $\int \dfrac{dx}{x^2 + 2x + 5}$. (Q.E.)

5. $\int \dfrac{\cos^3 \theta \, d\theta}{\sin^2 \theta}$.

6. $\int \dfrac{dx}{2x^2 + 2x + 5}$.

7. $\int \dfrac{dx}{x^2\sqrt{(4 + x^2)}}$ (Hint, put $x = 2 \tan t$).

8. $\int (\sec^2 \theta - 1) \sec^2 \theta \, d\theta$. (L.U.)

9. $\int \sin (x - \pi/3) \cos (x + \pi/3) dx$.

10. $\int x\sqrt{(x + 2)} dx$ (Hint, put $x + 2 = t^2$).

Evaluate the following definite integrals:—

11. $\int_0^{\pi/2} \sin 5x \cos x \, dx$. (L.U.)

12. $\int_1^4 \dfrac{dx}{(5x + 2)^3}$. (L.U.)

13. $\int_0^\pi \sin^3 \theta \, d\theta$.

14. $\int_{-\alpha}^\alpha \sin^3 \theta \, d\theta$.

15. $\int_0^{\pi/2} \sin^5 x \cos^2 x \, dx$.

16. $\int_0^{\pi/4} (\theta + \cos \theta) \sin \theta \, d\theta$.

17. $\int_0^1 \dfrac{x \, dx}{(2 + x^2)^2}$.

18. $\int_0^{\pi/2} \dfrac{\sec^2 \theta \, d\theta}{4 + \tan^2 \theta}$.

19. $\int_0^{\pi/2} x \sin^2 x \, dx$.

20. $\int_0^a x^3(a^2 - x^2)^{3/2} \, dx$.

21. By writing $x = a\cos^2\theta + b\sin^2\theta$ show that

$$\int_a^b \frac{dx}{(x-a)^{1/2}(b-x)^{1/2}} = \pi.$$

22. If $\displaystyle\int_0^{\pi/2} x\sin x\, dx = \int_0^2 (ax^2 + 2x)dx$, find the value of a.

23. Find substitutions to show that each of the integrals

$$\text{(i)}\ \int_0^1 \frac{x^2\, dx}{(1+x^2)^2} \quad \text{and} \quad \text{(ii)}\ \int_0^{1/\sqrt2} \frac{x^2\, dx}{\sqrt{(1-x^2)}}$$

is equal to $\displaystyle\int_0^{\pi/4} \sin^2\theta\, d\theta$.

24. Use Simpson's rule and seven ordinates to find the area bounded by the curve $xy = 12$, the x-axis and ordinates at $x = 1$ and $x = 4$.

25. The coordinates of a curve are given in the table below. Find the area between the curve and the x-axis.

x	0	5	10	15	20	25	30	35	40	45	50
y	0	10	18	20	19	20	17	7	3	1	0

(Q.E.)

CHAPTER 12

SOME APPLICATIONS OF THE INTEGRAL CALCULUS

12.1. Introduction

A few geometrical and dynamical applications of the integral calculus were given in Chapter 10. More elaborate examples of these applications can be undertaken now that the student has studied some methods of integration. Further applications of integration are discussed in this chapter. These occur in finding mean values, centres of mass, moments of inertia, lengths of arcs and areas of surfaces of revolution.

12.2. Further examples of the calculation of area

Equation (10.13) gives the area bounded by the curve $y = f(x)$ the x-axis and ordinates $x = a$, $x = b$ as $\int_a^b y\, dx$. If the curve lies below the x-axis, y is negative and the area obtained from the definite integral appears as a negative quantity (see Example 11 of Chapter 10). If therefore we require the *total* area enclosed by a curve which crosses the x-axis at points between the two abscissae in which we are interested, we appeal to a diagram and divide the range of integration into appropriate sub-ranges. Typical cases of this procedure are given in Examples 1 and 2 below.

Example 1. *Find the whole area enclosed by the curve $y = \sin x$ and the x-axis between $x = 0$ and $x = 2\pi$.*

A sketch of the curve is shown in Fig. 73 and it is clear that

$$\text{area } OAB = \int_0^\pi \sin x\, dx = \left[-\cos x \right]_0^\pi$$
$$= -\cos \pi + \cos 0 = 2,$$
$$\text{area } BCD = \int_\pi^{2\pi} \sin x\, dx = \left[-\cos x \right]_\pi^{2\pi}$$
$$= -\cos 2\pi + \cos \pi = -2.$$

The area required is therefore 4 units, and the working could have been set out

$$\text{whole area} = \int_0^\pi \sin x\, dx - \int_\pi^{2\pi} \sin x\, dx$$
$$= \left[-\cos x \right]_0^\pi - \left[-\cos x \right]_\pi^{2\pi} = 4.$$

If the range of integration had not been sub-divided at $x = \pi$, the result

212

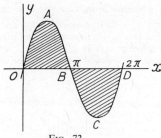

FIG. 73

would have been

$$\text{area} = \int_0^{2\pi} \sin x \, dx = \left[-\cos x \right]_0^{2\pi} = -\cos 2\pi + \cos 0 = 0,$$

and although this is a correct value of the definite integral $\int_0^{2\pi} \sin x \, dx$, it is not a correct interpretation in terms of area.

Example 2. *Find the area enclosed by the curve $y = x(x - 1)(x - 2)$ and the x-axis between $x = 0$ and $x = 4$.*

A rough graph of the curve (Fig. 74) shows that the area required is

$$\int_0^1 y \, dx - \int_1^2 y \, dx + \int_2^4 y \, dx.$$

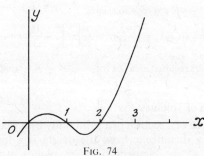

FIG. 74

Now $y = x(x - 1)(x - 2) = x^3 - 3x^2 + 2x$, so the area is

$$\left[\frac{x^4}{4} - x^3 + x^2 \right]_0^1 - \left[\frac{x^4}{4} - x^3 + x^2 \right]_1^2 + \left[\frac{x^4}{4} - x^3 + x^2 \right]_2^4$$

$$= \tfrac{1}{4} - 1 + 1 - \{4 - 8 + 4 - \tfrac{1}{4} + 1 - 1\} + 64 - 64 + 16 - 4 + 8 - 4$$

$$= 16 \cdot 5 \text{ units.}$$

In all our examples so far, the x-axis has been a bounding line of the area. Sometimes the area enclosed between two different curves is required and in such cases the result may be obtained as the difference between two areas each bounded partly by the x-axis. A typical example follows.

Example 3. *Find the area bounded by the curves $y^2 = 4x$ and $x^2 = 4y$.*

In Fig. 75, the curve $y^2 = 4x$ is shown by a full line and the curve $x^2 = 4y$ by a dotted line. The two curves intersect at the origin O and at the point B where $x = 4$. The required area is $OA'BA$ and this is clearly the difference between the two areas $OCBA$ and $OCBA'$, BC being the ordinate at $x = 4$.

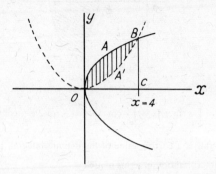

FIG. 75

For the curve $y^2 = 4x$, $y = 2\sqrt{x}$ and

$$\text{area } OCBA = \int_0^4 2\sqrt{x}\, dx = \frac{4}{3}\left[x^{3/2}\right]_0^4 = \frac{32}{3} \text{ units.}$$

For the curve $x^2 = 4y$, $y = x^2/4$ so that

$$\text{area } OCBA' = \int_0^4 \frac{x^2}{4}\, dx = \frac{1}{12}\left[x^3\right]_0^4 = \frac{16}{3} \text{ units.}$$

The required area $OA'BA = \frac{32}{3} - \frac{16}{3} = \frac{16}{3}$ units.

12.3. Calculation of volumes

Suppose the area of a section of a solid body by a plane perpendicular to the x-axis at a distance x from the origin is a function $S(x)$ of x. The volume δV of the element between planes at distances x and $x + \delta x$ from the origin lies between $S(x)\delta x$ and $S(x + \delta x)\delta x$.* Following the method of § 10.5 we therefore have, for the volume V of the solid,

$$V = \lim_{\delta x \to 0} \Sigma\, S(x)\delta x = \int_a^b S(x)dx, \tag{12.1}$$

where a and b are the end values of x for the solid under discussion.

For equation (12.1) to apply, it is not necessary for the solid to be a figure of revolution about the x-axis. If it is such a solid, $S(x)$ is replaced by πy^2 where $y = f(x)$ is the generating curve. In this case,

---

* See § 20.2.

(12.1) gives

$$V = \pi \int_a^b y^2 \, dx,$$

as given in equation (10.14). Volumes of solids of revolution generated by the rotation about the x-axis of the area enclosed between two intersecting curves can be found as the difference of two other volumes in a similar way to that shown in Example 3 above. The second example below illustrates the procedure.

Example 4. *The top and bottom of a large cistern are horizontal rectangles whose longer and shorter sides are respectively parallel. The top rectangle measures 16 m by 10 m and the bottom rectangle 9 m by 6 m. The sides of the cistern are plane and the vertical depth is 8 m. Find an expression for the area of a horizontal cross-section at a vertical height of x m from the bottom. Find, by integration, the volume of the cistern.* (L.U.)

In Fig. 76, $ABCD$ is a side elevation with $AD = 16$ m, $BC = 9$ m and $ABEF$ is an end elevation with $AF = 10$ m, $BE = 6$ m. XY is the longer side and XZ the shorter side of the rectangular cross-section at height x m from the

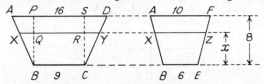

SIDE ELEVATION   END ELEVATION

FIG. 76

bottom of the cylinder. BP and CS are vertical lines through B and C respectively intersecting the lines AD, XY at P,S and Q,R as shown. From the first diagram it is clear that $AP = SD$ and hence $2AP = 16 - 9$ giving $AP = 3{\cdot}5$ m. Also, from the similar triangles XBQ, ABP,

$$\frac{XQ}{BQ} = \frac{AP}{BP} \quad \text{or} \quad \frac{XQ}{x} = \frac{3{\cdot}5}{8},$$

so that $XQ = 7x/16$. Also $XQ = RY$, so that $XY = 2XQ + QR$ giving

$$XY = \left(\frac{7x}{8} + 9\right) \text{ m.}$$

Working in precisely the same way from the second diagram we find that

$$XZ = \left(\frac{x}{2} + 6\right) \text{ m.}$$

The area $S(x)$ of the horizontal cross-section at height x m is therefore given by

$$S(x) = XY.XZ = \left(\frac{7x}{8} + 9\right)\left(\frac{x}{2} + 6\right)$$

$$= \left(\frac{7x^2}{16} + \frac{39x}{4} + 54\right) \text{ square metres.}$$

The volume of the cistern

$$= \int_0^8 S(x)dx = \int_0^8 \left(\frac{7x^2}{16} + \frac{39x}{4} + 54\right)dx$$

$$= \left[\frac{7x^3}{48} + \frac{39x^2}{8} + 54x\right]_0^8 = 818\cdot7 \text{ m}^3.$$

Example 5. *The area included between the parts of the two curves*

$$x^2 + y^2 = 1 \quad and \quad 4x^2 + y^2 = 4$$

for which y is positive is rotated about the x-axis. Find the volume of the solid thus formed. (O.C.)

A sketch of the curves shows that the first (full line) meets the x-axis at A and B (abscissae -1 and 1 respectively) and the y-axis at C ($y = 1$), while the second curve (dotted line) passes through A and B and meets the y-axis at D ($y = 2$).

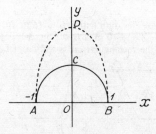

FIG. 77

The required volume is therefore the difference in volumes of the solids formed by rotating the areas ABD, ABC about the x-axis. For the curve ABD, $4x^2 + y^2 = 4$ giving $y^2 = 4(1 - x^2)$, while for ABC, $x^2 + y^2 = 1$ giving $y^2 = 1 - x^2$. Hence the volume required

$$= \pi \int_{-1}^1 4(1 - x^2)dx - \pi \int_{-1}^1 (1 - x^2)dx = 3\pi \int_{-1}^1 (1 - x^2)dx$$

$$= 3\pi \left[x - \frac{x^3}{3}\right]_{-1}^1 = 4\pi \text{ units.}$$

EXERCISES 12 (a)

1. Find the area enclosed by the curve $y = 4x - x^2$, the x-axis and ordinates at $x = 0$ and $x = 6$

2. Find the area bounded by $y^2 = 16x$ and $y = 3x$.

3. Find the area between the curves $a^7y = x^8$ and $a^7x = y^8$. (L.U.)

4. Find the area enclosed by the line $y = 2$ and the curve $y = x(3 - x)$. (L.U.)

5. The area enclosed by the two curves $y^2 = x^3$ and $y^2 = 8(x - 1)^3$ is rotated about the axis of x. Find the volume of the solid thus formed. (O.C.)

6. The area enclosed by the curves $y = 2 + x^2$, $y = 3 + x^2$ and ordinates at $x = 0$, $x = a$ is rotated about the x-axis. Find the volume of the solid so formed.

7. Water is poured into a hemispherical bowl whose axis is vertical and radius is a. Show that at the instant when the depth measured from the water's surface to the lowest point of the bowl is x, the area of the water's surface is given by

$$S(x) = \pi x(2a - x).$$

Find, by integration, the volume of the water at this instant.

8. A regular pyramid has a square of side a for its base and its height is h. Show that the area of a plane section at depth x below the vertex is $(ax/h)^2$. Prove also that the volume of the pyramid is $a^2h/3$.

12.4. Mean values

Let y be a function $\phi(x)$ of x and suppose that the range from $x = a$ to $x = b$ is divided into n equal sub-ranges each of width δx. Let $y_1, y_2, y_3, \ldots, y_n$ be the values of y at the middle points of each sub-range. The arithmetic mean of these n values of y is

$$\frac{1}{n}(y_1 + y_2 + y_3 + \ldots + y_n),$$

and, since $n\,\delta x = b - a$, this can be written

$$\frac{(y_1 + y_2 + y_3 + \ldots + y_n)\delta x}{b - a}.$$

If as $n \to \infty$ or $\delta x \to 0$, the expression has a limiting value, the limit is

$$\frac{\int_a^b y\,dx}{b - a}, \qquad\qquad (12.2)$$

and this is called the "*mean value*" of y over the range $(b - a)$.

Figure 78 shows the curve $y = \phi(x)$ and PA, QB are ordinates at $x = a$ and $x = b$. If we construct on AB a rectangle of area equal to that enclosed by the curve, the x-axis and PA, QB, its height H will

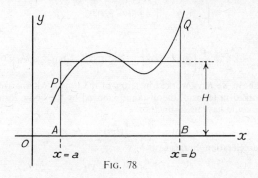

FIG. 78

be given by

$$H(b-a) = \int_a^b y\, dx.$$

Hence
$$H = \frac{\int_a^b y\, dx}{b-a},$$

which is the mean value of y over the range $(b-a)$. Thus in the geometrical representation, the mean value is the altitude of the rectangle on base $(b-a)$ whose area is equal to that included between the curve $y = \phi(x)$, the extreme ordinates and the x-axis.

Example 6. *Find the mean value of $\sin\theta$ over the range 0 to π.*

The required mean value

$$= \frac{\int_0^\pi \sin\theta\, d\theta}{\pi - 0} = \frac{1}{\pi}\left[-\cos\theta\right]_0^\pi = \frac{2}{\pi} = 0\cdot637.$$

Sometimes a quantity whose mean value is required can be expressed in terms of one or other of more than one variable. In such cases it is important to state which is the variable whose range is sub-divided in calculating the mean. Example 7 is an illustrative example.

Example 7. *A body has an initial velocity of 80 m/s and it is subjected to a retardation of 32 m/s². Find the mean value of the velocity of the body during its forward motion.*

If v is the velocity at time t we have $dv/dt = -32$, so that

$$v = \int (-32)dt = -32t + C,$$

where C is a constant. But $v = 80$ when $t = 0$, so that $C = 80$ and

$$v = 80 - 32t.$$

Forward motion ceases when $v = 0$, i.e., when $t = 80/32 = 2\cdot5$ seconds, and hence the mean velocity with respect to *time*

$$= \frac{\int_0^{2\cdot5} v\, dt}{2\cdot5} = \frac{\int_0^{2\cdot5} (80 - 32t)dt}{2\cdot5}$$

$$= \frac{1}{2\cdot5}\left[80t - 16t^2\right]_0^{2\cdot5} = \frac{2}{5}(200 - 100) = 40 \text{ m/s}.$$

In the above we have worked in terms of the time t. Alternatively we could have worked in terms of the distance s moved by the body. Instead of using dv/dt we could have used $v\,dv/ds$ or

$$\frac{d}{ds}(\tfrac{1}{2}v^2)$$

for the acceleration. In this case

$$\frac{d}{ds}(\tfrac{1}{2}v^2) = -32,$$

so that $\frac{1}{2}v^2 = -32s + C'$,

and, since $v = 80$ when $s = 0$, $C' = 3200$. Hence

$$v^2 = 6400 - 64s,$$

and forward motion ceases when $s = 6400/64 = 100$ m. Thus the mean velocity with respect to *space* is

$$\frac{1}{100}\int_0^{100} \sqrt{(6400 - 64s)}ds$$

$$= \frac{8}{100}\int_0^{100} \sqrt{(100 - s)}ds$$

$$= \frac{8}{100}\left[-\frac{2}{3}(100 - s)^{3/2}\right]_0^{100}$$

$$= \frac{8}{100}\cdot\frac{2}{3}\cdot 1000 = 53\cdot3 \text{ m/s.}$$

EXERCISES 12 (b)

1. Find the mean value of $\sin^2\phi$ for values of ϕ between 0 and π.

2. The pressure p and volume v of a quantity of steam are related by the law $pv^{1\cdot2} = 500$. Find the mean pressure as the volume of steam increases from 3 to 8.

3. Find the mean height of the curve $y^2 = 2x$ between $x = 1$ and $x = 3$.

4. Find the mean value of $x\sin x$ as x varies from 0 to π.

5. The quantities v, s, t are connected by the relations

$$v = n\sqrt{(a^2 - s^2)}, \quad s = a\sin nt.$$

    Show that the mean value of v considered as a function of t between $t = 0$ and $t = \pi/2n$ is $(2an)/\pi$.   (O.C.)

6. A stone is projected vertically upwards with velocity u and, on its upward journey, its retardation is g (constant). Show that the mean values of its velocity on the upward journey are (i) with respect to time, $\frac{1}{2}u$, (ii) with respect to space, $\frac{2}{3}u$.

12.5. Centres of mass

Let m_1, m_2, m_3, ... denote the masses of a system of particles situated at points P_1, P_2, P_3, ... whose abscissae are x_1, x_2, x_3, ... The forces acting on the particles are proportional to m_1, m_2, m_3, ... and they act through the centre of the earth's gravitational field. If the particles are distributed over a region whose linear dimensions are small compared with the distance to the centre of the gravitational field, the directions of the forces can be considered as being parallel. The resultant of the force system is a single force proportional to $(m_1 + m_2 + m_3 + ...)$ acting at a point whose abscissa $\bar{x}$ is given by

$$(m_1 + m_2 + m_3 + ...)\bar{x} = m_1x_1 + m_2x_2 + m_3x_3 + ..., \quad (12.3)$$

and the point at which this resultant acts is known as the *centre of mass* or *centre of gravity*.

Writing $m_1 + m_2 + m_3 + \ldots = M$, the total mass of the system, and $m_1x_1 + m_2x_2 + m_3x_3 + \ldots = N$, sometimes called the x-*moment* or *first moment* with respect to x of the system, equation (12.3) can be written

$$M\bar{x} = N, \qquad (12.4)$$

where $$M = \Sigma m \quad \text{and} \quad N = \Sigma mx, \qquad (12.5)$$

the sign Σ denoting summation over all the masses of the system.

To obtain an extension from a system of particles to a continuous body, it is natural to replace the particles by elements of the body and to use limiting sums (i.e., integrals) in place of summations.

Consider first a rod of length l situated along the x-axis and with one end at the origin. Let the line-density of the rod at a point of abscissa x be ρ; this may be a function of x and, for simplicity, we shall here consider that ρ increases (or at least does not decrease) with x. If δM is the mass of the element of the rod for points whose abscissae lie between x and $x + \delta x$,

$$\rho\,\delta x \leqslant \delta M \leqslant (\rho + \delta\rho)\delta x.$$

Following the method of § 10.5, these inequalities lead to

$$\frac{dM}{dx} = \rho,$$

and the total mass M of the rod is therefore given by

$$M = \int_0^l \rho\,dx. \qquad (12.6)$$

The x-moment δN of this element lies between $x\rho\,\delta x$ and $(x + \delta x)(\rho + \delta\rho)\delta x$, for lower and upper bounds for x are x and $x + \delta x$, and for the mass the bounds are $\rho\,\delta x$ and $(\rho + \delta\rho)\delta x$. Hence

$$x\rho\,\delta x < \delta N < (x + \delta x)(\rho + \delta\rho)\delta x$$

leading to

$$\frac{dN}{dx} = x\rho,$$

and, for the total x-moment of the rod,

$$N = \int_0^l \rho x\,dx. \qquad (12.7)$$

The abscissa $\bar{x}$ of the centre of mass of the rod is given by (12.4), viz.

$$\bar{x} = N/M, \qquad (12.8)$$

but now the values $M = \int_0^l \rho\,dx$, $N = \int_0^l x\rho\,dx$ for the rod replace those given in (12.5) for the particle system.

For the case of a rod of *uniform* density, ρ is independent of x and can therefore be taken outside the signs of integration in (12.6), (12.7). Hence, in this case,

$$M = \rho \int_0^l dx = \rho l, \quad N = \rho \int_0^l x \, dx = \tfrac{1}{2}\rho l^2,$$

and (12.8) gives $\bar{x} = \tfrac{1}{2}l$, showing that the centre of mass of a uniform rod is at its mid-point.

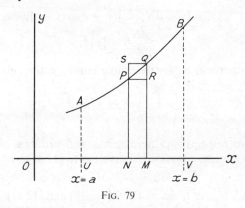

FIG. 79

Now consider a lamina $AUVB$ (Fig. 79) bounded by the curve $y = \phi(x)$, the x-axis and ordinates at $x = a$ and $x = b$. For simplicity we shall here take the surface-density ρ to be *uniform* but the analysis is easily modified when the surface-density varies from point to point of the lamina. The total mass M of the lamina is simply ρ times its area, so that

$$M = \rho \int_a^b y \, dx. \tag{12.9}$$

$PNMQ$ is a typical element bounded by ordinates PN, QM at abscissae x and $x + \delta x$. PR and QS are drawn parallel to the x-axis to intersect QM and PN respectively in R and S. The x-moment δN_x of the element lies between $x\rho y \, \delta x$ and $(x + \delta x)\rho(y + \delta y)\delta x$ for $\rho y \, \delta x$ and $\rho(y + \delta y)\delta x$ are respectively the masses of the rectangles $PNMR$ and $SNMQ$. Hence

$$x\rho y \, \delta x < \delta N_x < (x + \delta x)\rho(y + \delta y)\delta x,$$

leading in the usual way to

$$\frac{dN_x}{dx} = x\rho y.$$

The total x-moment N_x of the whole lamina is therefore given by

$$N_x = \int_a^b x\rho y \, dx = \rho \int_a^b xy \, dx, \tag{12.10}$$

since ρ is here assumed to be constant.

The mass of the rectangle $PNMR$ is $\rho y \, \delta x$ and its centre of mass is at a distance $\frac{1}{2}y$ from Ox. The corresponding quantities for the rectangle $SNMQ$ are $\rho(y + \delta y)\delta x$ and $\frac{1}{2}(y + \delta y)$. Hence the mass of the element $PNMQ$ lies between $\rho y \, \delta x$ and $\rho(y + \delta y)\delta x$ and the height of its centre of mass between $\frac{1}{2}y$ and $\frac{1}{2}(y + \delta y)$. If δN_y is the y-moment of the element

$$\tfrac{1}{2}y\rho y \, \delta x < \delta N_y < \tfrac{1}{2}(y + \delta y)\rho(y + \delta y)\delta x,$$

leading to
$$\frac{dN_y}{dx} = \tfrac{1}{2}\rho y^2.$$

The total y-moment N_y for the whole lamina is thus given by

$$N_y = \int_a^b \tfrac{1}{2}\rho y^2 \, dx = \tfrac{1}{2}\rho \int_a^b y^2 \, dx. \tag{12.11}$$

If $\bar{x}, \bar{y}$ are respectively the abscissa and ordinate of the centre of mass of the lamina, equations equivalent to (12.4) give

$$M\bar{x} = N_x, \quad M\bar{y} = N_y, \tag{12.12}$$

where M, N_x, N_y are given by (12.9), (12.10) and (12.11). Solving (12.12) for $\bar{x}, \bar{y}$ and inserting the expressions for M, N_x, N_y, we find that the coordinates of the centre of mass of the lamina $AUVB$ (Fig. 79) are given by

$$\bar{x} = \frac{\displaystyle\int_a^b xy \, dx}{\displaystyle\int_a^b y \, dx}, \quad \bar{y} = \frac{\frac{1}{2}\displaystyle\int_a^b y^2 \, dx}{\displaystyle\int_a^b y \, dx}. \tag{12.13}$$

Finally consider a solid body of uniform volume-density ρ. If the area of a cross-section by a plane perpendicular to the x-axis at distance x from the origin is $S(x)$ and if the solid is bounded by planes perpendicular to the x-axis at distances a and b from the origin, the volume of the solid is (see § 12.3) $\displaystyle\int_a^b S(x)dx$ and its total mass M is given by

$$M = \rho \int_a^b S(x)dx. \tag{12.14}$$

If δN is the x-moment of the element bounded by planes distant x and $x + \delta x$ from the origin,

$$x\rho S(x)\delta x < \delta N < (x + \delta x)S(x + \delta x)\delta x,$$

giving
$$\frac{dN}{dx} = x\rho S(x).$$

The total x-moment for the whole solid is therefore given by

$$N = \rho \int_a^b xS(x)dx. \tag{12.15}$$

The abscissa $\bar{x}$ of the centre of mass is then given by

$$M\bar{x} = N \tag{12.16}$$

where M and N are given by (12.14), (12.15).

In the case of the solid of revolution formed by the rotation about the x-axis of the area bounded by the curve $y = \phi(x)$, the x-axis and ordinates at $x = a$, $x = b$,

$$S(x) = \pi y^2.$$

The expressions for M and N become

$$M = \pi\rho \int_a^b y^2 \, dx, \quad N = \pi\rho \int_a^b xy^2 \, dx,$$

and (12.16) gives in this case,

$$\bar{x} = \frac{\displaystyle\int_a^b xy^2 \, dx}{\displaystyle\int_a^b y^2 \, dx}. \tag{12.17}$$

Example 8. *The density of a rod AB varies as the distance from the end A. Find the position of the centre of mass of the rod if its total length is 2 m.*

If we take the rod to lie along the x-axis with the end A at the origin, the density ρ at a point with abscissa x is kx where k is constant. The total mass M of the rod is given by

$$M = \int_0^2 \rho \, dx = k \int_0^2 x \, dx = k\left[\tfrac{1}{2}x^2\right]_0^2 = 2k.$$

The x-moment N is given by

$$N = \int_0^2 x\rho \, dx = k \int_0^2 x^2 \, dx = k\left[\frac{x^3}{3}\right]_0^2 = \frac{8k}{3}.$$

The abscissa of the centre of mass is therefore

$$\bar{x} = \frac{N}{M} = \frac{(8k/3)}{2k} = \frac{4}{3},$$

so that the centre of mass is 1·333 m from the end A.

Example 9. *Find the coordinates of the centre of mass of the lamina of uniform density bounded by the curve $y = (1 - x)^3$, the x-axis and ordinates for which $x = 0$ and $x = 1$.* (O.C.)

If ρ is the surface density, the total mass M of the lamina is given by

$$M = \rho \int_0^1 y \, dx = \rho \int_0^1 (1 - x)^3 \, dx$$

$$= \rho\left[-\frac{(1 - x)^4}{4}\right]_0^1 = \frac{\rho}{4}.$$

The x-moment N_x is

$$N_x = \rho \int_0^1 xy \, dx = \rho \int_0^1 x(1 - x)^3 \, dx$$

$$= \rho \int_0^1 (x - 3x^2 + 3x^3 - x^4) dx$$

$$= \rho \left[\frac{x^2}{2} - x^3 + \frac{3x^4}{4} - \frac{x^5}{5} \right]_0^1 = \frac{\rho}{20}.$$

For the y-moment, N_y,

$$N_y = \tfrac{1}{2}\rho \int_0^1 y^2 \, dx = \tfrac{1}{2}\rho \int_0^1 (1 - x)^6 \, dx$$

$$= \tfrac{1}{2}\rho \left[-\frac{(1 - x)^7}{7} \right]_0^1 = \frac{\rho}{14}.$$

The coordinates $\bar{x}$, $\bar{y}$ of the centre of mass are therefore given by

$$\bar{x} = N_x/M = \left(\frac{\rho}{20}\right) \Big/ \left(\frac{\rho}{4}\right) = \frac{1}{5}.$$

$$\bar{y} = N_y/M = \left(\frac{\rho}{14}\right) \Big/ \left(\frac{\rho}{4}\right) = \frac{2}{7}.$$

Example 10. *Find the position of the centre of gravity of a solid right circular cone of uniform density.*

As in Example 14 of Chapter 10, the cone can be regarded as the solid formed by the rotation of the line $y = x \tan \alpha$ about the x-axis for values of x from 0 to h, α being the semi-vertical angle and h the height of the cone.

The application of formula (12.17) gives, for the distance $\bar{x}$ of the centre of gravity from the vertex of the cone,

$$\bar{x} = \frac{\displaystyle\int_0^h x(x \tan \alpha)^2 \, dx}{\displaystyle\int_0^h (x \tan \alpha)^2 \, dx} = \frac{\displaystyle\int_0^h x^3 \, dx}{\displaystyle\int_0^h x^2 \, dx}$$

$$= \frac{\left[x^4/4 \right]_0^h}{\left[x^3/3 \right]_0^h} = \frac{3h}{4},$$

and, from symmetry, the centre of gravity lies on the axis of the cone.

EXERCISES 12 (c)

1. The density of a rod varies as the square of the distance from one end. If the length of the rod is l, find the distance of the centre of mass from this end.

2. From the point P where $x = 18$, $y = 12$ on the curve $y^2 = 8x$, PN is drawn perpendicular to the x-axis: find the distances from the axes of the centre of mass of a lamina of uniform density bounded by PN, the x-axis and the curve.

(O.C.)

3. Find the position of the centre of mass of a lamina of uniform density bounded by the curve $y^2 = x^3$, the x-axis and ordinates at $x = 1$ and $x = 4$.

4. Find the abscissa $\bar{x}$ of the centre of mass of a lamina of uniform density bounded by the curve $y = 1 + 10x - 2x^2$, the x-axis and ordinates for which $x = 1$ and $x = 5$.

   Verify that

$$\bar{x} = (y_1 + 12y_2 + 5y_3)/(y_1 + 4y_2 + y_3),$$

where y_1, y_2, y_3 are the ordinates of the curve at $x = 1, 3$ and 5.   (O.C.)

5. Find the volume cut-off from the solid obtained by rotating the curve $y^2 = 4ax$ about the x-axis by a plane at distance $5a$ from the origin.

   If the solid is of uniform density, find also the position of the centre of gravity of this part of the solid.   (O.C.)

6. Find the position of the centre of gravity of the solid within the surfaces formed by the revolution of the curve $y = x^4$ and the straight line $y = 6$ about the axis of y.   (O.C.)

12.6. Moments of inertia

Let $m_1, m_2, m_3, \ldots$ denote the masses of a system of particles situated at points $P_1, P_2, P_3, \ldots$ whose perpendicular distances from a given straight line are $r_1, r_2, r_3, \ldots$ Then the sum of the products of each mass and the *square* of its distance from the given line is called the *moment of inertia* of the system with respect to the given line. This moment, sometimes also called the *second moment*, is conveniently denoted by I, so that

$$I = m_1 r_1^2 + m_2 r_2^2 + m_3 r_3^2 + \ldots$$

If we imagine the total mass $M(= m_1 + m_2 + m_3 + \ldots)$ of the system to be concentrated at a point at distance k from the given line such that this single mass has the same moment of inertia about the given line as the particle system,

$$Mk^2 = I = m_1 r_1^2 + m_2 r_2^2 + m_3 r_3^2 + \ldots \qquad (12.18)$$

The distance k, calculated from this equation, is known as the *radius of gyration* about the given line.

As in the last section, an extension from a system of particles to a continuous body is obtained by replacing the particles by elements of the body and using integrals in place of summations. We give below a few examples of the application of the integral calculus to the calculation of moments of inertia but the scope of this book does not admit an exhaustive treatment.

Let a rod of length l be situated along the x-axis with one end at the origin and let its line-density at a point of abscissa x be ρ, here considered to be a non-decreasing function of x. The moment of inertia δI about the y-axis of the element of the rod between points with abscissae x and $x + \delta x$ satisfies the inequalities

$$x^2 \rho \delta x < \delta I < (x + \delta x)^2 (\rho + \delta \rho) \delta x,$$

so that

$$\frac{dI}{dx} = x^2 \rho.$$

The total moment of inertia I is therefore given by

$$I = \int_0^l \rho x^2 \, dx. \tag{12.19}$$

Since, from (12.6), the total mass M of the rod is given by $M = \int_0^l \rho \, dx$, its radius of gyration (k), from the equation $Mk^2 = I$, is given by

$$k^2 = \frac{\displaystyle\int_0^l \rho x^2 \, dx}{\displaystyle\int_0^l \rho \, dx}. \tag{12.20}$$

In the case of a uniform rod, ρ is constant and can be taken outside the signs of integration, so that

$$k^2 = \frac{\displaystyle\int_0^l x^2 \, dx}{\displaystyle\int_0^l dx} = \frac{\left[x^3/3\right]_0^l}{\left[x\right]_0^l} = \tfrac{1}{3}l^2. \tag{12.21}$$

Now consider the lamina $AUVB$ of Fig. 79. If the lamina is supposed to be of *uniform* density, the moment of inertia δI_y about the y-axis of the typical element $PNMQ$ satisfies

$$x^2 \rho y \, \delta x < \delta I_y < (x + \delta x)^2 \rho (y + \delta y) \delta x.$$

Hence

$$\frac{dI_y}{dx} = x^2 \rho y,$$

and the total moment of inertia I_y is given by

$$I_y = \rho \int_a^b x^2 y \, dx. \tag{12.22}$$

The mass of the rectangle $PNMR$ is $\rho y \, \delta x$ and its radius of gyration (from equation (12.21)) is $y/\sqrt{3}$. Its moment of inertia about the x-axis is $\rho y \, \delta x \times (y/\sqrt{3})^2$ or $\tfrac{1}{3}\rho y^3 \, \delta x$. Similarly the moment of inertia about Ox of the rectangle $SNMQ$ is $\tfrac{1}{3}\rho(y + \delta y)^3 \, \delta x$. Hence, if δI_x is the moment of inertia about the x-axis of the element $PNMQ$,

$$\tfrac{1}{3}\rho y^3 \, \delta x < \delta I_x < \tfrac{1}{3}\rho(y + \delta y)^3 \, \delta x,$$

leading to

$$\frac{dI_x}{dx} = \tfrac{1}{3}\rho y^3,$$

and
$$I_x = \tfrac{1}{3}\rho \int_a^b y^3 \, dx. \tag{12.23}$$

From (12.9), the mass M of the lamina is given by $M = \rho \int_a^b y \, dx$ and, if k_x, k_y denote the radii of gyration with respect to the x- and y-axes, (12.22) and (12.23) give

$$k_x{}^2 = \frac{\tfrac{1}{3}\displaystyle\int_a^b y^3 \, dx}{\displaystyle\int_a^b y \, dx}, \quad k_y{}^2 = \frac{\displaystyle\int_a^b x^2 y \, dx}{\displaystyle\int_a^b y \, dx}. \tag{12.24}$$

For a circular lamina of uniform surface-density ρ, the moment of inertia about an axis OP through its centre O and perpendicular to its plane can be found by considering the ring element bounded by circles

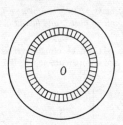

FIG. 80

of radii r and $r + \delta r$ shown shaded in Fig. 80. The mass of the ring is approximately $2\pi \rho r \, \delta r$ and all points within the ring lie at distances from the axis OP of between r and $r + \delta r$. Hence if δI is the moment of inertia of the elementary ring

$$r^2(2\pi \rho r \, \delta r) < \delta I < (r + \delta r)^2(2\pi \rho r \, \delta r).$$

Hence
$$\frac{dI}{dr} = 2\pi \rho r^3$$

and, if a is the radius of the lamina,

$$I = 2\pi \rho \int_0^a r^3 \, dr = \tfrac{1}{2}\pi \rho a^4.$$

Since the total mass M of the ring is $\pi \rho a^2$, this can be written $I = \tfrac{1}{2}Ma^2$.

This result can be used to find the moment of inertia (I) about its axis of symmetry of the solid formed by the revolution about the x-axis of the area bounded by the curve $y = \phi(x)$, the x-axis and ordinates at $x = a$, $x = b$. It is left as an exercise to the reader to show that, if ρ is the (uniform) volume-density of the solid

$$I = \tfrac{1}{2}\pi \rho \int_a^b y^4 \, dx. \tag{12.25}$$

Example 11. *The line-density of a rod varies as the distance from one end. Find its radius of gyration about an axis through this end perpendicular to the rod.*

If the rod lies along the x-axis with one end at the origin, the line-density $\rho = \lambda x$ where λ is a constant. If l is the length of the rod, the radius of gyration k is given by (12.20) as

$$k^2 = \frac{\int_0^l \rho x^2 \, dx}{\int_0^l \rho \, dx} = \frac{\int_0^l \lambda x^3 \, dx}{\int_0^l \lambda x \, dx}$$

$$= \frac{\left[x^4/4\right]_0^l}{\left[x^2/2\right]_0^l} = \frac{l^2}{2},$$

and $k = l/\sqrt{2}$.

Example 12. *A lamina of mass M is of uniform surface-density ρ and is in the form of a right-angled triangle of base b and height h. Show that the moment of inertia about the base is $\frac{1}{6}Mh^2$.*

The lamina (see Fig. 81) can be considered as the area bounded by the line $y = hx/b$, the x-axis and an ordinate at $x = b$. By (12.23), the moment of inertia

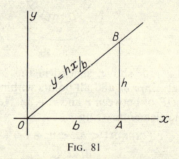

FIG. 81

about its base (the x-axis) is

$$\tfrac{1}{3}\rho \int_0^b y^3 \, dx = \tfrac{1}{3}\rho \int_0^b (hx/b)^3 \, dx$$

$$= \tfrac{1}{3}\rho \frac{h^3}{b^3}\left[\frac{x^4}{4}\right]_0^b = \tfrac{1}{12}\rho h^3 b.$$

But the area of the lamina is $\frac{1}{2}hb$, hence $M = \frac{1}{2}\rho hb$, and the moment of inertia can be written $\frac{1}{6}(\frac{1}{2}\rho hb).h^2$ or $\frac{1}{6}Mh^2$.

Example 13. *Show that the moment of inertia of a uniform solid right circular cone whose radius of base is r and mass M is $\frac{3}{10}Mr^2$.*

As in Example 14 of Chapter 10, the cone can be regarded as the solid formed by the rotation of the line $y = x \tan \alpha$ about the x-axis for values of x from 0 to h, α being the semi-vertical angle and h the height of the cone.

From (12.25) the moment of inertia I about the axis of symmetry is given by

$$I = \tfrac{1}{2}\pi\rho \int_0^h y^4 \, dx = \tfrac{1}{2}\pi\rho \int_0^h (x\tan\alpha)^4 \, dx$$

$$= \tfrac{1}{2}\pi\rho \tan^4\alpha \left[\frac{x^5}{5}\right]_0^h = \tfrac{1}{10}\pi\rho h^5 \tan^4\alpha,$$

where ρ is the volume-density of the cone.

Since the volume of the cone is (Example 14, Chapter 10) $\tfrac{1}{3}\pi h^3 \tan^2\alpha$,

$$M = \tfrac{1}{3}\pi\rho h^3 \tan^2\alpha.$$

By division, $I/M = \tfrac{3}{10}h^2 \tan^2\alpha,$

giving $I = \tfrac{3}{10}Mr^2$, for $r = h\tan\alpha$.

EXERCISES 12 (d)

1. Find the moment of inertia of a rod in which the line-density varies as the square of the distance from one end about an axis through that end perpendicular to the rod.

2. Find the moment of inertia about the y-axis of the lamina (surface-density ρ) bounded by the curve $ay = x(a - x)$ and the x-axis.

3. Find the moment of inertia of a uniform rectangular lamina of sides a and b about an axis parallel to the sides a and distant $\tfrac{1}{3}b, \tfrac{2}{3}b$ from them respectively.

4. Find the moment of inertia of a uniform lamina in the form of an isosceles triangle of height h about a line through its vertex parallel to its base.

5. Find the square of the radius of gyration about its axis of symmetry of the solid (of uniform density) obtained by rotating about the x-axis the area between the curve $y^2 = ax$ and an ordinate at $x = a$.

6. A uniform solid is formed by the rotation of a rectangle of sides a and b about a line in its plane distant $d\,(>\tfrac{1}{2}b)$ from its centre and parallel to the sides a. Find the square of the radius of gyration about the axis of symmetry of the solid so formed.

12.7. Length of arc

Fig. 82 shows the arc of a curve $y = \phi(x)$ whose end points A and B have abscissae $x = a$ and $x = b$. The tangent to the curve at a point P makes an angle ψ with the x-axis and, for simplicity, we shall only consider here arcs for which ψ is acute and increasing with x. Suppose $P_1, P_2, \ldots, P_{n-1}$ are points on the curve with abscissae $x_1, x_2, \ldots, x_{n-1}$ and let the tangents at these points make angles $\psi_1, \psi_2, \ldots, \psi_{n-1}$ with the x-axis. If we consider the typical element of arc between P_r and P_{r+1}, and if ψ' represents the angle between the line joining P_r, P_{r+1} and the x-axis, the right-angled triangle P_rMP_{r+1} of the subsidiary diagram shows that

$$P_rP_{r+1} = (x_{r+1} - x_r)\sec\psi'.$$

Now ψ' is greater than the angle ψ_r at P_r and less than the angle ψ_{r+1} at P_{r+1} and it follows that $AP_1 + P_1P_2 + \ldots + P_{n-1}B$ lies between $\Sigma(x_{r+1} - x_r)\sec\psi_r$ and $\Sigma(x_{r+1} - x_r)\sec\psi_{r+1}$. As the number of

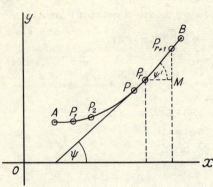

FIG. 82

points P_1, P_2, ..., P_{n-1} between A and B increases, each of these quantities approaches the same limit $\int_a^b \sec \psi \, dx$. Since

$$\sec^2 \psi = 1 + \tan^2 \psi \quad \text{and} \quad \tan \psi = dy/dx,$$

the length of arc, which is defined to be this limit, is given by

$$\text{arc } AB = \int_a^b \sqrt{\left\{1 + \left(\frac{dy}{dx}\right)^2\right\}} dx. \tag{12.26}$$

Example 14. *Find the length of the arc of the curve $6xy = 3 + x^4$ between the points whose abscissae are 1 and 4.*

Here $y = (3 + x^4)/6x = 1/(2x) + x^3/6$ and hence

$$\frac{dy}{dx} = -\frac{1}{2x^2} + \frac{x^2}{2}.$$

$$1 + \left(\frac{dy}{dx}\right)^2 = 1 + \left(\frac{x^2}{2} - \frac{1}{2x^2}\right)^2$$

$$= 1 + \frac{x^4}{4} - \frac{1}{2} + \frac{1}{4x^4}$$

$$= \frac{1}{4}\left(x^4 + 2 + \frac{1}{x^4}\right)$$

$$= \left[\frac{1}{2}\left(x^2 + \frac{1}{x^2}\right)\right]^2.$$

Therefore $\quad\quad \sqrt{\left\{1 + \left(\frac{dy}{dx}\right)^2\right\}} = \frac{1}{2}\left(x^2 + \frac{1}{x^2}\right),$

and $\quad\quad \text{arc} = \int_1^4 \frac{1}{2}\left(x^2 + \frac{1}{x^2}\right) dx$

$$= \frac{1}{2}\left[\frac{x^3}{3} - \frac{1}{x}\right]_1^4 = \frac{1}{2}\left\{\frac{64}{3} - \frac{1}{4} - \frac{1}{3} + 1\right\}$$

$$= 10\frac{7}{8}.$$

12.8. Areas of surfaces of revolution

Consider first the frustum of a right circular cone.* If similar and similarly situated polygons of n sides are inscribed within the circular ends, by joining corresponding vertices PQ, $P'Q'$ we obtain a series of n trapezia of which $PP'Q'Q$ is typical (Fig. 83). If M and N are the mid-points of the sides PP', QQ'

$$\text{area } PP'Q'Q = \tfrac{1}{2}(PP' + QQ')MN.$$

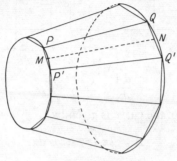

FIG. 83

As the number n of sides of the polygons increases, the surface area of the solid formed by the trapezia tends to a limit which is defined as the surface area of the conical frustum. The sum of the areas of the trapezia is

$$\tfrac{1}{2}(\text{sum of perimeters of the polygons})MN,$$

and since the perimeters of the polygons tend to the circumferences of the end circles of the frustum and MN tends to the limit PQ, the surface area of the frustum

$= \tfrac{1}{2}(\text{sum of circumferences of circular ends}) \times PQ$
$= \text{mean circumference of the frustum} \times \text{its slant height}.$

Now consider the solid formed by the rotation of the area bounded by the curve $y = \phi(x)$, the x-axis and ordinates at $x = a$, $x = b$ about the x-axis. Let P and Q be of abscissae x and $x + \delta x$. If PQ be joined by a straight line, the rotation of this line will form a frustum of a cone of mean circumference $2\pi(y + \tfrac{1}{2}\delta y)$ and slant height PQ. The slant height PQ (as in § 12.7) will lie between $\delta x \sec \psi$ and $\delta x \sec (\psi + \delta \psi)$ where ψ is the angle between the tangent to the curve $y = \phi(x)$ at P and the x-axis. Thus if δA is the area of the conical frustum generated by the rotation of the line PQ

$$2\pi(y + \tfrac{1}{2}\,\delta y)\,\delta x \sec \psi < \delta A < 2\pi(y + \tfrac{1}{2}\,\delta y)\delta x \sec (\psi + \delta \psi).$$

The surface area A of the solid of revolution is defined as the limit to which the sum of such areas tends as δx tends to zero.

---

* See Chapters 19 and 20 for the necessary definitions.

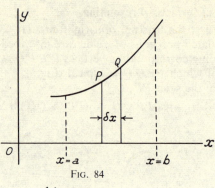

FIG. 84

Hence
$$\frac{dA}{dx} = 2\pi y \sec \psi,$$

and the total area of the curve is given by

$$A = 2\pi \int_a^b y \sec \psi \, dx.$$

Since $\sec^2 \psi = 1 + \tan^2 \psi = 1 + (dy/dx)^2$, this can be recast into the form

$$A = 2\pi \int_a^b y\sqrt{\{1 + (dy/dx)^2\}}dx. \tag{12.27}$$

Example 15. *Find the area of the surface formed by the rotation of the curve* $y^2 = 4x$ *about the x-axis, from the origin to* $x = 3$.

Here $y = 2\sqrt{x}$, $dy/dx = 1/\sqrt{x}$, so that

$$\text{area required} = 2\pi \int_0^3 2\sqrt{x}.\sqrt{\{1 + (1/\sqrt{x})^2\}}dx$$

$$= 4\pi \int_0^3 \sqrt{(x + 1)}dx$$

$$= 4\pi \left[\tfrac{2}{3}(x + 1)^{3/2}\right]_0^3$$

$$= \frac{8\pi}{3}\left[4^{3/2} - 1^{3/2}\right]$$

$$= \frac{56\pi}{3}.$$

EXERCISES 12 (e)

1. Find the length of the arc of the curve $y^2 = 8x^3$ between $x = 1$ and $x = 3$.

2. Find the length of the arc of the curve $x^2 + y^2 = r^2$ between $x = 0$ and $x = r$ (this curve is a quadrant of a circle of radius r, see § 16.1).

3. Find the length of the arc of the curve $x^{2/3} + y^{2/3} = 4$ between $x = 0$ and $x = 8$.

4. Find the surface area of the solid generated by the rotation about the x-axis of the area bounded by the curve $y = \sqrt{(r^2 - x^2)}$, the x-axis and ordinates at $x = 0$, $x = r$.

5. Find the area of the surface formed by the revolution about the x-axis of that part of the curve $x^{2/3} + y^{2/3} = 4$ which lies between $x = 0$ and $x = 8$.

   (Hint, the integral expressing the area can be evaluated by writing $x^{2/3} = t$.)

6. The curve $3y = (3 - x)\sqrt{x}$ between $x = 0$ and $x = 3$ rotates about the x-axis. Find the surface area of the solid so formed.

EXERCISES 12 (f)

1. Sketch the curve $y = x(3 - x)$ for values of x between -2 and $+5$. Find the area contained between the curve, the x-axis and ordinates at $x = 0$ and $x = 5$. Explain why this area is not $\int_0^5 x(3 - x)dx$.

2. Find the area enclosed between the two curves $y^2 = x$ and $x^2 = 8y$.
   (L.U.)

3. Find the area enclosed between the two curves $y^2 = 8(6 - x)$ and $y^2 = 12x$.

4. Find the area of the segment cut off from the curve $y^2 = 4ax$ by the line $y = x$. Find also the volume of the solid obtained by rotating this segment about the x-axis.
   (L.U.)

5. The curve $y = 1 + \cos x$ is rotated about the axis of x. Find the volume contained by the surface of revolution and the planes $x = +\pi/2$ and $x = -\pi/2$.
   (Q.E.)

6. The capping of a stone pillar is a solid with every horizontal cross-section a square. The centres of these squares lie on a vertical axis and their corners lie on the surface of a sphere of radius 0·1 m whose centre is on the axis, 0·05 m above the plane base of the solid. Calculate the volume of the capping.
   (L.U.)

7. Find the mean value of $\sin (n\phi + \alpha)$ over the ranges:—
   $$\text{(i)} \quad \phi = -\alpha/n \text{ to } \phi = 2m\pi - (\alpha/n),$$
   $$\text{(ii)} \quad \phi = -\alpha/n \text{ to } \phi = (\pi/2n) - (\alpha/n),$$
   where m and n are positive integers.

8. The horizontal range R of a stone projected with velocity V at angle α to the horizontal is given by
   $$R = (V^2/g) \sin 2\alpha,$$
   where g is a constant. Show that the mean range for all angles of projection from $0°$ to $90°$ is $2V^2/\pi g$.

9. Show that, if a is a positive constant, the mean ordinate of that part of the curve $y = ax - x^2$ which lies in the first quadrant is two-thirds of the maximum ordinate.

10. A number n is divided at random into two parts. Show that the mean value of the products of these parts is $n^2/6$.

11. Find the position of the centre of mass of a lamina of uniform density bounded by the coordinate axes and that part of the curve $(x^2/a^2) + (y^2/b^2) = 1$ which lies in the first quadrant.

12. Find the position of the centre of mass of a lamina of uniform density bounded by the curve $y = a \sin(x/a)$ and the x-axis between $x = 0$ and $x = a\pi$.

13. Find the area of the loop of the curve $y^2 = 4x^2(1 - x)$. Find also the position of the centre of mass of a uniform lamina bounded by this loop. (Hint, put $1 - x = t^2$ to evaluate the integrals.)

14. Find the position of the centre of gravity of the solid (of uniform density) formed by the rotation of the curve $x^2 + y^2 = a^2$ from $x = 0$ to $x = a$ about the x-axis.

15. A solid of uniform density is formed by rotating the portion of the curve $y = x^2 - 3x$ which is cut off by the x-axis about that axis. Find the position of its centre of gravity.

16. Find the moment of inertia about one of its sides of a square lamina of uniform density ρ and side a.

17. Find the square of the radius of gyration about the x-axis of a lamina of uniform density bounded by the curve $y = \sin x$ and the x-axis from $x = 0$ to $x = \pi$.

18. Show that the moment of inertia about the x-axis of a uniform lamina of mass M bounded by the curve $y^2 = 4ax$, the x-axis and an ordinate at $x = a$ is $\frac{4}{3}Ma^2$.

19. A uniform solid is formed by rotating the curve $x^2 + y^2 = a^2$ from $x = -a$ to $x = a$ about the x-axis. Show that the square of the radius of gyration about the axis of symmetry of the solid is $2a^2/5$.

20. The surface-density of a circular lamina of radius a varies as the distance from the centre. If the total mass of the lamina is M, find its moment of inertia about an axis through its centre and perpendicular to its plane.

21. Find the length of the arc of the curve $x^2 + y^2 = a^2$ between points where $x = a \cos \alpha$ and $x = a \cos \beta$.

22. The slope of a curve at a point whose abscissa is x is $2\sqrt{(x + x^2)}$. Find the length of the arc of the curve from $x = 1$ to $x = 10$.

23. Find the length of the arc of the curve $x^{2/3} - y^{2/3} = a^{2/3}$ for points between $x = a$ and $x = b$.
    (Hint, put $x^{2/3} = t$ to evaluate the integral.)

24. Find the area of the surface formed by the revolution of the curve $9y^2 = x(3 - x)^2$ from $x = 0$ to $x = 3$ about the x-axis.

25. Find the area of the surface generated by the revolution about the x-axis of that part of the curve $y = x^3$ which lies between $x = 0$ and $x = 1$.

CHAPTER 13

THE LOGARITHMIC AND EXPONENTIAL FUNCTIONS

13.1. Introduction

In reading Chapters 10 and 11 the student will probably wonder why the result

$$\int x^n \, dx = \frac{x^{n+1}}{n+1} + C$$

is invalid when $n = -1$ and will have noticed that a discussion of $\int x^{-1} \, dx$ was postponed. We commence the present chapter by considering this integral and then show how it leads to two functions of great importance in mathematics (and in some of its applications to physics and mechanics). Once this integral has been established, many more functions can be integrated and the gaps in § 11.2 can be filled in.

13.2. The area below the curve $y = 1/x$

Fig. 85 shows the graph of $y = 1/x$ for positive values of x. Suppose we wish to find the area enclosed by the curve, the x-axis and ordinates

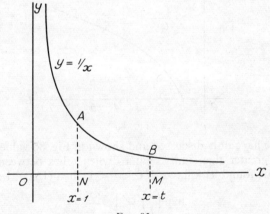

FIG. 85

at $x = 1$ and $x = t$. It can be inferred from the diagram that the area, or the definite integral $\int_1^t x^{-1} \, dx$, exists. The area or integral could in fact be found approximately by counting squares or by using one of the approximate methods of integration given in § 11.7, and the

235

student who has worked Exercise 11 (e), 2 will have carried out this procedure for the case of $t = 2$.

We can write

$$\int_1^t \frac{dx}{x} = \text{area } ANMB, \tag{13.1}$$

and it should be clear that as t increases the ordinate BM moves to the right and the area and integral are increasing functions of t. The curve $y = 1/x$ is (see § 17.17) a hyperbola and a convenient notation for the area bounded by the curve, the x-axis and ordinates at $x = 1$ and $x = t$ is hyp (t). With this notation, equation (13.1) can be rewritten

$$\int_1^t \frac{dx}{x} = \text{area } ANMB = \text{hyp}(t). \tag{13.2}$$

It should be noticed at this stage that when $t = 1$, the ordinates AN, BM coincide and the area $ANMB$ vanishes so that

$$\text{hyp}(1) = 0. \tag{13.3}$$

A graph of hyp (t) calculated by giving t various values and performing the integration by one of the approximate methods would appear, for values of t greater than unity, as shown in Fig. 86.

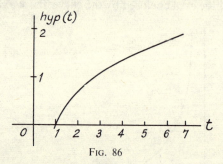

FIG. 86

So far we have only discussed and shown in Fig. 86 values of hyp (t) when t is greater than unity. Values when t lies between zero and unity can be found as follows. If $0 < t < 1$, then $(1/t) > 1$ and

$$\text{hyp}\left(\frac{1}{t}\right) = \int_1^{1/t} \frac{dx}{x}.$$

Now change the variable in this integral from x to z where $x = 1/z$ by the method of § 11.5. The limits of integration correspond to values of z of 1 and t, and, since $(dx/dz) = -(1/z^2)$, the integral can be written

$$\int_1^t z\left(-\frac{1}{z^2}\right)dz, \quad \text{or} \quad -\int_1^t \frac{dz}{z}.$$

This is $-\text{hyp}(t)$ and we have the important result that

$$\text{hyp}\left(\frac{1}{t}\right) = -\text{hyp}(t). \tag{13.4}$$

This formula enables the graph of $\text{hyp}(t)$ to be completed for values of t between zero and unity. For example, if $t = 2$, formula (13.4) gives $\text{hyp}(1/2) = -\text{hyp}(2) = -0.69$ approximately, if we read off $\text{hyp}(2)$ from Fig. 86. Other values of $\text{hyp}(t)$ for the range $0 < t < 1$ can be found in the same way and a completed graph of the function is shown in Fig. 87.

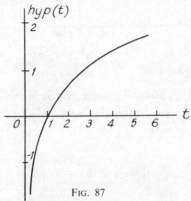

FIG. 87

13.3. Some important properties of the function hyp (t)

From the definition $\text{hyp}(t) = \displaystyle\int_1^t x^{-1}\,dx$ it should be clear that

$$\frac{d}{dt}\{\text{hyp}(t)\} = \frac{1}{t},$$

and this result can also be inferred from the diagram shown in Fig. 88. This shows the curve $y = 1/x$ and AN, BM, CP are ordinates at $x = 1$, $x = t$ and $x = t + \delta t$. BD, CE are drawn parallel to the x-axis to meet CP and BM at D and E respectively. The figure shows that the area $BMPC$ lies between the areas of the rectangles $EMPC$ and $BMPD$. Since $BM = 1/t$, $CP = 1/(t + \delta t)$ and $MP = \delta t$, this can be expressed as

$$\frac{\delta t}{t + \delta t} < \text{area } BMPC < \frac{\delta t}{t}. \tag{13.5}$$

Since $\text{hyp}(t) = \text{area } ANMB$ and $\text{hyp}(t + \delta t) = \text{area } ANPC$, we have, by subtraction,

$$\text{hyp}(t + \delta t) - \text{hyp}(t) = \text{area } ANPC - \text{area } ANMB$$
$$= \text{area } BMPC. \tag{13.6}$$

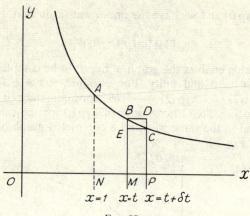

$$x=1 \quad x=t \quad x=t+\delta t$$

FIG. 88

Combining (13.5), (13.6) and dividing by δt, we have

$$\frac{1}{t + \delta t} < \frac{\text{hyp}\,(t + \delta t) - \text{hyp}\,(t)}{\delta t} < \frac{1}{t},$$

showing that

$$\lim_{\delta t \to 0} \left\{ \frac{\text{hyp}\,(t + \delta t) - \text{hyp}\,(t)}{\delta t} \right\} = \frac{1}{t},$$

or,

$$\frac{d}{dt}\{\text{hyp}\,(t)\} = \frac{1}{t}. \tag{13.7}$$

Using (13.7) and the formula for the differentiation of a function of a function, if c is a constant,

$$\frac{d}{dt}\{\text{hyp}\,(ct)\} = \frac{1}{ct} \times c = \frac{1}{t}.$$

Hence

$$\frac{d}{dt}\{\text{hyp}\,(ct) - \text{hyp}\,(t)\} = \frac{1}{t} - \frac{1}{t} = 0,$$

showing that

$$\text{hyp}\,(ct) - \text{hyp}\,(t) = \text{constant}.$$

The value of the constant can be found by putting $t = 1$, and since from (13.3) hyp (1) = 0, the constant is hyp (c). Hence

$$\text{hyp}\,(ct) - \text{hyp}\,(t) = \text{hyp}\,(c). \tag{13.8}$$

Writing $c = a$, $t = b$ in (13.8) we have

$$\text{hyp}\,(ab) = \text{hyp}\,(a) + \text{hyp}\,(b). \tag{13.9}$$

Similarly putting $c = a/b$, $t = b$, we can show that

$$\text{hyp}\left(\frac{a}{b}\right) = \text{hyp}\,(a) - \text{hyp}\,(b). \tag{13.10}$$

One other property of the function is required. Writing $t = a^n$ in (13.2),

$$\text{hyp}\,(a^n) = \int_1^{a^n} \frac{dx}{x}.$$

Changing the variable in the integral by the substitution $x = u^n$, the limits of integration become 1 and a, and since $(dx/du) = nu^{n-1}$,

$$\text{hyp}\,(a^n) = \int_1^a \frac{1}{u^n} . nu^{n-1}\,du$$

$$= n \int_1^a \frac{du}{u}$$

$$= n\,\text{hyp}\,(a). \qquad (13.11)$$

13.4. The logarithmic function

The properties

$$\text{hyp}\,(ab) = \text{hyp}\,(a) + \text{hyp}\,(b),$$

$$\text{hyp}\,\left(\frac{a}{b}\right) = \text{hyp}\,(a) - \text{hyp}\,(b),$$

$$\text{hyp}\,(a^n) = n\,\text{hyp}\,(a),$$

and $\qquad\qquad \text{hyp}\,(1)\ = 0,$

of the function $\text{hyp}\,(t)$ suggest that there is a connection with the logarithmic function. That $\text{hyp}\,(t)$ is not exactly the same as $\log_{10} t$ can be seen from the graph of Fig. 87 and a table of common logarithms. Thus the graph gives $\text{hyp}\,(2)$ and $\text{hyp}\,(3)$ as approximately 0·69 and 1·1 while a table of logarithms shows that $\log_{10} 2 = 0·301$ and $\log_{10} 3 = 0·477$.

The details of the connection between the "hyp" and logarithmic functions can be established as follows. From the graph of Fig. 87 it may be inferred that there is a value of t (which we shall denote by e) which makes the value of $\text{hyp}\,(t)$ unity: the graph shows that the value of e is about 2·7. Hence

$$\text{hyp}\,(e) = 1,$$

and writing $a = e, n = x$ in (13.11),

$$\text{hyp}\,(e^x) = x\,\text{hyp}\,(e) = x.$$

Putting $e^x = y$, this gives $\text{hyp}\,(y) = x$ and a combination of these two equations shows that

$$y = e^{\,\text{hyp}\,(y)}. \qquad (13.12)$$

Thus $\text{hyp}\,(y)$ is the power to which the number e must be raised to make it equal to y, and hence, by the definition of a logarithm, the "hyp" function is the logarithm to base e, or

$$\text{hyp}\,(y) \equiv \log_e y.$$

The logarithm to base e is called a *natural* or Napierian logarithm and an alternative notation for $\log_e y$ is $\ln y$, the "n" signifying the word "natural".

A summary of the contents of the first four sections of this chapter is contained in the two important results

$$\frac{d}{dx}\{\log_e x\} = \frac{1}{x}, \tag{13.13}$$

and its inverse,

$$\int \frac{dx}{x} = \log_e x + C. \tag{13.14}$$

The variable in these results has for convenience been here written as x and it should be noted that the function $\log_e x$ has been defined for *positive* values of x only. The identity between the "hyp" and logarithmic functions means that the graph of Fig. 87 gives also a graph of the logarithmic function. The relation (13.13) shows that the slope of the curve is very gentle when the independent variable is very large and that it is very steep when the variable is very small.

Example 1. *Show that common and natural logarithms are connected by the relation* $\log_{10} x = \log_e x \times \log_{10} e$.

This follows immediately from Example 3 of Chapter 2 (page 27 by writing $N = x, a = 10, b = e$.

Example 2. *Differentiate $\log_e \sec x$ with respect to x and hence find* $\int \tan x \, dx$.

If $y = \log_e \sec x$, we can write

$$y = \log_e \left(\frac{1}{\cos x}\right) = \log_e 1 - \log_e \cos x = -\log_e \cos x,$$

since $\log_e 1 = 0$. Writing $u = \cos x$, the result (13.13) and the formula for differentiating a function of a function gives

$$\frac{dy}{dx} = \frac{dy}{du} \times \frac{du}{dx}$$

$$= \frac{d}{du}\{-\log_e u\} \times \frac{d}{dx}(\cos x)$$

$$= -\frac{1}{u} \times -\sin x = \frac{\sin x}{\cos x} = \tan x.$$

Since $\dfrac{d}{dx}(\log_e \sec x) = \tan x$, the inverse relation is

$$\int \tan x \, dx = \log_e \sec x + C.$$

EXERCISES 13 (a)

1. Differentiate with respect to x:—

(i) $x^2 \log_e x$,   (ii) $\log_e (1/x)$.

2. Differentiate $x(\log_e x - 1)$ with respect to x and hence find $\int \log_e x \, dx$.

3. Taking $e = 2.718$, use the result of Example 1 of this Chapter, viz.

$$\log_{10} x = \log_e x \times \log_{10} e,$$

to draw the graph of $y = \log_e (x + 3)$ between $x = -2.5$ and $x = 3$. With the same axes and scales draw the graph of $5y = x^3$ and use these graphs to solve approximately the equation

$$(x + 3)^5 = e^{x^3}. \tag{L.U.}$$

4. Differentiate with respect to x:—

(i) $\log_e \sqrt{\left(\dfrac{x^2 - 1}{x^2 + 1} \right)}$,   (Q.E.),   (ii) $\log_e (\operatorname{cosec} x + \cot x)$.   (L.U.)

5. Find dy/dx when $y = \log_e \{x + \sqrt{(x^2 + 1)}\}$, and hence evaluate $\displaystyle\int \frac{dx}{\sqrt{(1 + x^2)}}$.
(L.U.)

6. A submarine telegraph cable consists of a copper core with a concentric sheath of nonconducting material. The ratio of the radius of the core to the thickness of the sheath is x and it is known that the speed of signalling is equal to $kx^2 \log_e (1/x)$, where k is a constant.

Show that the greatest speed of signalling is reached in a cable for which $x = 1/\sqrt{e}$.

13.5. The exponential function

If
$$x = \log_e t, \tag{13.15}$$

x is given uniquely for positive values of t. It may be inferred from the graph of Fig. 87 that for any assigned value of x, there is one and only one value of t and that value is positive. We may regard t as a function of x and this function is single-valued and everywhere positive. We could write t as antilog$_e (x)$ but it is more usual to write the inverse relation to (13.15) in the form

$$t = e^x, \tag{13.16}$$

or, in a notation which is particularly useful when x is replaced by a complicated expression,

$$t = \exp (x). \tag{13.17}$$

t is called the *exponential* function of x and this function is of great importance in mathematics and its application to physical science.

The graph of the function $t = e^x$ can be obtained from that of $x = \log_e t$ by interchanging the axes of Fig. 87. This is shown in Fig. 89. The function $t = e^{-x}$ has many physical applications and its graph is easily obtained from that of e^x for $e^{-x} = 1/e^x$. A sketch of the graph of e^{-x} for positive values of x is shown in Fig. 90.

The exponential function can be expressed in the form of a limit as follows. Since

$$\frac{d}{dt}\{\log_e t\} = \frac{1}{t},$$

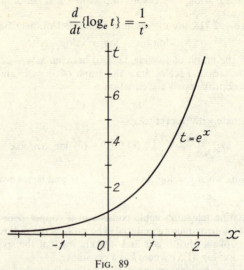

FIG. 89

it follows that as h tends to zero

$$\frac{1}{h}\{\log_e (t + h) - \log_e t\}$$

tends to $1/t$. Putting $t = 1$, since $\log_e 1 = 0$, we have that

$$\frac{1}{h}\log_e (1 + h)$$

tends to unity as h tends to zero. Writing $h = x/n$, this can be expressed by

$$\frac{n}{x}\log_e \left(1 + \frac{x}{n}\right) \to 1 \text{ as } n \to \infty,$$

showing that

$$\log_e \left\{\left(1 + \frac{x}{n}\right)^n\right\} \to x \text{ as } n \to \infty.$$

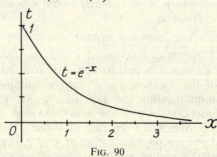

FIG. 90

This can be written in the alternative form

$$\left(1 + \frac{x}{n}\right)^n \to e^x \text{ as } n \to \infty$$

or,

$$e^x = \lim_{n \to \infty} \left(1 + \frac{x}{n}\right)^n. \tag{13.18}$$

13.6. Differentiation and integration of e^x

If $y = e^x$, then $x = \log_e y$ and (13.13) gives

$$\frac{dx}{dy} = \frac{1}{y} = \frac{1}{e^x}.$$

Since from (8.11),

$$\frac{dy}{dx} = 1 \Big/ \left(\frac{dx}{dy}\right)$$

we have

$$\frac{dy}{dx} = e^x,$$

giving the important result that

$$\frac{d}{dx}(e^x) = e^x. \tag{13.19}$$

This shows that the slope of the curve $y = e^x$ at a point whose abscissa is x is equal to the ordinate at this point.

The formula for the differential coefficient of a function of a function then gives, if a is a constant

$$\frac{d}{dx}(e^{ax}) = ae^{ax}, \tag{13.20}$$

and the inverse relation is

$$\int e^{ax}\, dx = \frac{1}{a}e^{ax} + C. \tag{13.21}$$

Suppose now that a is a positive constant and that $y = a^x$. Then $\log_e y = \log_e (a^x) = x \log_e a$, so that

$$y = e^{x \log_e a}.$$

It follows from (13.20) that

$$\frac{d}{dx}(a^x) = \frac{d}{dx}(e^{x \log_e a})$$

$$= (\log_e a)(e^{x \log_e a})$$

$$= y \log_e a$$

$$= a^x \log_e a. \tag{13.22}$$

Example 3. *If $y = xe^{-x}$ show that $\dfrac{d^2y}{dx^2} + 2\dfrac{dy}{dx} + y = 0$.* (L.U.)

By the formula for the differential coefficient of a product

$$\frac{dy}{dx} = \frac{d}{dx}(x).e^{-x} + x\frac{d}{dx}(e^{-x}) = e^{-x} - xe^{-x}$$

$$= (1 - x)e^{-x} = e^{-x} - y. \tag{13.23}$$

Hence
$$\frac{d^2y}{dx^2} = -e^{-x} - \frac{dy}{dx},$$

giving
$$\frac{d^2y}{dx^2} + \frac{dy}{dx} = -e^{-x}. \tag{13.24}$$

The required result follows immediately from the addition of (13.23), (13.24).

Example 4. *Show that* $e^{\log_e x} = x$ *and that* $e^{-2\log_e x} = 1/x^2$.

If $y = e^{\log_e x}$, then $\log_e y = \log_e x . \log_e e = \log_e x$, since $\log_e e = 1$.
Hence $y = x$.

If $z = e^{-2\log_e x}$, then

$$\log_e z = -2 \log_e x . \log_e e = -2 \log_e x = \log_e (1/x^2)$$

so that $z = 1/x^2$.

EXERCISES 13 (b)

1. Differentiate with respect to x:—
$$\text{(i) } e^{2x} \sin 3x, \quad \text{(ii) } e^{x^2}.$$

2. If $y = e^{-2x} \cos 4x$, prove that $\dfrac{d^2y}{dx^2} + 4\dfrac{dy}{dx} + 20y = 0$.

3. If $y = x^n e^{ax}$, show that $\dfrac{dy}{dx} - ay = \dfrac{ny}{x}$. (L.U.)

4. If $y = e^x \sin x$, show that
$$\frac{dy}{dx} = \sqrt{2}.e^x \sin (x + \pi/4), \quad \frac{d^2y}{dx^2} = 2e^x \sin (x + \pi/2). \quad \text{(O.C.)}$$

5. Evaluate the following definite integrals:—
$$\text{(i) } \int_0^1 e^{-3x} dx, \quad \text{(ii) } \int_0^1 (e^x - e^{-x})^2 dx, \quad \text{(iii) } \int_0^2 (x - 1)e^x dx. \quad \text{(L.U.)}$$

6. Show that the length of arc of the curve $2y = e^x + e^{-x}$ from $x = 0$ to $x = a$ is $\frac{1}{2}(e^a - e^{-a})$.

7. Find the following by the method of integration by parts:—
$$\text{(i) } \int x^2 e^x dx, \quad \text{(ii) } \int e^x \cos x dx.$$

8. Find the maximum ordinate of the curve $y = xe^{-x}$, and draw a rough sketch of the curve.

13.7. Some integrals depending on $\int x^{-1} dx$

Now that the integral

$$\int x^{-1} dx = \log_e x + C \tag{13.25}$$

is available, some methods of integration which were omitted from the discussion of Chapter 11 can be given. We first note that an important extension of the integral occurs when x is replaced by the linear expression $(ax + b)$ where a and b are constants. By (11.2), this is

$$\int \frac{dx}{ax + b} = \frac{1}{a} \log_e (ax + b) + C, \tag{13.26}$$

and for the particular case of $a = 1$, this reduces to

$$\int \frac{dx}{x + b} = \log_e (x + b) + C. \tag{13.27}$$

Now consider the integral

$$\int \frac{f'(x)}{f(x)} . dx,$$

in which the numerator of the integrand is the derivative with respect to x of the denominator. To evaluate this integral write $f(x) = t$, so that $f'(x) . (dx/dt) = 1$ and $(dx/dt) = 1/f'(x)$. The rule (11.7) for integration by change of variable then gives

$$\int \frac{f'(x)}{f(x)} . dx = \int \frac{f'(x)}{t} . \frac{1}{f'(x)} . dt$$

$$= \int \frac{dt}{t}$$

$$= \log_e t + C$$

$$= \log_e f(x) + C. \tag{13.28}$$

Hence the *integral of a fraction in which the numerator is the differential coefficient of the denominator is simply* $\log_e$ *(denominator)*.

The important rule of (13.28) enables the integrals of certain trigonometrical functions to be written down. Thus

$$\int \tan x \, dx = \int \frac{\sin x}{\cos x} dx$$

$$= - \int - \frac{\sin x}{\cos x} dx$$

$$= - \int \frac{\frac{d}{dx}(\cos x)}{\cos x} dx = - \log_e \cos x + C. \tag{13.29}$$

Similarly

$$\int \cot x \, dx = \int \frac{\cos x}{\sin x} dx$$

$$= \int \frac{\frac{d}{dx}(\sin x) dx}{\sin x}$$

$$= \log_e \sin x + C. \tag{13.30}$$

Also,
$$\int \operatorname{cosec} x \, dx = \int \frac{dx}{\sin x}$$

$$= \int \frac{dx}{2 \sin \tfrac{1}{2}x \cos \tfrac{1}{2}x}$$

$$= \int \frac{dx}{2 \tan \tfrac{1}{2}x \cos^2 \tfrac{1}{2}x}$$

$$= \int \frac{\tfrac{1}{2} \sec^2 \tfrac{1}{2}x}{\tan \tfrac{1}{2}x} dx$$

$$= \int \frac{\dfrac{d}{dx}(\tan \tfrac{1}{2}x)}{\tan \tfrac{1}{2}x} dx$$

$$= \log_e \tan \tfrac{1}{2}x + C. \qquad (13.31)$$

The integral of $\sec x$ can then be found by noting that
$$\cos x = \sin \left(\tfrac{1}{2}\pi + x\right),$$

so that
$$\int \sec x \, dx = \int \operatorname{cosec} \left(\tfrac{1}{2}\pi + x\right) dx$$

$$= \log_e \tan \left(\tfrac{1}{4}\pi + \tfrac{1}{2}x\right) + C, \qquad (13.32)$$

by using (13.31) and (11.2).

Some other instances of the important result (13.28) will be found in the following examples.

Example 5. *Evaluate* $\displaystyle\int_0^1 \frac{x^2 \, dx}{x^3 + 1}$. $\qquad\qquad$ (L.U.)

Since $\dfrac{d}{dx}(x^3 + 1) = 3x^2$, the integral can be written

$$\frac{1}{3} \int_0^1 \frac{\dfrac{d}{dx}(x^3 + 1)}{x^3 + 1} dx$$

$$= \frac{1}{3} \left[\log_e(x^3 + 1) \right]_0^1$$

$$= \frac{1}{3} \log_e 2,$$

since $\log_e 1 = 0$.

Example 6. *Find* $\displaystyle\int \frac{(4x + 5)dx}{x^2 + 2x + 2}$.

The derivative with respect to x of the denominator is $2x + 2$, and if we write the numerator as $2(2x + 2) + 1$, the integral can be expressed as the sum

$$2 \int \frac{(2x + 2)dx}{x^2 + 2x + 2} + \int \frac{dx}{x^2 + 2x + 2}.$$

The first integral is $2 \log_e (x^2 + x + 2)$ and the second integral can be written as

$$\int \frac{dx}{(x + 1)^2 + 1}$$

or by § 11.2, $\tan^{-1}(x + 1)$. Hence finally

$$\int \frac{(4x + 5)dx}{x^2 + 2x + 2} = 2 \log_e (x^2 + 2x + 2) + \tan^{-1}(x + 1) + C.$$

The integration of rational algebraical fractions (i.e., fractions in which the numerator and denominator contain only positive integral powers of the variable and constant coefficients) can often be made to depend on the integral

$$\int \frac{dx}{ax + b} = \frac{1}{a} \log_e (ax + b) + C. \tag{13.33}$$

We consider below a few of the simpler cases. In all cases, if the degree of the numerator is equal to or greater than that of the denominator, the numerator must first be divided by the denominator until the remainder is of lower degree than the denominator.

(a) Let the denominator be of the *first* degree. In this case the remainder after the division will be independent of the variable and the integral will be given as a sum of terms involving powers of the variable and a logarithmic term.

Example 7. *Find* $\int \dfrac{x^3\, dx}{x - 1}$.

By division of x^3 by $(x - 1)$ we find

$$\frac{x^3}{x - 1} \equiv x^2 + x + 1 + \frac{1}{x - 1},$$

the quotient being $x^2 + x + 1$ and the remainder unity.

Hence

$$\int \frac{x^3\, dx}{x - 1} = \int (x^2 + x + 1)dx + \int \frac{dx}{x - 1}$$

$$= \tfrac{1}{3}x^3 + \tfrac{1}{2}x^2 + x + \log_e (x - 1) + C.$$

Example 8. *Integrate* $\dfrac{3 - 2x}{2x - 1}$ *with respect to x.*

Here the division process gives

$$\frac{3 - 2x}{2x - 1} \equiv -1 + \frac{2}{2x - 1},$$

so that

$$\int \frac{3 - 2x}{2x - 1}dx = \int (-1)dx + 2\int \frac{dx}{2x - 1}$$

$$= -x + 2.\tfrac{1}{2} \log_e (2x - 1) + C$$

$$= -x + \log_e (2x - 1) + C.$$

(b) Let the denominator be of the *second* degree and let it break up into a pair of linear factors. In this case, we can split the integrand by resolving it into partial fractions and each partial fraction can be integrated by (13.33).

Thus to find $\int \dfrac{dx}{a^2 - x^2}$ we first write the integrand as

$$\frac{A}{a - x} + \frac{B}{a + x},$$

where A, B are two constants to be found. The usual method (§ 2.8) for resolution into partial fractions then gives as the identity from which A and B are to be determined

$$A(a + x) + B(a - x) \equiv 1.$$

By letting $x = \pm a$ in turn we find

$$A = B = \frac{1}{2a},$$

so that

$$\int \frac{dx}{a^2 - x^2} = \frac{1}{2a} \int \frac{dx}{a - x} + \frac{1}{2a} \int \frac{dx}{a + x}$$

$$= -\frac{1}{2a} \log_e (a - x) + \frac{1}{2a} \log_e (a + x) + C$$

$$= \frac{1}{2a} \log_e \left(\frac{a + x}{a - x}\right) + C.$$

Example 9. *Find* $\int \dfrac{x^3 \, dx}{x^2 + x - 20}$.

Here the numerator of the integrand is of higher degree than the denominator, and division until a remainder of lower degree is obtained shows that

$$\frac{x^3}{x^2 + x - 20} \equiv x - 1 + \frac{21x - 20}{x^2 + x - 20}.$$

The last term on the right is resolved into partial fractions by writing

$$\frac{21x - 20}{x^2 + x - 20} \equiv \frac{A}{x + 5} + \frac{B}{x - 4},$$

so that the identity for the determination of A and B is

$$A(x - 4) + B(x + 5) \equiv 21x - 20.$$

By letting $x = -5$ and 4 respectively we find that

$$A = \frac{125}{9}, \quad B = \frac{64}{9}.$$

Hence

$$\int \frac{x^3 \, dx}{x^2 + x - 20} = \int (x - 1)dx + \frac{125}{9} \int \frac{dx}{x + 5} + \frac{64}{9} \int \frac{dx}{x - 4}$$

$$= \frac{1}{2}x^2 - x + \frac{125}{9} \log_e (x + 5) + \frac{64}{9} \log_e (x - 4) + C.$$

(c) When the denominator is of higher degree than the second, the method of resolution into partial fractions is still often useful. A few instances will be found among the example and exercises which follow but a complete discussion is outside the scope of this book.

Example 10. *Find* $\int \dfrac{dx}{(x-1)^2(x^2+1)}$.

Let, according to the usual rules for resolution into partial fractions,

$$\frac{1}{(x-1)^2(x^2+1)} \equiv \frac{A}{x-1} + \frac{B}{(x-1)^2} + \frac{Cx+D}{x^2+1},$$

then $\quad A(x-1)(x^2+1) + B(x^2+1) + (Cx+D)(x-1)^2 \equiv 1.$

By letting $x = 1$, we find $B = 1/2$. Equating the coefficients of x^2 and the terms independent of x gives respectively

$$-A + B - 2C + D = 0 \quad \text{and} \quad -A + B + D = 1.$$

Subtraction of these two relations then gives $2C = 1$ or $C = 1/2$. A comparison of the coefficients of x^3, shows that $A + C = 0$, so that $A = -C = -1/2$. Finally, putting $A = -1/2$, $B = 1/2$ in the relation $-A + B + D = 1$ shows that $D = 0$.

Hence

$$\int \frac{dx}{(x-1)^2(x^2+1)} = -\frac{1}{2} \int \frac{dx}{x-1} + \frac{1}{2} \int \frac{dx}{(x-1)^2} + \frac{1}{2} \int \frac{x\,dx}{x^2+1}$$

$$= -\frac{1}{2}\log_e(x-1) - \frac{1}{2(x-1)} + \frac{1}{4}\log_e(x^2+1) + C,$$

the last integral being evaluated by writing

$$\frac{1}{2} \int \frac{x\,dx}{x^2+1} = \frac{1}{4} \int \frac{(2x)\,dx}{x^2+1}$$

$$= \frac{1}{4} \int \frac{\dfrac{d}{dx}(x^2+1)}{x^2+1} dx.$$

The method of integration by parts is often useful in cases where the integrand contains the function $\log_e x$. If this function is taken as the "u" in formula (11.10), i.e.

$$\int u\frac{dv}{dx}dx = uv - \int v\frac{du}{dx}dx,$$

then du/dx in the integral on the right is simply $1/x$ and this integral can then often be found easily. As a simple example, the integral $\int \log_e x\,dx$ can be found by taking

$$u = \log_e x, \quad \frac{dv}{dx} = 1,$$

so that $\qquad \dfrac{du}{dx} = \dfrac{1}{x}, \qquad v = x,$

and we have $\displaystyle\int \log_e x\, dx = x \log_e x - \int x.\frac{1}{x}.dx$

$$= x \log_e x - \int dx$$

$$= x \log_e x - x + C.$$

Another example which depends on the work of this section is given below.

Example 11. *Find* $\displaystyle\int \tan^{-1} x\, dx$.

By writing $u = \tan^{-1} x$, $dv/dx = 1$, so that $du/dx = 1/(1 + x^2)$ and $v = x$, in the formula for integration by parts, we have

$$\int \tan^{-1} x\, dx = x \tan^{-1} x - \int \frac{x}{1 + x^2}dx.$$

$$= x \tan^{-1} x - \frac{1}{2}\int \frac{\frac{d}{dx}(x^2 + 1)}{x^2 + 1}dx$$

$$= x \tan^{-1} x - \tfrac{1}{2} \log_e (x^2 + 1) + C.$$

EXERCISES 13 (c)

Evaluate the following indefinite and definite integrals:—

1. $\displaystyle\int \frac{x\, dx}{1 + x^2}.$

2. $\displaystyle\int \frac{(6x^2 + x - 1)dx}{4x^3 + x^2 - 2x + 3}.$

3. $\displaystyle\int_1^{10} \frac{3t + 1}{3t^2 + 2t}dt.$

4. $\displaystyle\int \frac{x + a}{x^2 + a^2}dx.$

5. $\displaystyle\int \frac{e^x - e^{-x}}{e^x + e^{-x}}dx.$

6. $\displaystyle\int \frac{\sec^2 x}{3 + 4 \tan x}dx.$

7. $\displaystyle\int \frac{dt}{t \log_e t}.$

8. $\displaystyle\int \frac{(8t - 3)dt}{2t^2 + 2t + 1}.$

Integrate the following with respect to x:—

9. $\displaystyle\frac{x}{2x - 3}.$

10. $\displaystyle\frac{2x + 3}{x + 2}.$

11. $\displaystyle\frac{x^3}{4 - 2x}.$

12. $\displaystyle\frac{5}{x^2 + x - 6}.$

13. $\displaystyle\frac{6x}{(x + 1)(x - 2)}.$

14. $\displaystyle\frac{x^3 + 2}{x^2 - 1}.$

15. $\displaystyle\frac{4x + 3}{(x - 3)^2}.$

16. $\displaystyle\frac{7x + 5}{(x - 3)(x^2 + 4)}.$

Evaluate the following definite integrals:—

17. $\int_0^1 \dfrac{dx}{(x + 1)(x + 2)}$.

18. $\int_1^2 \dfrac{dt}{t^2(1 + t)}$.

Use the method of integration by parts to find the following integrals:—

19. $\int x^2 \log_e x \, dx$.

22. $\int x \log_e (x + 4) dx$.

20. $\int x^2 \tan^{-1} x \, dx$.

23. $\int \dfrac{\log_e x \, dx}{(x + 1)^2}$.

21. $\int x \sec^2 x \, dx$.   (Q.E.)

24. $\int \sec^2 x \log_e \tan x \, dx$.

13.8. Logarithmic differentiation

When a function consists of a number of factors it is often convenient to take the logarithm before differentiating. Thus, suppose that

$$y = \frac{u_1 u_2 u_3 \ldots}{v_1 v_2 v_3 \ldots},$$

where u_1, u_2, u_3, ..., v_1, v_2, v_3, ... are functions of x. We have

$$\log_e y = \log_e u_1 + \log_e u_2 + \log_e u_3 + \ldots$$
$$- \log_e v_1 - \log_e v_2 - \log_e v_3 - \ldots$$

so that

$$\frac{1}{y}\frac{dy}{dx} = \frac{1}{u_1}\frac{du_1}{dx} + \frac{1}{u_2}\frac{du_2}{dx} + \frac{1}{u_3}\frac{du_3}{dx} + \ldots$$

$$- \frac{1}{v_1}\frac{dv_1}{dx} - \frac{1}{v_2}\frac{dv_2}{dx} - \frac{1}{v_3}\frac{dv_3}{dx} - \ldots$$

A similar method can be applied to find the differential coefficient of the function

$$y = u^v,$$

where u and v are functions of x. Taking the logarithm we have

$$\log_e y = v \log_e u,$$

so that

$$\frac{1}{y}\frac{dy}{dx} = \frac{dv}{dx}\log_e u + \frac{v}{u}\frac{du}{dx}.$$

Example 12. *Find the differential coefficients with respect to x of*

$$(i) \ y = \sqrt{\left(\frac{1 + x}{1 - x}\right)}, \quad (ii) \ y = x^x.$$

From (i), $\log_e y = \frac{1}{2} \log_e (1 + x) - \frac{1}{2} \log_e (1 - x)$,

so that

$$\frac{1}{y}\frac{dy}{dx} = \frac{1}{2(1 + x)} + \frac{1}{2(1 - x)} = \frac{1}{1 - x^2}.$$

Hence
$$\frac{dy}{dx} = \frac{y}{1 - x^2}$$

$$= \frac{1}{(1 - x^2)\sqrt{}}\sqrt{\left(\frac{1 + x}{1 - x}\right)}$$

$$= \frac{1}{(1 - x)^{3/2}(1 + x)^{1/2}}.$$

From (ii), $\log_e y = x \log_e x$, so that

$$\frac{1}{y}\frac{dy}{dx} = 1 + \log_e x,$$

using the usual formula for the differentiation of the product $x \log_e x$.

Hence
$$\frac{dy}{dx} = y(1 + \log_e x) = x^x(1 + \log_e x).$$

13.9. Successive approximations and Maclaurin's series

Consider first the function $1/(1 - x)$ for values of x less than unity. It is easy to verify by ordinary algebraical processes that

$$\frac{1}{1 - x} = 1 + \frac{x}{1 - x},$$

$$\frac{1}{1 - x} = 1 + x + \frac{x^2}{1 - x},$$

$$\frac{1}{1 - x} = 1 + x + x^2 + \frac{x^3}{1 - x},$$

and so on. Hence

$$1, \quad 1 + x, \quad 1 + x + x^2, \ldots$$

are successive approximations to the function $1/(1 - x)$ for the respective errors are

$$\frac{x}{1 - x}, \quad \frac{x^2}{1 - x}, \quad \frac{x^3}{1 - x}, \ldots,$$

and, for values of x less than unity, these errors become progressively smaller. It should be noted that the successive approximations

$$1, \quad 1 + x, \quad 1 + x + x^2, \ldots$$

are all equal for $x = 0$. From $(1 + x)$ onwards, they all have the same first derivative at $x = 0$; from $(1 + x + x^2)$ onwards, they all have the same second derivative for this value of x and so on.

This suggests the following method of approximating to a function $f(x)$. Let

$$f(x) \doteqdot a_0 + a_1 x + a_2 x^2 + a_3 x^3 + \ldots + a_n x^n.$$

Choose $a_0, a_1, a_2, a_3, \ldots, a_n$ so that for a certain value of x, say $x = 0$, the function $f(x)$ and its first n derivatives (assumed to exist) are the same as the values at $x = 0$ of the polynomial and its first n derivatives. By this procedure, the polynomial may well be a successively better approximation to the function as the number $(n + 1)$ of terms in the polynomial increases.

The first n derivatives of the polynomial are

$$a_1 + 2a_2x + 3a_3x^2 + \ldots + na_nx^{n-1},$$
$$2a_2 + 6a_3x + \ldots + n(n-1)a_nx^{n-2},$$
$$6a_3 + \ldots + n(n-1)(n-2)a_nx^{n-3},$$

and so on. At $x = 0$, the values of the polynomial and its first n derivatives are

$$a_0, \quad a_1, \quad 2a_2, \quad 6a_3, \ldots, (n)!\, a_n.$$

Equating these to the values at $x = 0$ of $f(x)$ and its first n derivatives, we find

$$a_0 = f(0), \qquad a_3 = \tfrac{1}{6}f'''(0),$$
$$a_1 = f'(0), \qquad \ldots \ldots$$
$$a_2 = \tfrac{1}{2}f''(0), \qquad a_n = \frac{1}{(n)!}f^{(n)}(0).$$

Thus we may expect to be able to write

$$f(x) \doteqdot f(0) + xf'(0) + \frac{x^2}{2!}f''(0) + \frac{x^3}{3!}f'''(0) + \ldots + \frac{x^n}{(n)!}f^{(n)}(0). \quad (13.34)$$

The above procedure is only satisfactory if (i) $f(x)$ and its first n derivatives all exist and are continuous at $x = 0$, and (ii) the difference between $f(x)$ and the polynomial on the right of (13.34) decreases as n increases. Thus the method would fail entirely for the case $f(x) = 1/x$, for none of $f(x)$ nor its derivatives exist at $x = 0$. The method would fail also for the case $f(x) = 1/(1 - x)$ when $x > 1$ since, for such values of x, the errors in the polynomial representations would increase as the number of terms in them increased.

In the example $f(x) = 1/(1 - x)$ it was easy to calculate the errors of the approximations at each stage. It is not so easy to do this for the general function $f(x)$. It is, in fact, beyond the scope of the present book to attempt to do more than point out that there are many functions $f(x)$ for which $f(x)$ and all its derivatives exist and are continuous at $x = 0$ and for which the series

$$f(0) + xf'(0) + \frac{x^2}{2!}f''(0) + \ldots + \frac{x^n}{(n)!}f^{(n)}(0) + \ldots$$

converges. Under such conditions $f(x)$ is the limit of the sum of this series and we can write

$$f(x) = f(0) + xf'(0) + \frac{x^2}{2!}f''(0) + \ldots + \frac{x^n}{(n)!}f^{(n)}(0) + \ldots, \quad (13.35)$$

the series on the right of (13.35) being known as *Maclaurin's series* for $f(x)$.

13.10. Series for e^x and $\log_e(1 + x)$

As examples of the expansion of functions in their Maclaurin's series we consider here the series for e^x and $\log_e(1 + x)$.

Firstly, if $f(x) = e^x$, since $\dfrac{d}{dx}(e^x) = e^x$,

$$f(x) = f'(x) = f''(x) = \ldots = f^{(n)}(x) = \ldots = e^x,$$
and $\qquad f(0) = f'(0) = f''(0) = \ldots = f^{(n)}(0) = \ldots = e^0 = 1.$

Hence (13.35) gives

$$e^x = 1 + x + \frac{x^2}{2!} + \frac{x^3}{3!} + \ldots + \frac{x^n}{(n)!} + \ldots \qquad (13.36)$$

It is clear that e^x and all its derivatives exist and are continuous at $x = 0$ and it can be shown (but we shall not attempt to do so here) that the difference between e^x and the first $(n + 1)$ terms of the series of (13.36) tends to zero as $n \to \infty$ for all values of x. The series of (13.36) is therefore a valid representation of the function e^x for all x.

The above series for e^x is known as the *exponential series* and has many useful applications. Here we shall consider only its use in evaluating e^x numerically as a function of x. For example, if $x = 1$, we have

$$e = 1 + 1 + \frac{1}{2!} + \frac{1}{3!} + \frac{1}{4!} + \frac{1}{5!} + \frac{1}{6!} + \ldots$$
$$= 1{\cdot}0000 + 1{\cdot}0000 + 0{\cdot}5000 + 0{\cdot}1667 + 0{\cdot}0417 + 0{\cdot}0083$$
$$+ 0{\cdot}0014 + 0{\cdot}0002 + \ldots$$
$$= 2{\cdot}718 \ldots$$

To obtain similar accuracy, for values of x less than unity less terms of the series would be required, but for larger values of x it would be necessary to retain more terms.

Now take $f(x) = \log_e(1 + x)$ so that $f(0) = \log_e 1 = 0$. Then

$$f'(x) = \frac{1}{1 + x}, \quad f''(x) = -\frac{1}{(1 + x)^2}, \quad f'''(x) = \frac{2}{(1 + x)^3},$$

$$f^{iv}(x) = -\frac{3!}{(1 + x)^4}, \quad \ldots\ldots\ldots\ldots, \quad f^{(n)}(x) = \frac{(-1)^{n-1}(n - 1)!}{(1 + x)^n},$$

and

$$f'(0) = 1, \quad f''(0) = -1, \quad f'''(0) = 2, \quad f^{iv}(0) = -3!, \ldots$$
$$f^{(n)}(0) = (-1)^{n-1}(n - 1)!$$

Maclaurin's series for $\log_e(1 + x)$ is therefore

$$\log_e (1 + x) = x - \frac{x^2}{2} + \frac{x^3}{3} - \frac{x^4}{4} + \ldots + \frac{(-1)^{n-1} x^n}{n} + \ldots \qquad (13.37)$$

Again $\log_e (1 + x)$ and all its derivatives exist and are continuous at $x = 0$ but it can be shown that the above expansion is valid in the sense that the difference between $\log_e (1 + x)$ and the first n terms of its series tends to zero as $n \to \infty$ *only when* $-1 < x \leqslant 1$.

The reader may wonder why we have given a series for $\log_e (1 + x)$ rather than for $\log_e x$. The reason is that $\log_e x$ and its derivatives do not exist at $x = 0$ and $\log_e x$ cannot therefore be represented by a series in ascending powers of x.

The *logarithmic series* (13.37) is only useful for calculating natural logarithms for small values of x; in fact, the series is not valid when $x > 1$ and even for values of x approaching unity many terms would have to be retained to obtain reasonable accuracy. Some algebraical manipulation of the series can, however, permit the logarithms of larger numbers to be calculated and a typical example is given below.

Example 13. *If the absolute magnitude of x is greater than unity, show that*

$$log_e \left(\frac{x + 1}{x - 1}\right) = 2\left(\frac{1}{x} + \frac{1}{3x^3} + \frac{1}{5x^5} + \ldots\right),$$

and use this series to calculate $log_e 2$ *to three places of decimals.*

We can write

$$\log_e \left(\frac{x + 1}{x - 1}\right) = \log_e \left(\frac{1 + 1/x}{1 - 1/x}\right) = \log_e \left(1 + \frac{1}{x}\right) - \log_e \left(1 - \frac{1}{x}\right).$$

If x is greater than unity, x can be replaced by $1/x$ and $-1/x$ in the series (13.37) so that

$$\log_e \left(\frac{x + 1}{x - 1}\right) = \frac{1}{x} - \frac{1}{2x^2} + \frac{1}{3x^3} - \frac{1}{4x^4} + \frac{1}{5x^5} - \ldots$$

$$- \left\{-\frac{1}{x} - \frac{1}{2}\left(-\frac{1}{x}\right)^2 + \frac{1}{3}\left(-\frac{1}{x}\right)^3 - \frac{1}{4}\left(-\frac{1}{x}\right)^4 + \frac{1}{5}\left(-\frac{1}{x}\right)^5 - \ldots\right\}$$

$$= 2\left(\frac{1}{x} + \frac{1}{3x^3} + \frac{1}{5x^5} + \ldots\right).$$

Putting $x = 3$, we have

$$\log_e \left(\frac{4}{2}\right) = 2\left(\frac{1}{3} + \frac{1}{3 \times 3^3} + \frac{1}{5 \times 3^5} + \ldots\right)$$

giving $\log_e 2 = 2(0 \cdot 3333 + 0 \cdot 0124 + 0 \cdot 0008 + \ldots) = 0 \cdot 693 \ldots$

EXERCISES 13 (d)

Use the method of logarithmic differentiation to find dy/dx when:—

1. $y = (1 - 2x)(1 - 4x)(1 - 6x)$.   3. $y = 10^x$.

2. $y = \dfrac{\sqrt{(1 - 3x)}}{(1 - x)^3}$.   4. $y = (\log_e x)^x$.

Assuming the expansions given below are convergent, show that:—

5. $\sin x = x - \dfrac{x^3}{3!} + \dfrac{x^5}{5!} - \ldots$

6. $\cos x = 1 - \dfrac{x^2}{2!} + \dfrac{x^4}{4!} - \ldots$

7. $\tan(x + \theta) = \tan\theta + x\sec^2\theta + x^2\sec^2\theta\tan\theta + \ldots$

8.   Show that $\dfrac{1}{2}\left(e + \dfrac{1}{e}\right) = 1 + \dfrac{1}{2!} + \dfrac{1}{4!} + \dfrac{1}{6!} + \ldots + \dfrac{1}{(2n)!} + \ldots$

9.   If α is so small that its cube and higher powers may be neglected find the values of the constants A, B and C so that
$$e^{x+\alpha} = Ae^x + Be^{x+\frac{1}{2}\alpha} + Ce^{x+\frac{1}{4}\alpha}.$$

10.   Show that $\log_e(2 + 3x) = \log_e 2 + \frac{3}{2}x - \frac{9}{8}x^2 + \frac{9}{8}x^3 - \ldots$, and state the limits between which x must lie for the expansion to be valid.

11.   By writing $1 + x + x^2 = (1 - x^3)/(1 - x)$ show that, if $-1 \leqslant x < 1$, $\log_e(1 + x + x^2) = x + \frac{1}{2}x^2 - \frac{2}{3}x^3 + \frac{1}{4}x^4 + \frac{1}{5}x^5 - \frac{1}{3}x^6 + \ldots$

12.   If λ, β are the roots of the quadratic equation $x^2 - px + q = 0$, show that for suitable values of x,
$$\log_e(1 + px + qx^2) = (\lambda + \beta)x - \tfrac{1}{2}(\lambda^2 + \beta^2)x^2 + \tfrac{1}{3}(\lambda^3 + \beta^3)x^3 - \ldots$$

EXERCISES 13 (e)

1.   Find dy/dx when:—

     (i) $y = \log_e \tan x$,   (ii) $y = \log_e(\sec x + \tan x)$.   (L.U.)

2.   Find the maximum value of $(\log_e x)/x$.

3.   Differentiate the following with respect to x:—

     (i) $\log_e\left(\dfrac{x}{1 + x}\right)$,   (ii) $\log_e\{\sqrt{(x + 1)} + \sqrt{(x - 1)}\}$.

4.   If $y = \cos(\log_e x)$, prove that
$$x^2\frac{d^2y}{dx^2} + x\frac{dy}{dx} + y = 0.$$

5.   If $y = (e^{2x} - 1)/(e^{2x} + 1)$, show that
$$x = \frac{1}{2}\log_e\left(\frac{1 + y}{1 - y}\right).$$

6.   Evaluate the following:—

     (i) $\displaystyle\int\left(e^{3x} + \frac{2}{x - 1}\right)dx$,   (ii) $\displaystyle\int_0^{\pi/4}(e^{-2x} + 2\tan x)dx$.

7.   If $y = \exp(\tan^{-1} x)$, show that
$$(1 + x^2)\frac{d^2y}{dx^2} + (2x - 1)\frac{dy}{dx} = 0.$$

8.   Show that if α and β are suitably chosen, $y = e^{\alpha x}\cos\beta x$ satisfies the

equation

$$\frac{d^2y}{dx^2} + A\frac{dy}{dx} + By = 0$$

and find α, β in terms of A and B.

9. If ψ is the angle of inclination to the x-axis of the tangent at a point of the curve $y = \frac{1}{2}(e^x + e^{-x})$, show that

(i) $y = \sec \psi$,   (ii) $x = \log_e (\tan \psi + \sec \psi)$.

10. The arc of the curve $y = e^{x/2} + e^{-x/2}$ from $x = -2$ to $x = 2$ rotates about the x-axis. Find the area of the surface so generated.

11. Integrate the following with respect to x:—

(i) $\dfrac{e^x + e^{-x}}{e^x - e^{-x}}$,   (ii) $\dfrac{x^{e-1} + e^{x-1}}{x^e + e^x}$.

12. Show that

$$\int \frac{x\,dx}{2x^2 - x + 1} = \frac{1}{4}\log_e (2x^2 - x + 1) + \frac{1}{2\sqrt{7}}\tan^{-1}\left(\frac{4x - 1}{\sqrt{7}}\right) + C.$$

13. Integrate the following with respect to x:—

(i) $\dfrac{x + 1}{x + 2}$,   (ii) $\dfrac{x^2 - 3x + 1}{x + 1}$.

14. Evaluate the following:—

(i) $\displaystyle\int_0^1 \frac{dt}{4 + t - 3t^2}$,   (ii) $\displaystyle\int \frac{(2t + 1)dt}{t^3 + 2t^2 - 3t}$.

15. Integrate the following with respect to x:—

(i) $\dfrac{1 - 3x^2}{3x - x^3}$,   (ii) $\dfrac{1}{(x^2 - 1)^2}$.

16. Find the following:—

(i) $\displaystyle\int \frac{x^2\,dx}{x^2 + 7x + 10}$,   (ii) $\displaystyle\int \frac{(10x^2 + 13x + 9)dx}{(x - 2)(2x + 1)^2}$.

17. Evaluate:—

(i) $\displaystyle\int_1^2 \frac{dt}{t^2 + 4t}$,   (ii) $\displaystyle\int_0^3 \frac{2t\,dt}{(1 + t^2)(3 + t^2)}$.

18. Find the area enclosed by the curve $(1 + 3x + 2x^2)y = 1$, the x-axis and ordinates at $x = 0$ and $x = 5$.

19. Sketch the curve $y^2(x + a) + x(x - a) = 0$, where a is a positive constant, and show that y is a maximum or a minimum when $x = \pm a\sqrt{2} - a$. Prove that the volume enclosed by rotating about the axis of x the part of the curve which lies between $x = 0$ and $x = a$ is $\pi a^2(1\cdot 5 - 2\log_e 2)$. (Q.E.)

20. Show that

(i) $\displaystyle\int_0^{\pi/4} \theta \sec^2 \theta \,d\theta = \frac{\pi}{4} - \frac{1}{2}\log_e 2$,

(ii) $\displaystyle\int_1^2 (x - 1)^2 \log_e x \,dx = \frac{2}{3}\log_e 2 - \frac{5}{18}$.

21. Find dy/dx when

$$\text{(i) } y = \frac{x(1 + x^2)}{\sqrt{(1 - x^2)}}, \quad \text{(ii) } y = 5^{1 + \sin^2 x}. \qquad \text{(Q.E.)}$$

22. Show that

$$\frac{d}{dx}(\sec x \tan^n x) = \sec x \{n \tan^{n-1} x + (n + 1) \tan^{n+1} x\}.$$

Use this result to show that if $y = \sec x$,

$$\frac{d^2 y}{dx^2} = \sec x(1 + 2 \tan^2 x), \quad \frac{d^3 y}{dx^3} = \sec x(5 \tan x + 6 \tan^3 x),$$

$$\frac{d^4 y}{dx^4} = \sec x(5 + 28 \tan^2 x + 24 \tan^4 x).$$

Hence show that the first three terms of Maclaurin's series for $\sec x$ are given by

$$\sec x = 1 + \frac{x^2}{2} + \frac{5x^4}{24} + \dots$$

23. Find the coefficient of x^5 in the expansion of $e^x/(1 + 2x^2)$ in ascending powers of x.   (Q.E.)

24. In the equation $x^{2 + \lambda} = e^2$, λ is a small quantity whose third and higher powers may be neglected; prove that

$$\text{(i) } x = e. e^{-\frac{\lambda}{2 + \lambda}},$$
$$\text{(ii) } x = e(1 - \tfrac{1}{2}\lambda + \tfrac{3}{8}\lambda^2).$$

25. Find the coefficient of x^n in the expansion of

$$\log_e \left(\frac{6 - x}{3 - x}\right)$$

in ascending powers of x; it can be assumed that x lies within the range for which the expansion is convergent.

CHAPTER 14

COORDINATES. LENGTHS OF LINES. AREAS OF TRIANGLES. LOCI. INTERSECTION OF CURVES

14.1. Systems of coordinates

In elementary graphical work, it is customary to specify the position of a point P in a plane by its perpendicular distances OM, MP from two fixed perpendicular lines Oy, Ox (Fig. 91). The lines Ox, Oy are

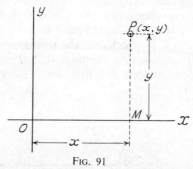

Fig. 91

called the axes, O the origin and the distances OM, MP are referred to as the abscissa and ordinate of the point P. A convenient method of referring shortly to the position of a point whose abscissa is x and ordinate y is to say that the point has *coordinates* x and y and to use the symbol (x, y) to designate the position of such a point.

The student is assumed to be familiar with the usual sign conventions used in graphical work. Briefly these are as follows—for points to the right of the axis Oy the abscissa is positive and for points to the left of Oy the abscissa is negative, while the ordinate is positive or

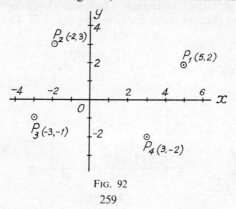

Fig. 92

259

negative according as the point is above or below the axis Ox. Thus, in Fig. 92, the points P_1, P_2, P_3, P_4 would be denoted by $(5, 2), (-2, 3), (-3, -1)$ and $(3, -2)$ respectively.

The above method of specifying the position of a point in a plane by its distances from two straight lines (the axes) was first introduced by the philosopher Des Cartes. It is not essential for the axes to be mutually perpendicular as in Figs. 91, 92, but it is often convenient to use, and in this book we shall only deal with, axes at right angles. Such a system of coordinates is known as the *Cartesian* system and, when the axes are at right angles to one another as in Fig. 91, x and y are referred to as the *rectangular Cartesian coordinates* of the point P. Other coordinate systems, for example that outlined below, exist but the Cartesian system is by far the most important.

Another method of specifying the position of a point in a plane is by its polar coordinates. Suppose that O is a fixed point, called the *origin or pole*, and Ox is a fixed line, called the *initial line*. Then

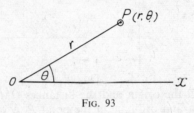

FIG. 93

in Fig. 93, the position of a point P is known when the angle $xOP\ (=\theta)$ and the length $OP\ (=r)$ are given. The two quantities r and θ are called the *polar coordinates* of P and the position of P is conveniently denoted in this system of coordinates by the symbol (r, θ).

14.2. The relation between Cartesian and polar coordinates

Suppose that the rectangular Cartesian coordinates of a point P are (x, y) and that its polar coordinates are (r, θ). Then, from Fig. 94,

$$\left.\begin{array}{l} x = OM = r\cos\theta, \\ y = MP = r\sin\theta, \end{array}\right\} \tag{14.1}$$

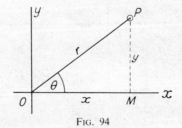

FIG. 94

and x, y can be found when r, θ are given. Conversely, if x, y are given, we have

$$r^2 = x^2 + y^2, \Big\} \atop \tan \theta = y/x. \quad\quad (14.2)$$

Equations (14.2) do not determine r, θ uniquely, for $r = \pm\sqrt{(x^2 + y^2)}$ and θ can take an indefinite number of different values. To obtain an unique correspondence, we take $r = +\sqrt{(x^2 + y^2)}$ and determine θ as the angle which lies between $-\pi$ and $+\pi$ satisfying the two equations $\cos \theta = x/r$, $\sin \theta = y/r$.

Example 1. *Find (i) the Cartesian coordinates of the point whose polar coordinates are* $(5, \pi/4)$ *and (ii) the polar coordinates of the point whose Cartesian coordinates are* $(-1, 1)$.

(i) $r = 5$, $\theta = \pi/4$, so that

$$x = 5 \cos (\pi/4) = 5/\sqrt{2} = 3.536,$$
$$y = 5 \sin (\pi/4) = 5/\sqrt{2} = 3.536.$$

(ii) $x = -1$, $y = 1$, so that

$$r = \sqrt{(1^2 + 1^2)} = \sqrt{2} = 1.414,$$
$$\cos \theta = -1/\sqrt{2}, \quad \sin \theta = 1/\sqrt{2} \quad \text{giving } \theta = 3\pi/4.$$

14.3. The distance between two points with given rectangular coordinates

Let P_1, P_2 be the two given points and let their coordinates be respectively (x_1, y_1), (x_2, y_2). Draw (Fig. 95) P_1M_1, P_2M_2 perpendicular to Ox, and draw P_2R parallel to Ox to meet P_1M_1 in R. Then

$$P_2R = M_2M_1 = OM_1 - OM_2 = x_1 - x_2,$$

and $\quad RP_1 = M_1P_1 - M_1R = M_1P_1 - M_2P_2 = y_1 - y_2.$

From the right-angled triangle P_2RP_1,

$$P_1P_2 = \sqrt{(P_2R^2 + RP_1{}^2)}$$
$$= \sqrt{\{(x_1 - x_2)^2 + (y_1 - y_2)^2\}}. \quad\quad (14.3)$$

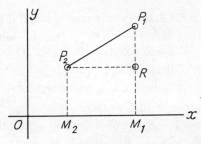

FIG. 95

The proof of formula (14.3) has been given only for the case in which the coordinates of both the points are all positive. When due regard is paid to the usual sign conventions for the coordinates, it will be found to remain true for any positions of the two given points.

Example 2. *Find the distance between the points* $(4, -7)$ *and* $(-1, 5)$.

By writing $x_1 = 4$, $y_1 = -7$, $x_2 = -1$, $y_2 = 5$ in formula (14.3), or (preferably) working from first principles from a diagram showing the given points, it will be found that the required distance

$$= \sqrt{\{(4 + 1)^2 + (-7 - 5)^2\}} = \sqrt{(25 + 144)} = \sqrt{(169)} = 13.$$

14.4. A proof of the addition formulae of trigonometry

A compact method of obtaining a formula for $\cos (A - B)$ where A and B are angles of any magnitude can now be given. In Fig. 96, diagrams are given for $90° < A < 180°, 0 < B < 90°$ and $0 < A < 90°$,

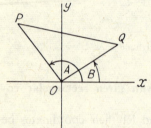 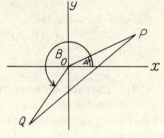

Fig. 96

$180° < B < 270°$ and corresponding diagrams can be constructed for angles of any size. If OP and OQ are each of unit length, the Cartesian coordinates of P are $(\cos A, \sin A)$ and of Q $(\cos B, \sin B)$. In both cases $\cos POQ = \cos (A - B)$, and the cosine formula for the triangle POQ gives

$$PQ^2 = OP^2 + OQ^2 - 2OP.OQ \cos POQ$$
$$= 1 + 1 - 2 \cos (A - B)$$
$$= 2 - 2 \cos (A - B).$$

But, from formula (14.3) for the distance between P and Q,

$$PQ^2 = (\cos A - \cos B)^2 + (\sin A - \sin B)^2$$
$$= 2 - 2 \cos A \cos B - 2 \sin A \sin B,$$

when use is made of the relations

$$\cos^2 A + \sin^2 A = 1, \cos^2 B + \sin^2 B = 1.$$

Equating these values of PQ^2 we have

$$\cos (A - B) = \cos A \cos B + \sin A \sin B.$$

The formulae for $\cos (A + B)$, $\sin (A + B)$, $\sin (A - B)$ can then be deduced by replacing B by $-B$, $90° - B$ and $B - 90°$.

EXERCISES 14 (a)

1. Plot the points (i) whose rectangular coordinates are $(2, -3)$ and $(-4, -1)$, (ii) whose polar coordinates are $(2, 2\pi/3)$ and $(2, -2\pi/3)$.

2. (i) Find the Cartesian coordinates of the points whose polar coordinates are:—

    $$(a)\ (4, \pi/3),\quad (b)\ (5, -\pi/4).$$

    (ii) Find the polar coordinates of the points whose Cartesian coordinates are:—

    $$(a)\ (1, 1),\quad (b)\ (-5, 12).$$

3. Show that the distance between two points whose polar coordinates are (r_1, θ_1) and (r_2, θ_2) is

    $$\sqrt{\{r_1{}^2 + r_2{}^2 - 2r_1 r_2 \cos(\theta_1 - \theta_2)\}}.$$

4. Find the distances between the following pairs of points:—
    (i) $(0, 0)$ and (a, b),
    (ii) $(a, 0)$ and $(0, b)$,
    (iii) $(p + q, q + r)$ and $(q + r, r + p)$.

5. Find the lengths of the sides of a triangle whose vertices are the points $(5, -6), (-3, -2)$ and $(1, -3)$.

6. Show that the triangle whose vertices are the points $(-2, 2), (2, 3)$ and $(-1, -2)$ is isosceles.

7. Show that one of the angles of the triangle whose vertices are the points $(5, 1), (-3, 7)$ and $(8, 5)$ is a right angle.

8. The points A and C have coordinates $(4, 1)$ and $(1, 4)$ respectively. The point B is constructed by drawing AB equal and parallel to OC, where O is the origin. Find the coordinates of the point B and show that the figure $OABC$ is a rhombus.

14.5. The coordinates of a point which divides the join of two given points in a given ratio

In Fig. 97, let P_1 and P_2 be the points (x_1, y_1), (x_2, y_2) respectively and suppose we require to find the coordinates of a point P which

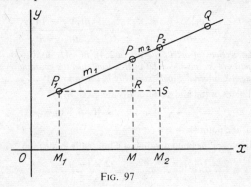

FIG. 97

divides the line P_1P_2 in the ratio $m_1:m_2$. Draw P_1M_1, P_2M_2, PM perpendicular to the axis Ox and draw P_1S parallel to Ox to meet PM, P_2M_2 in R and S respectively. Then, if P is the point (x, y),

$$OM_1 = x_1, \quad OM_2 = x_2, \quad OM = x.$$

From the construction of Fig. 97 it should be clear that

$$P_1R = M_1M = OM - OM_1 = x - x_1,$$
$$RS = MM_2 = OM_2 - OM = x_2 - x.$$

Since PR, P_2S are parallel,

$$\frac{P_1R}{RS} = \frac{P_1P}{PP_2} = \frac{m_1}{m_2},$$

so that

$$\frac{x - x_1}{x_2 - x} = \frac{m_1}{m_2}.$$

Solving for x we find

$$x = \frac{m_1x_2 + m_2x_1}{m_1 + m_2}.$$

By drawing P_1N_1, PN, P_2N_2 perpendicular to Oy, we could show in the same way that the ordinate of P is given by

$$y = \frac{m_1y_2 + m_2y_1}{m_1 + m_2}.$$

Hence the coordinates (x, y) of a point P which divides the join of the points $P_1 (x_1, y_1)$ and $P_2 (x_2, y_2)$ *internally* in the ratio $m_1:m_2$ are given by

$$x = \frac{m_1x_2 + m_2x_1}{m_1 + m_2}, \quad y = \frac{m_1y_2 + m_2y_1}{m_1 + m_2}. \tag{14.4}$$

These are called Joachimsthal's section-formulae. As a special case, the coordinates of the middle point of the line joining (x_1, y_1) to (x_2, y_2) are $\frac{1}{2}(x_1 + x_2)$ and $\frac{1}{2}(y_1 + y_2)$ for here $m_1 = m_2$.

If the point Q divides the line P_1P_2 *externally* in the ratio $m_1:m_2$, i.e., if $P_1Q:QP_2::m_1:m_2$, its coordinates would be found in a similar way to be

$$x = \frac{m_1x_2 - m_2x_1}{m_1 - m_2}, \quad y = \frac{m_1y_2 - m_2y_1}{m_1 - m_2}. \tag{14.5}$$

Example 3. *The vertices of a triangle are $A(2, 4)$, $B(-4, -6)$, $C(6, 0)$. Write down the coordinates of X, the mid-point of BC. Find also the coordinates of the points which divide AX internally and externally in the ratio $2:1$.*

The coordinates of the mid-point of BC are given by

$$x = \tfrac{1}{2}(-4 + 6) = 1, \quad y = \tfrac{1}{2}(-6 + 0) = -3,$$

so that X is the point $(1, -3)$.

The coordinates of the point dividing AX internally in the ratio $2:1$ are given by

$$x = \frac{2 \times 1 + 1 \times 2}{2 + 1} = \frac{4}{3}, \quad y = \frac{2 \times (-3) + 1 \times 4}{2 + 1} = -\frac{2}{3}.$$

The coordinates of the point dividing AX externally in the ratio $2:1$ are given by

$$x = \frac{2 \times 1 - 1 \times 2}{2 - 1} = 0, \quad y = \frac{2 \times (-3) - 1 \times 4}{2 - 1} = -10.$$

14.6. The area of a triangle whose vertices have given coordinates

In Fig. 98, let the vertices of the triangle ABC be the points $A(x_1, y_1)$, $B(x_2, y_2)$ and $C(x_3, y_3)$. Draw AL, BM, CN perpendicular

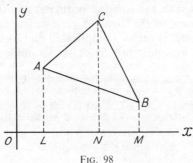

FIG. 98

to the axis Ox. Then

area $\triangle ABC$

$$= \text{area trapezium } ALNC + \text{area trapezium } CNMB$$
$$\qquad\qquad - \text{area trapezium } ALMB$$
$$= \tfrac{1}{2}LN(LA + NC) + \tfrac{1}{2}NM(NC + MB) - \tfrac{1}{2}LM(LA + MB)$$
$$= \tfrac{1}{2}(x_3 - x_1)(y_1 + y_3) + \tfrac{1}{2}(x_2 - x_3)(y_3 + y_2)$$
$$\qquad\qquad - \tfrac{1}{2}(x_2 - x_1)(y_1 + y_2).$$

After a slight rearrangement of the right-hand side, we have

$$\text{area } \triangle ABC = \tfrac{1}{2}\{x_1(y_2 - y_3) + x_2(y_3 - y_1) + x_3(y_1 - y_2)\}. \quad (14.6)$$

By taking the point C to be at the origin O ($x_3 = y_3 = 0$), the area of the triangle OAB with vertices $O(0, 0)$, $A(x_1, y_1)$ and $B(x_2, y_2)$ is given by

$$\text{area } \triangle OAB = \tfrac{1}{2}(x_1 y_2 - x_2 y_1). \quad (14.7)$$

It should be noted that for the formula (14.6) to give a positive value for the area, it is necessary for the points A, B and C to be taken in a special order. This is such that in starting from the point A and proceeding round the perimeter of the triangle in the order A, B and C, the area of the triangle must always be on the *left*. In a specific example it is probably better to draw a rough diagram and work from first principles (see Example 4).

The above formula for the area of a triangle can be used to obtain the condition that three points shall be collinear (i.e., that they shall lie on the same straight line). If the three points (x_1, y_1), (x_2, y_2) and (x_3, y_3) lie on the same straight line, the area of the triangle formed by them vanishes and, from (14.6) the condition for collinearity is therefore

$$x_1(y_2 - y_3) + x_2(y_3 - y_1) + x_3(y_1 - y_2) = 0. \qquad (14.8)$$

The area of a quadrilateral (or other plane polygon) with given vertices can be obtained in a similar way. Perpendiculars are dropped from the vertices on to the axis Ox (or on to a line parallel to Ox through the vertex with the smallest ordinate) and the area can then be expressed in terms of the areas of various trapezia (or triangles). The resulting formula is rather complicated and again it is probably preferable to work from first principles (see Example 6).

Example 4. *Find the area of the triangle formed by the points* $(-2, 3)$, $(-7, 5)$, $(3, -5)$.

Denoting the points $(-2, 3)$, $(-7, 5)$ and $(3, -5)$ by A, B and C, we drop perpendiculars AL, BM on to a line through C parallel to the axis Ox. Then,

$$\begin{aligned}
\text{area } \triangle ABC &= \text{area trapezium } BMLA + \text{area } \triangle ALC - \text{area } \triangle BMC \\
&= \tfrac{1}{2}(5)(10 + 8) + \tfrac{1}{2}(5)(8) - \tfrac{1}{2}(10)(10) \\
&= 15 \text{ units.}
\end{aligned}$$

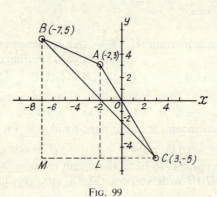

FIG. 99

Alternatively, since in proceeding round the perimeter of the triangle in the order, A, B, C, the area of the triangle is always on the left, we could obtain the same result by writing $x_1 = -2$, $x_2 = -7$, $x_3 = 3$, $y_1 = 3$, $y_2 = 5$, $y_3 = -5$ in formula (14.6).

Example 5. *Find the relation between x and y if the point (x, y) lies on the line joining the points $(2, 3)$ and $(5, 4)$.*

The condition (14.8) for collinearity of the three given points gives

$$x(3 - 4) + 2(4 - y) + 5(y - 3) = 0,$$

i.e.,

$$x - 3y + 7 = 0.$$

Example 6. *Find the area of the quadrilateral whose vertices are the points* $(2, 1)$, $(3, 5)$, $(-3, 4)$ *and* $(-2, -2)$.

In Fig. 100, the four points $(2, 1)$, $(3, 5)$, $(-3, 4)$, $(-2, -2)$ are denoted by A, B, C and D. Perpendiculars AL, BM, CN are dropped from A, B and C on

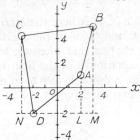

FIG. 100

to a line through D (the point with the smallest ordinate) parallel to Ox. Then

area $ABCD$ = area trapezium $CNMB$ − area $\triangle CND$

$\qquad\qquad\qquad\qquad$ −area $\triangle DLA$ − area trapezium $ALMB$

$\qquad = \frac{1}{2}(6)(6 + 7) - \frac{1}{2}(1)(6) - \frac{1}{2}(4)(3) - \frac{1}{2}(1)(3 + 7)$

$\qquad = 25$ units.

EXERCISES 14 (b)

1. Find the coordinates of the points dividing the line joining the point $(7, -5)$ to the point $(-2, 7)$ internally in the ratio $5:4$ and externally in the ratio $3:2$.

2. Find the distance between the points $(-2, 3)$, $(3, -1)$: find also the co-ordinates of the point of trisection that lies nearer to $(-2, 3)$.

3. The coordinates of the angular points A, B, C, D of a quadrilateral $ABCD$ are respectively $(2, 5)$, $(8, 7)$, $(10, 3)$ and $(0, 1)$. E, F, G, H are the middle points of the sides AB, BC, CD, DA respectively. Show that the middle points of EG and FH coincide.

4. If x and y are the coordinates of the middle point of the line joining the points $(2, 3)$ and $(3, 4)$, show that $x - y + 1 = 0$.

5. Find the areas of the triangles whose vertices are:—

$\qquad$ (i) $(0, 4)$, $(3, 6)$ and $(-8, -2)$,

$\qquad$ (ii) $(a, c + a)$, (a, c) and $(-a, -c - a)$.

6. Show that the four points $(2, 9)$, $(-3, 12)$, $(-8, 15)$ and $(7, 6)$ all lie on the same straight line.

7. Find the area of the quadrilateral whose angular points are $(1, 1)$, $(3, 5)$, $(-2, 4)$ and $(-1, -5)$.

8. Find the area of the triangle whose vertices are the points with polar coordinates $(1, 30°)$, $(2, 60°)$ and $(3, 90°)$.

14.7. The equation to a locus

If a point moves so that it always satisfies a given condition, or conditions, the path traced out by it is called its *locus*. For example, if a point always lies on the straight line joining the points (2, 3) and (5, 4) we have seen in Example 5 of this chapter that its coordinates (x, y) always satisfy the equation

$$x - 3y + 7 = 0. \qquad (14.9)$$

The locus of the point (x, y) is here the straight line joining the two given points and equation (14.9) is the *equation of the locus*. (It will be shown in the next chapter that when the locus of the moving point is a straight line, its equation is of the first degree in x and y.)

Again, suppose that a point P moves in a plane so that the lines AP, BP joining it to two fixed points A and B in the same plane are

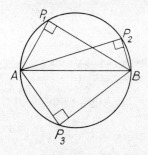

FIG. 101

always at right angles to each other. Then we know that P lies on the circumference of a circle on AB as diameter (see Fig. 101 in which P_1, P_2, P_3 show three possible positions of the point P). The circumference of this circle is the locus of P when its motion is restricted so as to fulfil the above condition.

If a point moves in a plane so as to satisfy some condition such as the above, it will describe in general a definite curve or locus. An equation can in general be found which relates the coordinates (x, y) of the point and this relation will be true only for points lying on the locus. Conversely, to every equation relating x and y there is, in general, a definite geometrical locus. Thus equation (14.9) above represents a straight line. Many other examples will occur to the reader who will already be familiar with the concept of an equation representing a curve from his previous work on graphs and curve sketching.

A few examples on the method of formation of the equation to a locus are appended.

Example 7. *Find the equation to the locus of a point which is always equidistant from the two points (0, 0) and (3, 4).*

In Fig. 102, O is the point $(0, 0)$ and A the point $(3, 4)$. P_1, P_2, P_3 are three possible positions of a point P on the required locus. Let $P(x, y)$ be any point

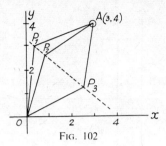

FIG. 102

on the locus. Then

$$OP^2 = x^2 + y^2,$$
$$PA^2 = (x - 3)^2 + (y - 4)^2.$$

Since OP is always to be equal to PA,

$$x^2 + y^2 = (x - 3)^2 + (y - 4)^2.$$

This equation reduces to

$$6x + 8y = 25,$$

which is the required equation to the locus of P. On geometrical grounds it is clear that the locus is the perpendicular bisector of the line joining the two points O, A and the equation $6x + 8y = 25$ therefore represents this bisector.

Example 8. *Find the equation to the locus of a point which is always at distance a from the origin.*

If P is the point (x, y) and O is the origin

$$OP^2 = x^2 + y^2.$$

Since $OP = a$, the required equation to the locus of P is

$$x^2 + y^2 = a^2,$$

the equation representing a circle centre the origin and radius a.

Example 9. *Find the equation to the locus of a point which moves so that its distance from a line parallel to the axis Oy through the point $(-1, 0)$ is equal to its distance from the point $(1, 0)$.*

In Fig. 103, AB is the line through $C(-1, 0)$ parallel to the axis Oy. S is the point $(1, 0)$ and P the point (x, y). PM is the distance of P from the line AB and is clearly equal to the sum of the abscissa of P and the length CO. Hence

$$PM = x + 1.$$

Also, since P and S are respectively the points (x, y) and $(1, 0)$,

$$PS^2 = (x - 1)^2 + y^2.$$

Since $PS = PM$, we therefore have

$$(x - 1)^2 + y^2 = (x + 1)^2,$$

or $$y^2 = 4x.$$

This is the required equation to the locus of P. The locus is shown dotted in

Fig. 103 and its exact shape could be obtained by plotting the graph of $y^2 = 4x$.

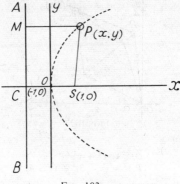

FIG. 103

14.8. The intersections of two curves whose equations are known

The coordinates of the points of intersection, or common points, of two curves will satisfy simultaneously the equations of both curves. The problem of finding the coordinates of the points of intersection of two curves therefore reduces to that of finding the solutions of a pair of simultaneous equations.

As an example, consider the points of intersection of the straight line $x - 3y + 7 = 0$ (equation (14.9)) with the curve

$$3x^2 + 3y^2 - 18x - 20y + 57 = 0.$$

The coordinates of the points of intersection are given by the solutions of the simultaneous equations

$$x - 3y + 7 = 0,$$
$$3x^2 + 3y^2 - 18x - 20y + 57 = 0.$$

Substituting the value $x = 3y - 7$, obtained from the first equation, in the second, the ordinates of the common points are given by

$$3(3y - 7)^2 + 3y^2 - 18(3y - 7) - 20y + 57 = 0.$$

This reduces to $3y^2 - 20y + 33 = 0,$

or, $(3y - 11)(y - 3) = 0.$

Hence $y = 11/3$ or 3, and from the relation $x = 3y - 7$, the corresponding values of x are 4 and 2. Hence the required coordinates of the points of intersection are $(4, 11/3)$ and $(2, 3)$. If the graphs of the line and curve are plotted and the coordinates of the points of intersection are read from the diagram, they will be found to agree with these values.

Example 10. *Find the length of the line joining the points of intersection of the straight line $x - y + 1 = 0$ and the curve $y = 2x^2 - x + 1$.*

From the first equation, $y = x + 1$ and substitution in the second gives for the determination of the abscissae of the points of intersection

$$x + 1 = 2x^2 - x + 1,$$

or, $2x^2 - 2x = 0.$

This can be written $x(x - 1) = 0,$

so that $x = 0$ or 1. The corresponding ordinates, found from $y = x + 1$, are therefore $y = 1$ and 2 and the points of intersection are therefore $(0, 1)$ and $(1, 2)$. The length of the line joining these points

$$= \sqrt{\{(0 - 1)^2 + (1 - 2)^2\}} = \sqrt{2}.$$

EXERCISES 14 (c)

1. Find the equation to the locus of a point which moves so that its distance from the point $(3, 4)$ is always 5 units.

2. A and B are respectively the points $(1, 0)$ and $(7, 8)$. A point P moves so that the angle APB is a right angle. Find the equation to the locus of P.

3. The coordinates of the points A and B are $(1, 5)$ and $(5, 3)$. Find the equation to the perpendicular bisector of the line AB.

4. Find the equation to the locus of a point P such that the areas of the triangles PAB, PCD are equal, where A, B, C, D are the points $(0, 1)$, $(0, 7)$, $(3, 0)$, $(5, 0)$.

5. Find the coordinates of the point of intersection of the two lines $2x - 3y = 5$ and $3x + y + 2 = 0$.

6. Find the coordinates of the points of intersection and length of the common chord of the two curves $y = 2x^2$ and $y^2 = 4x$.

7. Find the coordinates of the common points of the line $x - 2 = 0$ and the curve $x^2 + y^2 - 4x - 10y + 4 = 0$. Find also the coordinates of the middle point of the line joining the common points.

8. The equations to the sides AB, BC, CA of a plane triangle are respectively $y = 2x + 1$, $x - 3y + 4 = 0$ and $y = x$. D is the mid-point of BC. By actually finding the coordinates of A, B, C and D verify that

$$AB^2 + AC^2 = 2(AD^2 + BD^2).$$

EXERCISES 14 (d)

1. Find the perimeter of the triangle whose vertices are, in polar coordinates, the points $(0, 0)$, $(2, \pi/6)$, $(4, \pi/2)$.

2. Find the lengths of the sides of a triangle whose vertices are the points $(2, 3)$, $(4, -5)$, $(-3, -6)$.

3. Find the coordinates of a point equidistant from the three points $(2, 3)$, $(4, 5)$, $(6, 1)$.

4. Show that the four points $(-2, 3)$, $(0, -1)$, $(6, 7)$ and $(8, 3)$ are the angular points of a rectangle.

5. Find the coordinates of the centre of the circumscribed circle of the triangle whose vertices are the points $(-2, 2)$, $(1, -2)$, $(1, 3)$.

6. Find the radius of the circumscribed circle of the triangle the polar coordinates of whose vertices are $(0, 0)$, (r_1, θ_1), (r_2, θ_2).

7. The line joining the points $(-6, 8)$ and $(8, -6)$ is divided into four equal parts. Find the coordinates of each point of section.

8. A and B are respectively the points $(5, 6)$ and $(7, 2)$. If the line AB is produced to a point C such that $BC = \frac{1}{2}AB$, find the coordinates of C.

9. By finding the coordinates of the middle points of the lines joining the middle points of pairs of opposite sides of a quadrilateral the coordinates of whose angular points are given, prove that the lines joining the middle points of the opposite sides of any quadrilateral bisect one another.

10. The polar coordinates of three points A, B and C are given in the following table:

$$\theta = \pi/6, \quad \pi/3, \quad 3\pi/4,$$
$$r = 100, \quad 100, \quad 160.$$

Find the lengths of the sides and the area of the triangle ABC.   (Q.E.)

11. Find the area of the triangle whose vertices are the points $(2, 1)$, $(3, -2)$ and $(-4, -1)$.

12. Find the area of the quadrilateral whose vertices are the points $(1, 1)$, $(2, 3)$, $(3, 3)$ and $(4, 1)$.

13. A is the point $(2, 3)$ and B is the point $(0, -1)$. The angle BAC is a right angle and BC is 5 units in length. Find the coordinates of the two possible positions of C.   (L.U.)

14. A and B are the points $(-3, 2)$, $(5, 8)$ respectively. Find the equation to the locus of a point P which moves so that $PA^2 - PB^2 = 50$.   (O.C.)

15. Find the equation to the locus of a point which moves so that its distance from the point $(a, 0)$ is always equal to four times its distance from the y-axis.

16. A point moves so that the sum of the squares of its distances from the two fixed points $(a, 0)$ and $(-a, 0)$ is constant and equal to $2c^2$. Show that the equation to its locus is

$$x^2 + y^2 = c^2 - a^2.$$

17. The coordinates of two of the vertices of a plane triangle are $(3, 2)$ and $(5, 6)$. Find the locus of the third vertex if the area of the triangle is 12 units.

18. The coordinates of points A, B are respectively $(2, 5)$ and $(5, 8)$. P is a point such that $AP = 2BP$. Find the equation to the locus of P.   (O.C.)

19. Find the equation to the perpendicular bisector of the line joining the two points $(4, 1)$ and $(-2, 3)$.   (O.C.)

20. A variable point P moves so that the square of its distance from the origin is equal to the area of the triangle PAB where A and B are the points $(13, 0)$ and $(0, 1)$. Find the equation to the locus of P.

21. Find the lengths of the sides of the triangle formed by the three straight lines $x - 4y = 6$, $3x - 2y + 1 = 0$ and $x + y = 2$.

22. Find the coordinates of the middle point of the line joining the common points of the line $2x - 3y + 8 = 0$ and the curve $y^2 = 8x$.   (O.C.)

23. Find the values of a and b if the straight lines $ax + 5y = 7$ and $4x + by = 5$ intersect at the point $(2, -1)$. If the lines meet the x-axis at A and B, find the length of AB.   (O.C.)

24. Show that the coordinates of the common point of the line $y = mx + a/m$ and the curve $y^2 = 4ax$ are $(a/m^2, 2a/m)$.

25. Show that the length of the line joining the common points of the line $y = mx + c$ and the curve $y^2 = 4x$ is

$$\frac{4}{m^2}(1 + m^2)^{\frac{1}{2}}(1 - mc)^{\frac{1}{2}}.$$

CHAPTER 15

THE COORDINATE GEOMETRY OF THE STRAIGHT LINE

15.1. An equation of the first degree in x and y represents a straight line

The most general equation of the first degree in x and y is

$$Ax + By + C = 0, \tag{15.1}$$

where A, B and C are constants. Let $P_1(x_1, y_1)$, $P_2(x_2, y_2)$, $P_3(x_3, y_3)$ be any three points lying on the locus represented by the equation (15.1). Since P_1 lies on the locus, its coordinates (x_1, y_1) must satisfy equation (15.1) and hence

$$Ax_1 + By_1 + C = 0.$$

Similarly
$$Ax_2 + By_2 + C = 0,$$

and
$$Ax_3 + By_3 + C = 0.$$

Subtracting the second of these equations from the first and the third from the second we obtain

$$A(x_1 - x_2) + B(y_1 - y_2) = 0$$

and
$$A(x_2 - x_3) + B(y_2 - y_3) = 0.$$

By equating the values of the ratio A/B obtained from these two equations we have

$$\frac{y_1 - y_2}{x_1 - x_2} = -\frac{A}{B} = \frac{y_2 - y_3}{x_2 - x_3},$$

giving
$$(x_1 - x_2)(y_2 - y_3) = (x_2 - x_3)(y_1 - y_2).$$

This can be rearranged as

$$x_1(y_2 - y_3) + x_2(y_3 - y_1) + x_3(y_1 - y_2) = 0.$$

From formula (14.6) we see that the area of the triangle formed by the three points P_1, P_2 and P_3 is zero. Since the points P_1, P_2, P_3 are *any* three points on the locus represented by equation (15.1), the locus must be a straight line; for a curved line could not be such that the area of the triangle formed by joining *any* three points on it should be zero.

15.2. The equation to a straight line which is parallel to one of the coordinate axes

In Fig. 104, let AB be a line parallel to the axis Oy. Let AB meet the axis Ox in C and let $OC = c$. Let P be any point on the line AB and let its coordinates be (x, y).

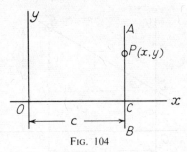

Fig. 104

Since the abscissa of the point P is always c, we have

$$x = c.$$

This relation is true for every point on the line AB and for no other point. The relation $x = c$ is therefore the equation to the line.

Similarly the equation to a line parallel to the axis Ox and at distance d from it is

$$y = d.$$

The coordinate axes are special cases of these lines for which $c = d = 0$. The equation of the axis Ox is therefore $y = 0$ and that of the axis Oy is $x = 0$.

15.3. Special forms of the equation to a straight line

It is often useful to be able to write the general equation

$$Ax + By + C = 0$$

of a straight line in a form in which the constants A, B, C, or rather the ratios of two of them to the third, are related to some geometrical properties of the line. Such properties might be the slope of the line and the coordinates of a point on it, the intercepts it makes on the coordinate axes and so on. Special forms of the equation to a line in terms of various properties possessed by it are developed in the paragraphs which follow.

(a) *The equation to a line in terms of its slope and its intercept on the y-axis*

Let the line have slope m and make an intercept c on the y-axis. Then in Fig. 105 where CP is the line, CR is parallel to the axis Ox and PP' is parallel to the axis Oy, $\tan \theta = m$ and the coordinates of the point C are $(0, c)$.

If P be any point on the line CP with coordinates (x, y), the figure shows that

$$CR = OP' = x,$$
$$PR = PP' - RP' = PP' - OC = y - c,$$

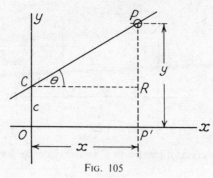

FIG. 105

$$m = \tan \theta = \frac{PR}{CR} = \frac{y - c}{x}.$$

Hence $y = mx + c$ (15.2)

is the relation between the coordinates of any point (x, y) on CP and this is the required equation to the line in terms of its slope m and intercept c.

By writing the general equation $Ax + By + C = 0$ in the form

$$y = -\frac{A}{B}x - \frac{C}{B},$$

and comparing it with (15.2) we see that the ratios A/B, C/B of the constants in the general equation can be expressed in terms of the slope m of the line and its intercept c on the y-axis by

$$\frac{A}{B} = -m, \quad \frac{C}{B} = -c.$$ (15.3)

For the line of slope m passing through the origin of coordinates O, $c = 0$ and the equation to such a line is

$$y = mx.$$

(b) *The equation to a line in terms of its slope and the coordinates of a point on it*

The equation to any line of slope m is, from (15.2),

$$y = mx + c.$$

If the point whose coordinates are (x_1, y_1) lies on the line, these values of x and y will satisfy the above equation so that

$$y_1 = mx_1 + c.$$

By subtraction we have

$$y - y_1 = m(x - x_1)$$ (15.4)

as the equation to the line of slope m which passes through the point (x_1, y_1).

Example 1. *Find the equation to the straight line which passes through the point* $(1, 2)$ *and makes an angle of* $45°$ *with the x-axis.*

Here $m = \tan 45° = 1$, $x_1 = 1$, $y_1 = 2$ and the required equation is

$$y - 2 = (1)(x - 1),$$

or,

$$x - y + 1 = 0.$$

(c) The equation to a line passing through two given points

If one of the given points has coordinates (x_1, y_1) the equation to a line passing through it is (equation (15.4))

$$y - y_1 = m(x - x_1).$$

If the second given point has coordinates (x_2, y_2) and the above line passes through it, its equation will be satisfied by $y = y_2$, $x = x_2$. Hence

$$y_2 - y_1 = m(x_2 - x_1).$$

By division, we have

$$\frac{x - x_1}{x_2 - x_1} = \frac{y - y_1}{y_2 - y_1}, \tag{15.5}$$

as the required equation to the line passing through the two points (x_1, y_1), (x_2, y_2).

Example 2. *If B and C are respectively the points* $(-1, -3)$ *and* $(5, -1)$ *find the equation to the line BC.*

From (15.5) the required equation is

$$\frac{x - (-1)}{5 - (-1)} = \frac{y - (-3)}{-1 - (-3)},$$

i.e.

$$\frac{x + 1}{6} = \frac{y + 3}{2},$$

giving

$$x - 3y = 8.$$

(d) The equation to a line in terms of its intercepts on the coordinate axes

Let the line AB (Fig. 106) make intercepts of lengths a and b on the coordinate axes Ox, Oy respectively. Then the line passes through

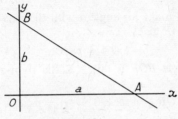

FIG. 106

the points $(a, 0)$ and $(0, b)$ and formula (15.5) gives for its equation

$$\frac{x - a}{0 - a} = \frac{y - 0}{b - 0},$$

or,

$$\frac{x - a}{-a} = \frac{y}{b}.$$

This, by cross multiplication, gives

$$bx - ab = -ay,$$

or, after division by ab and a slight rearrangement,

$$\frac{x}{a} + \frac{y}{b} = 1. \tag{15.6}$$

The general equation $Ax + By + C = 0$ can be written

$$\frac{x}{(-C/A)} + \frac{y}{(-C/B)} = 1,$$

and a comparison with equation (15.6) shows that the line $Ax + By + C = 0$ makes intercepts on the coordinate axes of $-C/A$ and $-C/B$.

(e) *The equation to a line in terms of the length of the perpendicular from the origin and the angle this perpendicular makes with the x-axis*

In Fig. 107, let the line cut the coordinate axes in A and B, let p be the length of the perpendicular OP drawn from the origin O on

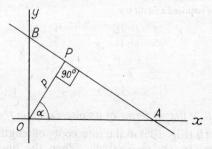

FIG. 107

to AB and let OP make an angle α with the axis Ox. Then the right-angled triangle OAP gives

$$OA = p \sec \alpha,$$

and, since the angle BOP is $90° - \alpha$, the right-angled triangle BOP gives

$$OB = p \sec (90° - \alpha) = p \operatorname{cosec} \alpha.$$

The line AB therefore makes intercepts $p \sec \alpha$ and $p \operatorname{cosec} \alpha$ on the coordinate axes and, from (15.6), its equation is

$$\frac{x}{p \sec \alpha} + \frac{y}{p \operatorname{cosec} \alpha} = 1,$$

or, $x \cos \alpha + y \sin \alpha = p.$ (15.7)

In working problems on the straight line, a correct choice of the form of the equation to the line can often reduce the algebraical work. We conclude this section with some illustrative examples.

Example 3. *Two parallel lines AP, BQ pass through the points $A(5, 0)$ and $B(-5, 0)$ respectively. Find the slope of these lines if they meet the line $4x + 3y = 25$ in points P and Q such that the distance PQ is 5 units.* (L.U.)

Let the slope of the parallel lines AP, BQ be m. Then, by (15.4), the equations of AP, BQ are respectively $y = m(x - 5)$ and $y = m(x + 5)$. The coordinates of P, the point of intersection of AP and the line $4x + 3y = 25$, are given by the values of x and y which satisfy the simultaneous equations

$$y = m(x - 5),$$
$$4x + 3y = 25.$$

Substitution of y from the first equation in the second gives

$$4x + 3m(x - 5) = 25$$

giving $x = \dfrac{25 + 15m}{4 + 3m}.$

Since $y = m(x - 5)$ the ordinate of the point P is

$$y = m\left(\frac{25 + 15m}{4 + 3m} - 5\right) = \frac{5m}{4 + 3m},$$

and the coordinates of P are therefore

$$\left(\frac{25 + 15m}{4 + 3m}, \frac{5m}{4 + 3m}\right).$$

The coordinates of Q are similarly found from the solution of the simultaneous equations $y = m(x + 5)$, $4x + 3y = 25$ to be

$$\left(\frac{25 - 15m}{4 + 3m}, \frac{45m}{4 + 3m}\right).$$

The distance PQ is therefore given by

$$PQ^2 = \left(\frac{25 + 15m}{4 + 3m} - \frac{25 - 15m}{4 + 3m}\right)^2 + \left(\frac{5m}{4 + 3m} - \frac{45m}{4 + 3m}\right)^2$$

$$= \frac{2500m^2}{(4 + 3m)^2}.$$

Since $PQ = 5$, we therefore have, for the determination of the slope m,

$$\frac{2500m^2}{(4 + 3m)^2} = 5^2,$$

giving $91m^2 - 24m - 16 = 0,$

or, $(7m - 4)(13m + 4) = 0.$

Hence the required values of the slope are $4/7$ or $-4/13$.

Example 4. *If O is the origin, A the point $(8, 0)$ and B the point $(0, 6)$, find the coordinates of the points P and Q, where the line $3x + 2y = c$ meets OA and AB respectively. If the area of the triangle OPQ is one-half that of the triangle OAB, find the value of c.* (L.U.)

The line OA is the x-axis and its equation is $y = 0$. The coordinates of P are therefore given by the solution of the simultaneous equations $y = 0$, $3x + 2y = c$. Hence P is the point $(c/3, 0)$.

The line AB makes intercepts of 8 and 6 units on the coordinate axes and, from (15.6), its equation is

$$\frac{x}{8} + \frac{y}{6} = 1,$$

or
$$3x + 4y = 24.$$

The coordinates of Q are given by the solution of the simultaneous equations $3x + 4y = 24$ and $3x + 2y = c$. These are easily found to be

$$x = \frac{2c}{3} - 8, \quad y = 12 - \frac{c}{2}.$$

Since, from (14.7), the area of the triangle formed by the points $(0, 0)$, (x_1, y_1), (x_2, y_2) is $\frac{1}{2}(x_1 y_2 - x_2 y_1)$, the area of the triangle OPQ is given by

$$\frac{1}{2}\left\{\frac{c}{3}\left(12 - \frac{c}{2}\right) - \left(\frac{2c}{3} - 8\right)(0)\right\} = 2c - \frac{c^2}{12}.$$

The area of the triangle OAB is that of a right-angled triangle of base 8 and height 6 and hence is 24 units. If $\triangle OPQ = \frac{1}{2} \triangle OAB$,

$$2c - \frac{c^2}{12} = \frac{24}{2},$$

or,
$$c^2 - 24c + 144 = 0,$$

giving $(c - 12)^2 = 0$ or $c = 12$.

Example 5. *Find the length of the perpendicular from the origin on to the straight line passing through the two points $(6, 4)$ and $(9, 8)$.*

By (15.5) the line through the two given points is

$$\frac{x - 6}{9 - 6} = \frac{y - 4}{8 - 4},$$

or, $4x - 3y = 12$.

This can be compared with the form $x \cos \alpha + y \sin \alpha = p$ by dividing through by $\sqrt{(4^2 + 3^2)}$ or 5, for 4/5 and $-3/5$ can be taken as the cosine and sine respectively of the angle $\tan^{-1}(-3/4)$. Thus if $\alpha = \tan^{-1}(-3/4)$ we can write the equation of the line as

$$x \cos \alpha + y \sin \alpha = 12/5,$$

showing that the required length of the perpendicular from the origin is 12/5.

EXERCISES 15 (a)

1. Find the equation to the straight line which:—

   (i) cuts off an intercept of 3 units from the negative y-axis and makes an angle of 120° with the x-axis,

(ii) makes intercepts of -5 and 3 respectively on the axes of x and y,

(iii) passes through the two points $(-2, 3)$ and $(4, -6)$.

2. Find the slope of the straight line cutting off intercepts of -3 and 4 from the axes of x and y respectively.

3. Find (i) the intercepts on the axes of coordinates, (ii) the slope and (iii) the perpendicular distance from the origin, of the line $5x - 12y = 65$.

4. Find the coordinates of the point of intersection P of the two straight lines $4x + 3y = 7$, $3x - 4y = -1$. Find also the equation to the line joining P to the point $(-2, 3)$.

5. Find the distance between the two parallel straight lines
$$2x + y = 4 \quad \text{and} \quad 4x + 2y = 2.$$

6. Find the equation to the straight line passing through the point of intersection of the two lines $2x + 3y = 4$ and $3x - 2y = 5$ and also through the point of intersection of $3x - 4y = 7$ and $2x + 5y = 2$.

7. Find the equation to a straight line which passes through the point $(3, 5)$ and makes equal intercepts on the coordinate axes.

8. Find the equation to a straight line which passes through the point $(4, 4)$ and forms with the coordinate axes in the fourth quadrant a triangle whose area is 4 units.

9. A straight line AB cuts the coordinate axes at A and B and $OA.OB = c^2$ where O is the origin. If AB is parallel to the line $x \cos \alpha + y \sin \alpha = p$, find its equation.

10. Find the condition that the straight line $(x/a) + (y/b) = 1$ should lie at unit distance from the origin.

11. Find the equations to the diagonals of the parallelogram whose sides have
   the equations:—
$$3x + y = 1, \quad 3y = 5x + 3,$$
$$3x + y = 15, \quad 3y = 5x - 11. \qquad \text{(L.U.)}$$

12. The straight line $y = m(x - 2a)$ through a fixed point $(2a, 0)$ meets the lines $x = a$ and $y = b$ in P and Q respectively. If O is the origin and A the point $(a, 0)$, find the equations to the lines OP, AQ and the coordinates of their point of intersection R. If m varies, show that the locus of the point R is the straight line $2bx - ay = ab$. (L.U.)

15.4. The angle between two straight lines

In Fig. 108, let the equations to the two straight lines AB, CD, intersecting at P be respectively $y = m_1 x + c_1$ and $y = m_2 x + c_2$. Then if AB, CD make angles θ_1, θ_2 with the axis Ox as shown,

$$\tan \theta_1 = m_1, \quad \tan \theta_2 = m_2. \qquad (15.8)$$

Since the external angle PBO of the triangle PDB is equal to the sum of the interior angles PDB and BPD, the angle APC between the lines is given by

$$\text{angle } APC = \text{angle } BPD$$
$$= \text{angle } PBO - \text{angle } PDB$$
$$= \theta_1 - \theta_2.$$

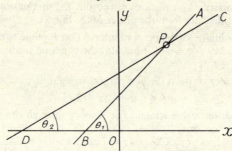

Fig. 108

Hence

$$\tan APC = \tan(\theta_1 - \theta_2)$$
$$= \frac{\tan\theta_1 - \tan\theta_2}{1 + \tan\theta_1\tan\theta_2}$$
$$= \frac{m_1 - m_2}{1 + m_1 m_2}$$

when use is made of the relations (15.8). The angle between the two lines $y = m_1 x + c_1$, $y = m_2 x + c_2$ is therefore

$$\tan^{-1}\left(\frac{m_1 - m_2}{1 + m_1 m_2}\right). \tag{15.9}$$

It should be noted that if the quantity in brackets in (15.9) is positive, it is the tangent of the acute angle between the lines (the angle APC of Fig. 108); if this quantity is negative, it is the tangent of the obtuse angle (the angle DPA).

If we require the angle between the two straight lines

$$A_1 x + B_1 y + C_1 = 0,$$
$$A_2 x + B_2 y + C_2 = 0,$$

their respective slopes m_1, m_2 are, from (15.3), given by

$$m_1 = -\frac{A_1}{B_1}, \quad m_2 = -\frac{A_2}{B_2}.$$

Substitution in (15.9) and a slight reduction shows that the angle between the two lines is

$$\tan^{-1}\left(\frac{A_2 B_1 - A_1 B_2}{A_1 A_2 + B_1 B_2}\right). \tag{15.10}$$

Finally, if the two lines are given in the form

$$x \cos \alpha + y \sin \alpha = p_1,$$
$$x \cos \beta + y \sin \beta = p_2,$$

we know that the perpendiculars from the origin on the two lines make angles α and β respectively with the x-axis. The angle between the lines is clearly equal to the angle (or its supplement) between the perpendiculars, so that the required angle is, in this case, either $\beta - \alpha$ or $\pi - (\beta - \alpha)$.

Example 6. *Find the angles between the following pairs of lines:*—

  (i) $y = 2x + 5$ *and* $3x + y = 7$,

  (ii) $3x - y + 7 = 0$ *and* $x - 3y + 8 = 0$.

(i) The slope m_1 of the first line is 2 and by writing the equation to second line in the form $y = -3x + 7$ we see that its slope $m_2 = -3$.

Hence, from (15.9) the angle between the lines is given by

$$\tan^{-1}\left\{\frac{2 - (-3)}{1 + 2(-3)}\right\}$$

$$= \tan^{-1}(-1) = 135°.$$

(ii) Comparing the given lines with $A_1 x + B_1 y + C_1 = 0$, $A_2 x + B_2 y + C_2 = 0$, $A_1 = 3$, $B_1 = -1$, $A_2 = 1$, $B_2 = -3$, so that by (15.10) the required angle is

$$\tan^{-1}\left\{\frac{(1)(-1) - (3)(-3)}{(3)(1) + (-1)(-3)}\right\}$$

$$= \tan^{-1}\left(\frac{4}{3}\right) = 53° \, 8'.$$

Example 7. *Find the equations to the lines through the point (2, 3) which make angles of 45° with the line $x - 2y = 1$.*   (L.U.)

In Fig. 109, P is the point (2, 3) and AB the line $x - 2y = 1$. This can be written $y = \frac{1}{2}x - \frac{1}{2}$ and its slope is therefore 1/2.

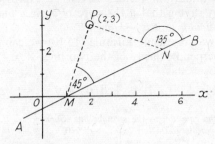

FIG. 109

It is clear from the diagram that there are two possible lines PM, PN which make angles of 45° with AB, and the tangents of these angles are respectively

$\tan 45° = 1$ and $\tan 135° = -1$. Hence, if m is the slope of PM, (15.9) gives

$$1 = \frac{m - (1/2)}{1 + (m/2)},$$

or $m = 3$. Similarly if m' is the slope of PN,

$$-1 = \frac{m' - (1/2)}{1 + (m'/2)},$$

or $m' = -1/3$.

The required lines are therefore those which pass through the point $(2, 3)$ and have slopes 3 and $-1/3$. By (15.4) the equations are

$$y - 3 = 3(x - 2) \quad \text{or} \quad 3x - y = 3,$$

and $\qquad\qquad y - 3 = -\tfrac{1}{3}(x - 2) \quad \text{or} \quad x + 3y = 11.$

15.5. Conditions for parallelism and perpendicularity

If the two lines $y = m_1 x + c_1$, $y = m_2 x + c_2$ are parallel, they have the same slope and hence

$$m_1 = m_2. \tag{15.11}$$

If the lines are given by $A_1 x + B_1 y + C_1 = 0$, $A_2 x + B_2 y + C_2 = 0$, their slopes are $-A_1/B_1$ and $-A_2/B_2$ and in this case the condition for parallelism becomes

$$\frac{A_1}{B_1} = \frac{A_2}{B_2}. \tag{15.12}$$

If the two straight lines $y = m_1 x + c_1$, $y = m_2 x + c_2$ are perpendicular, the angle between them is $90°$. Since the tangent of an angle of $90°$ is infinite, formula (15.9) shows that $1 + m_1 m_2 = 0$. Hence the condition for perpendicularity is

$$m_1 m_2 = -1, \tag{15.13}$$

i.e., *the product of the slopes of two perpendicular straight lines is -1*. This is a most important result—it can be expressed in a slightly different way by saying that *if the slope of a given line is m, the slope of a line perpendicular to it is $-1/m$.*

In a similar way, by using formula (15.10) we can deduce that the condition for perpendicularity of the two lines $A_1 x + B_1 y + C_1 = 0$, $A_2 x + B_2 y + C_2 = 0$ is

$$A_1 A_2 + B_1 B_2 = 0. \tag{15.14}$$

Example 8. *Find the equation to the straight line which passes through the point $(-2, 3)$ and is parallel to the line $7x - y - 6 = 0$.*

The equation to the given line can be written $y = 7x - 6$ so that its slope is 7. The equation to the required line will therefore be that to a line with slope 7 which passes through the point $(-2, 3)$. This is

$$y - 3 = 7(x + 2) \quad \text{or} \quad 7x - y + 17 = 0.$$

Example 9. *The coordinates of three points are $A(1, 2)$, $B(-1, -3)$, $C(5, -1)$. Find the equation to BC and the equation to the line through A perpendicular to BC.*
(L.U.)

The equation to BC is

$$\frac{x+1}{5+1} = \frac{y+3}{-1+3},$$

i.e., $x - 3y = 8$.

The slope of this line is $1/3$, so that the slope of a line perpendicular to it is -3. Thus we require the equation to a line of slope -3 which passes through the point $(1, 2)$. This is

$$y - 2 = -3(x - 1) \quad \text{or} \quad 3x + y = 5.$$

EXERCISES 15 (b)

1. At what angle are the lines whose equations are $ax + by + c = 0$ and $(a - b)x + (a + b)y + d = 0$ inclined to each other? (L.U.)

2. The vertices of a triangle are the points $A(1, 4)$, $B(5, 1)$, $C(-1, -1)$. Find the equations to its sides and the values of $\tan B$, $\tan C$. (L.U.)

3. Find the equations to the straight lines passing through the point $(3, -2)$ and making angles of $60°$ with the line $\sqrt{3}x + y = 1$.

4. The base of an isosceles triangle lies along the line $3x + 2y = 2$ and one of the equal sides lies along $y = 2x$. Find the equation to the other equal side if it too passes through the origin.

5. Write down the equation to the straight line which:

   (i) is parallel to the y-axis and passes through the point $(2, 3)$,
   (ii) is parallel to the line $4x + 3y + 8 = 0$ and passes through the point $(2, -3)$,
   (iii) is perpendicular to the line $4x + 3y + 8 = 0$ and passes through the origin.

6. A triangle ABC is formed by the lines $3x - 4y + 3 = 0$ (AB), $x + y - 3 = 0$ (BC) and $4x - 3y - 5 = 0$ (AC). Find the equation to the straight line through C perpendicular to AB. (L.U.)

7. Find the equation to the join of the points $(1, 2)$ and $(3, 4)$. Find also the coordinates of the middle point of the join and hence write down the equation to the perpendicular bisector of the join. (L.U.)

8. P, Q, R are three points with coordinates $(1, 0)$, $(2, -4)$, $(-5, -2)$ respectively. Determine

   (i) the equation to the line through P perpendicular to QR,
   (ii) the equation to the line through Q perpendicular to PR,
   (iii) the coordinates of the point of intersection of these lines. (L.U.)

9. Find the coordinates of the point of intersection of the perpendiculars from the vertices to the opposite sides of a triangle whose sides have equations $x - 3y = 0$, $4x - y = 0$ and $x + y = 20$.

10. Prove that the straight line joining any two of the four points

$$(am_1, a/m_1), \quad (am_2, a/m_2), \quad (am_3, a/m_3), \quad (am_4, a/m_4)$$

is perpendicular to the straight line joining the other two if $m_1 m_2 m_3 m_4 = -1$.

15.6. The perpendicular distance of a point from a straight line

In Fig. 110, AB is the straight line $Ax + By + C = 0$ meeting the coordinate axes in A and B. P is the point (h, k), PQ is perpendicular to AB and we require to find a formula for the distance $PQ = p$.

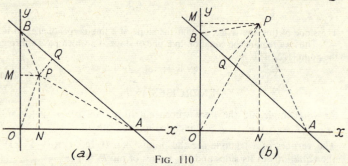

(a)   Fig. 110   (b)

Two cases, as shown in Figs. (a) and (b), arise according as P is or is not on the same side of the line AB as the origin. PN, PM are drawn perpendicular to the axes Ox, Oy respectively.

The equation to the line AB can be written in the form

$$\frac{x}{-C/A} + \frac{y}{-C/B} = 1,$$

showing that it makes intercepts on the coordinate axes given by

$$OA = -C/A, \quad OB = -C/B.$$

From the right-angled triangle OAB we have

$$AB = \sqrt{(OA^2 + OB^2)} = \sqrt{\{(-C/A)^2 + (-C/B)^2\}}$$
$$= C\sqrt{(A^2 + B^2)}/AB.$$

Since $PM = h$, $PN = k$, $PQ = p$,

$$\text{area} \triangle OAP = \tfrac{1}{2}OA \cdot PN = -\frac{Ck}{2A},$$

$$\text{area} \triangle BOP = \tfrac{1}{2}OB \cdot PM = -\frac{Ch}{2B},$$

$$\text{area} \triangle BPA = \tfrac{1}{2}AB \cdot PQ = \frac{Cp\sqrt{(A^2 + B^2)}}{2AB}$$

$$\text{area} \triangle OAB = \tfrac{1}{2}OA \cdot OB = \frac{C^2}{2AB}.$$

From Fig. 110(a),

$$\text{area} \triangle OAP + \text{area} \triangle BOP + \text{area} \triangle BPA = \text{area} \triangle OAB,$$

so that
$$-\frac{Ck}{2A} - \frac{Ch}{2B} + \frac{Cp\sqrt{(A^2 + B^2)}}{2AB} = \frac{C^2}{2AB}.$$

Solving for p we find

$$p = \frac{Ah + Bk + C}{\sqrt{(A^2 + B^2)}}.$$

From Fig. 110 (b), we should have

  area $\triangle OAP$ + area $\triangle BOP$ − area $\triangle$ area BPA = area $\triangle OAB$.

This would lead in a similar way to

$$p = -\frac{Ah + Bk + C}{\sqrt{(A^2 + B^2)}}.$$

To sum up, the perpendicular distance of the point $P(h, k)$ from the straight line $Ax + By + C = 0$ is given by

$$\pm \frac{Ah + Bk + C}{\sqrt{(A^2 + B^2)}}. \tag{15.15}$$

It is usual to quote only the magnitude of the distance irrespective of sign: if the positive root of $(A^2 + B^2)$ is always taken, a difference in sign in the distances of two points calculated from (15.15) without the alternative signs would indicate that the two points are on opposite sides of the given line. The formula is best remembered by observing that the numerator is obtained by writing the coordinates of the given point in the left-hand side of the equation to the line and the denominator is the square root of the sum of the squares of the coefficients of x and y in the equation.

Example 10. *Find the distance of the points* $(2, -1)$ *and* $(1, 1)$ *from the line* $3x + 4y = 6$.

Writing the line in the standard form $3x + 4y - 6 = 0$, the distance of the point $(2, -1)$ is given by

$$\frac{3(2) + 4(-1) - 6}{\sqrt{\{(3)^2 + (4)^2\}}} = -\frac{4}{5}.$$

The distance of the point $(1, 1)$ from the line is

$$\frac{3(1) + 4(1) - 6}{\sqrt{\{(3)^2 + (4)^2\}}} = \frac{1}{5}.$$

Thus the required distances would usually be quoted as 4/5 and 1/5 respectively: the calculated results show that the two points are on opposite sides of the line as can be seen from a plot of the given line and points.

15.7. The equations to the bisectors of the angles between two given lines

In Fig. 111 the given lines

$$A_1x + B_1y + C_1 = 0, \quad A_2x + B_2y + C_2 = 0$$

are shown as AB and CD, the point of intersection being R. If P is a point (x, y) on the bisector of the angle ARC, P will be equidistant from both lines and lie on opposite sides of the two lines from the origin. Hence from (15.15), the coordinates (x, y) of P will satisfy

$$\frac{A_1x + B_1y + C_1}{\sqrt{(A_1^2 + B_1^2)}} = -\frac{A_2x + B_2y + C_2}{\sqrt{(A_2^2 + B_2^2)}}.$$

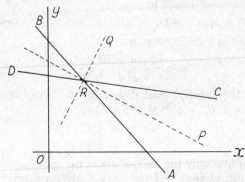

FIG. 111

If Q is the point (x, y) on the bisector of the angle ARD, Q will be equidistant from both lines and will lie on the same sides of the lines as the origin. Hence the coordinates (x, y) of Q will satisfy

$$\frac{A_1x + B_1y + C_1}{\sqrt{(A_1^2 + B_1^2)}} = \frac{A_2x + B_2y + C_2}{\sqrt{(A_2^2 + B_2^2)}}.$$

The equations to the bisectors of the angles between the two lines $A_1x + B_1y + C_1 = 0$, $A_2x + B_2y + C_2 = 0$ can therefore be written

$$\frac{A_1x + B_1y + C_1}{\sqrt{(A_1^2 + B_1^2)}} = \pm \frac{A_2x + B_2y + C_2}{\sqrt{(A_2^2 + B_2^2)}}. \qquad (15.16)$$

Example 11. *Write down the equations to the bisectors of the angles between the lines $3x + 4y = 12$ and $4x - 3y = 6$.*

Writing the equations to the lines in the standard forms

$$3x + 4y - 12 = 0 \quad \text{and} \quad 4x - 3y - 6 = 0,$$

the required equations to the bisectors are

$$\frac{3x + 4y - 12}{\sqrt{\{(3)^2 + (4)^2\}}} = \pm \frac{4x - 3y - 6}{\sqrt{\{(4)^2 + (-3)^2\}}}.$$

These reduce to $x - 7y + 6 = 0$ and $7x + y - 18 = 0$.

EXERCISES 15 (c)

1. (i) Find the distance of the point $(2, 3)$ from the line $5x - 12y + 39 = 0$.
    (ii) Find the distance of the point $(2, 1)$ from the line $3x + 2y = 8$ and explain the result.

2. Find the distance of the point (h, k) from the line $x \cos \alpha + y \sin \alpha = p$.
                                                                        (O.C.)

3. Show that the point $(1, 1)$ is equidistant from the lines $3x + 4y = 12$, $5x - 12y + 20 = 0, 4x - 3y = 6$.

By drawing a rough figure, decide whether the point is the centre of the inscribed or one of the escribed circles of the triangle formed by the above lines.   (O.C.)

4. Find the equation to the straight line which lies mid-way between the point $(2, -1)$ and the line $3x - 2y + 5 = 0$.   (O.C.)

5. Find the equation to the locus of a point which moves so that its perpendicular distance from the line $4x - 3y = 7$ is twice its perpendicular distance from the line $5x + 12y = 8$.

6. Find the equations to the bisectors of the angles between the lines $4x + 3y = 7$ and $24x + 7y = 31$.

7. Find the equations to the six bisectors of the angles between the lines $x + 7y = 3, 17x - 7y + 3 = 0, x - y + 1 = 0$ and show that three of them pass through the point $(1, 1)$.   (O.C.)

8. Q is the foot of the perpendicular from the point $P(h, k)$ on to a line RS whose equation is $Ax + By + C = 0$. Find the equations to the bisectors of the angles between the lines PQ and RS.

15.8. The equation to a straight line passing through the point of intersection of two given straight lines

Suppose the two given straight lines are represented by the equations $A_1x + B_1y + C_1 = 0$, $A_2x + B_2y + C_2 = 0$ and let P be the point of intersection of these two lines. Consider the equation

$$A_1x + B_1y + C_1 + \lambda(A_2x + B_2y + C_2) = 0, \qquad (15.17)$$

where λ is an arbitrary constant. The equation (15.17) represents some straight line for it is of the first degree in x and y. Further, the equation is satisfied by the coordinates of the common point of the two given lines since these coordinates simultaneously satisfy the equations $A_1x + B_1y + C_1 = 0$ and $A_2x + B_2y + C_2 = 0$. Hence equation (15.17) represents a straight line passing through the point of intersection of the given lines.

The method used above of deducing the equation to a line passing through the common point of two other lines is an example of a device of great use in coordinate geometry and should be studied with care. That there is an infinity of lines passing through the common point will be clear on geometrical grounds: it is also apparent from equation (15.17) in which the constant λ is quite arbitrary.

It will generally be required to find the equation to a line which, besides passing through the point of intersection of two given lines, satisfies a second condition such as passing through a second given point or having a given slope. This further condition will enable us to fix the value of λ and to pick out from the infinity of lines that one which fulfils all the required conditions. Some illustrative examples follow.

Example 12. *Find the equation to the line which passes through the point* (3, 2) *and the point of intersection of the lines* $2x + 3y - 1 = 0$ *and* $3x - 4y - 6 = 0$.

Any line through the point of intersection of the given lines is

$$2x + 3y - 1 + \lambda(3x - 4y - 6) = 0.$$

If this line passes through the point (3, 2),

$$2(3) + 3(2) - 1 + \lambda\{3(3) - 4(2) - 6\} = 0,$$

or, $5\lambda = 11$. Writing $\lambda = 11/5$ in the equation we have

$$2x + 3y - 1 + (11/5)(3x - 4y - 6) = 0,$$

or, $43x - 29y = 71$ as the required result.

Example 13. *Find the equation to the line through the intersection of the lines* $3x - 2y + 14 = 0$, $x + y = 6$ *and perpendicular to the line* $5x - 6y = 0$.

Writing the equation to the second given line in the standard form $x + y - 6 = 0$, the equation to any line through the intersection of the two given lines is

$$3x - 2y + 14 + \lambda(x + y - 6) = 0$$

or, $(3 + \lambda)x - (2 - \lambda)y + 14 - 6\lambda = 0.$

The slope of this line is $(3 + \lambda)/(2 - \lambda)$ and the slope of the line $5x - 6y = 0$ is $5/6$. Hence for the lines to be perpendicular

$$\frac{5}{6}\left(\frac{3 + \lambda}{2 - \lambda}\right) = -1,$$

leading to $\lambda = 27$. Inserting this value of λ, the required equation is

$$(3 + 27)x - (2 - 27)y + 14 - 6(27) = 0,$$

or, $30x + 25y = 148$.

15.9. The equation to a pair of straight lines

Consider the equation

$$3x^2 - 4xy + y^2 = 0. \tag{15.18}$$

This can be written

$$(x - y)(3x - y) = 0,$$

and the equation is satisfied by the coordinates of all points which make either $x - y = 0$ or $3x - y = 0$. In other words the equation is satisfied by the coordinates of points lying on either of the straight lines $y = x$ or $y = 3x$. Hence equation (15.18) represents two straight lines which pass through the origin and are inclined at angles of $45°$ and $\tan^{-1} 3$ respectively to the x-axis.

Consider now the more general equation

$$ax^2 + 2hxy + by^2 = 0. \tag{15.19}$$

Multiplying it by a, it can be written in the form

$$(a^2x^2 + 2ahxy + h^2y^2) - (h^2 - ab)y^2 = 0$$

or, $\{(ax + hy) + y\sqrt{(h^2 - ab)}\}\{(ax + hy) - y\sqrt{(h^2 - ab)}\} = 0.$

Thus the equation represents the two straight lines

$$ax + \{h + \sqrt{(h^2 - ab)}\}y = 0, \atop ax + \{h - \sqrt{(h^2 - ab)}\}y = 0,} \tag{15.20}$$

each of which passes through the origin: for equation (15.19) is satisfied by the coordinates of all the points which lie on the first of the lines given by (15.20) and also by those of all the points which lie on the second line.

It should be noted that the two lines given by (15.20) are real and different if $h^2 > ab$, they are real and coincident if $h^2 = ab$ and they are imaginary if $h^2 < ab$.

The angle between the two lines represented by equation (15.19) can be found as follows. Suppose the separate equations of the two lines are $y = m_1x$ and $y = m_2x$. Then equation (15.19) must be equivalent to

$$(y - m_1x)(y - m_2x) = 0$$

or,

$$m_1m_2x^2 - (m_1 + m_2)xy + y^2 = 0.$$

Comparing this with equation (15.19),

$$\frac{m_1m_2}{a} = -\frac{m_1 + m_2}{2h} = \frac{1}{b},$$

so that

$$m_1m_2 = a/b, \quad m_1 + m_2 = -2h/b.$$

By formula (15.9), the angle between the two lines is given by

$$\tan^{-1}\left(\frac{m_1 - m_2}{1 + m_1m_2}\right).$$

Since

$$(m_1 - m_2)^2 = (m_1 + m_2)^2 - 4m_1m_2$$
$$= (4h^2/b^2) - (4a/b)$$
$$= 4(h^2 - ab)/b^2,$$

the required angle is

$$\tan^{-1}\left\{\frac{2\sqrt{(h^2 - ab)}/b}{1 + (a/b)}\right\},$$

or,

$$\tan^{-1}\left\{\frac{2\sqrt{(h^2 - ab)}}{a + b}\right\}. \tag{15.21}$$

Since the tangent of an angle of 90° is infinite, formula (15.21) shows that the two straight lines given by the equation

$$ax^2 + 2hxy + by^2 = 0$$

are perpendicular if

$$a + b = 0. \tag{15.22}$$

The general equation of the second degree

$$ax^2 + 2hxy + by^2 + 2gx + 2fy + c = 0 \tag{15.23}$$

which contains terms in x, y and a constant as well as the second degree

terms of the left-hand side of equation (15.19) can, in certain circumstances, represent two straight lines. It will do so if the expression
on the left can be broken into two factors, each of the first degree.
It is, however, rather beyond the scope of the present book to obtain
the necessary relation between the constants a, b, c, f, g, h in equation
(15.23) for this equation to represent two straight lines and we restrict
the discussion to pairs of lines which pass through the origin.

Example 14. *Calculate the acute angle between the pair of straight lines represented
by the equation* $3x^2 - 10xy + 7y^2 = 0$. (L.U.)

Working from first principles, the equation can be written in the form

$$(x - y)(3x - 7y) = 0,$$

showing that the equation represents two lines through the origin of slopes 1
and 3/7. By (15.9), the angle between them is

$$\tan^{-1}\left\{\frac{1 - (3/7)}{1 + (3/7)}\right\} = \tan^{-1}\left(\frac{4}{10}\right) = 21° \, 48'.$$

Alternatively, since here $a = 3$, $2h = -10$, $b = 7$, formula (15.21) gives the
required angle as

$$\tan^{-1}\left\{\frac{2\sqrt{(25 - 21)}}{3 + 7}\right\} = \tan^{-1}\left(\frac{4}{10}\right) \text{ as before.}$$

EXERCISES 15 (d)

1. Two lines pass through the point of intersection of the lines $2x + 4y = 1$,
   $x + 3y = 2$. The first line passes through the point $(2, 1)$ and the second is
   parallel to the line $x - 2y + 3 = 0$. Find the equations to the two lines.

2. Find the equations to the two lines which pass through the point of intersection of the lines $x + y = 2$, $2x - 3y + 1 = 0$ and are at unit distance
   from the origin.

3. The sides AB, AC of a triangle are the lines $y = 0$ and $3x + y = 3$. B is the
   point $(5, 0)$ and C the point $(3, -6)$. Find

   (i) the equation to the line joining A to the middle point of BC,
   (ii) the equation to the line through A perpendicular to BC.

4. Find the equation to a line which cuts off equal intercepts from the coordinate
   axes and passes through the point of intersection of the lines $5x + y = 1$
   and $3x - 4y + 1 = 0$.

5. Find the equation to a straight line passing through the point of intersection
   of the lines $2x - y = 8$, $4x + y = 10$ and also through the point of intersection of $2x - 3y = 4$, $4x + y = 3$.

6. Find the angle between the two straight lines $2x^2 - 7xy + 3y^2 = 0$.

7. Show that the two straight lines $x^2 - 2xy \sec \theta + y^2 = 0$ make an angle θ
   with one another.

8. What two straight lines are represented by the equation

$$4x^2 - 24xy + 11y^2 = 0?$$

Show that the equation to the bisectors of the angles between these two lines can be written $12x^2 - 7xy - 12y^2 = 0$.

15.10. The determination of linear laws from experimental data

Relations between two varying quantities, such as the pressure and temperature of a given mass of air, can be shown graphically by plotting the results of experiments on squared paper. It is often possible to deduce formulae connecting the variable quantities from such results and formulae so derived are said to be *empirical*, this term meaning that the formulae depend on experiment.

If the points plotted from the experimental results lie approximately along a straight line, it is known from § 15.1 that the relation between the variables involved (say x and y) will be *linear*, that is, of the first degree. In such cases the relation can be assumed to be of the form

$$y = mx + c \qquad (15.24)$$

where m and c are constants. Further, the values of m and c can be deduced from the slope of the line and the length of the intercept made by it on the y-axis, or by substituting in equation (15.24) the coordinates (as read off from the graph) of two points on the line. The details are shown in Example 15 below.

Example 15. *In a certain experiment carried out at constant pressure, the volumes v cm^3 of gas at temperatures $\theta°C$ are given by the following table:—*

θ	10	21	33	50	65
v	103·7	107·7	112·0	118·4	123·9

Show that the approximate relation between v and θ is $v = 100 + 0.367\theta$ and find the volume of gas when the temperature is $30°C$.

The graph (Fig. 112) shows a plot of the given values of temperature and volume. This plot should be on as large a scale as possible and, if such a graph is looked at carefully, it will be found that the points do not lie exactly on a straight line but that it is possible to draw a line from which none of the points deviates greatly. In drawing the line, care should be taken to ensure that the plotted points are evenly distributed about it, some being above and some below. This can be done conveniently by moving a stretched thread or a transparent scale with a fine line etched on its lower surface and taking care that the points do not deviate systematically from the thread or line. In this example the points do lie very close to a straight line and we can assume that v and θ obey a law of the form $v = m\theta + c$. Reading from the graph, it can be seen that a rise in temperature of $70°C$ is accompanied by a rise of $125·7 - 100 = 25·7$ cm^3 in volume; the slope of the line is therefore given by $m = 25·7/70 = 0·367$. It can also be seen that the intercept made by the line on the v-axis is 100 so that $c = 100$ and the relation between v and θ is $v = 0·367\theta + 100$. Alternatively, reading off the coordinates of two points A and B (chosen to be well spaced to minimise errors in reading from the graph) we find that A is the point $(0, 100)$ and B is $(70, 125·7)$. Substituting these values in the equation $v = m\theta + c$, we have

$$100 = c, \quad 125·7 = 70\theta + c$$

and this pair of equations gives $m = 0.367$, $c = 100$. Finally, substituting $\theta = 30$ in the relation found

$$v = 100 + 0.367 \times 30 = 111 \text{ cm}^3.$$

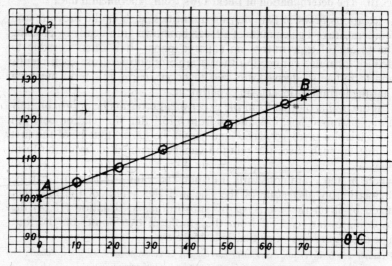

FIG. 112

It is worth noticing that if a law of the form $y = mx^n + c$ (where *the value of the index n is known*) is thought to be true, this can be checked by seeing if a plot of y against x^n yields a straight line. If it does, the values of the constants m and c can be deduced as before. An example in which the value of n is not known is given in Example 17.

Example 16. *Measurements of the coefficient of friction μ of steel tyres on a rail at speeds of V gave the following results:—*

V	30	38	49	55	60
μ	0·066	0·051	0·041	0·036	0·033

Show that these results follow approximately a law of the form $V = (m/\mu) + c$ and find the values of m and c.

We are here seeking to establish a linear relation between V and the reciprocal of μ so we first form the following table of $1/\mu$ and the corresponding values of V

$1/\mu$	15·0	19·6	24·4	27·8	30·0
V	30	38	49	55	60

and these values are plotted in Fig. 113.

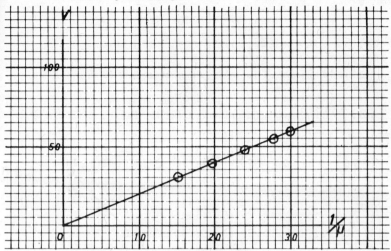

FIG. 113

Since the graph is a straight line, the law relating V and $1/\mu$ is of the form $V = (m/\mu) + c$ where m is the slope of the line and c is the intercept made on the V-axis. The latter is clearly zero and the former is $60/30 = 2$.

Measurements of two related sets of physical quantities frequently obey a "power" law of the form $y = Cx^n$ where the multiplier C and the index n are constants. This relation can be written

$$\log y = \log C + n \log x$$

and the linear relationship between $\log y$ and $\log x$ shows that the graph of these quantities is a straight line of slope n which makes an intercept of length $\log C$ on the axis of $\log y$. As it is possible, by using a stretched string or a transparent scale, to decide if a set of points do or do not lie along a straight line, the graph of $\log y$ against $\log x$ will show if the given values of the measurements x and y do or do not obey a power law of the form $y = Cx^n$. If they do, the values of n and $\log C$ are easily deduced from the graph and an example is given below.

Example 17. *The following table shows some values of θ obtained experimentally for the given values of t:—*

t	5	9	16	20	25
θ	0·45	0·60	0·80	0·89	1·00

By plotting $\log \theta$ against $\log t$ show that, allowing for small errors of observation, there is probably a relation between θ and t of the form $\theta = Ct^n$ and find approximate values for C and n.   (L.U.)

Tabulating $\log \theta$ against $\log t$ from the data

$\log t$	0·699	0·954	1·204	1·301	1·398
$\log \theta$	−0·347	−0·222	−0·097	−0·051	0·000

it should be remembered that the logarithms of numbers less than unity are negative so that

$$\log 0·45 = \bar{1}·653 = -1 + 0·653 = -0·347, \text{ etc.,}$$

and a graph of these quantities is shown in Fig. 114. This is a straight line so

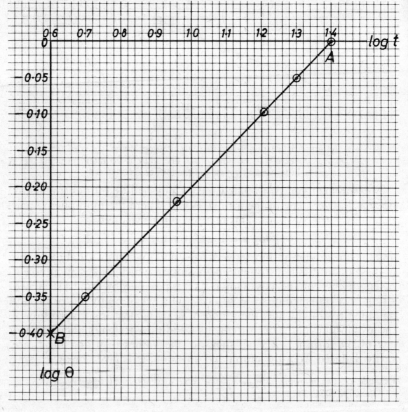

Fig. 114

that $\log \theta = \log C + n \log t$ and $\theta = Ct^n$. Substituting the coordinates $(1·398, 0)$, $(0·6, -0·399)$ of the points A and B we have

$$0 = \log C + 1·398n, \quad -0·399 = \log C + 0·6n.$$

These equations give $n = 0·5$ and $\log C = -0·699 = \bar{1}·301$ so that $n = \frac{1}{2}$ and $C = 0·2$.

EXERCISES 15 (e)

1. In a certain experiment, the recorded values of the load P lifted by an effort E were given by:—

P	100	200	300	400	500
E	5·60	8·65	11·85	14·88	18·00

By plotting these results, show that E and P are approximately related by an equation of the form $E = mP + c$ and find the values of the constants m and c.

2. Plot the points whose coordinates are given by:—

x	0	1	2	3	4	5	6
y	$-10·0$	$-7·7$	$-5·4$	$-3·1$	$-0·8$	1·5	3·8

Show that the relation between x and y is of the form $y = mx + c$ and deduce values for the constants m and c.

3. Values of the velocity v (metres per second) of a particle at time t seconds are given below:—

t	0·1	2·2	7·1	10·4	12·3
v	100·3	106·6	121·2	131·3	136·9

Show that the law relating v and t is of the form $v = u + at$ and determine values for the constants a and u. What was the initial velocity of the particle?

4. Corresponding values of V and r are given by:—

r	0·0	1·0	1·6	2·5	3·2
V	$-2·00$	$-1·50$	$-0·72$	1·13	3·12

Show that the relation between V and r is of the form $V + \alpha = \beta r^2$ and find values for the constants α and β.

5. Numerical values of two variables θ and t are given by:—

t	0·5	0·8	1·0	1·4	2·0
θ	4·08	4·21	4·33	4·65	5·33

By plotting θ against t^2, show that θ and t are approximately related by $\theta - c = \alpha t^2$ and find numerical values of α and c.

6. It is thought that two variables p and v are related by the formula

$$p = \frac{c}{v - b}$$

where b and c are constants. Show that the truth of this can be decided by testing if either the graph of v against $1/p$ or that of pv against p is a straight line.

7. The following table shows the values of a quantity p obtained experimentally for the given values of a quantity v:—

v	2	3	4	5
p	3·54	1·92	1·25	0·89

Show that the relation between p and v is of the type $p = Cv^n$ where C and n are constants; determine approximate values for C and n.

8. The coefficient of friction μ in a bearing running at a speed of V metres per second is given by a law of the form $\mu = CV^n$. Determine the constants C and n by using the *mean* results of each of two experiments in which corresponding measurements were:—

	Experiment 1		Experiment 2	
V	105	157	419	471
μ	0·0018	0·0021	0·0036	0·0040

9. The luminosity I of an electric lamp is thought to be related to the voltage V of the current in it by the law $I = kV^n$. Test if the following measurements obey this law and, if they do, find k and n.

V	20	35	50	70	100	120
I	0·08	0·75	3·13	12·01	50·00	103·70

10. The values of two variables y and z are given in terms of a third variable x by the table

x	$\frac{1}{2}$	$\frac{4}{3}$	$\frac{5}{2}$	6
y	1·061	1·732	2·371	3·673
x	1·250	1·667	2·250	4·000

Verify that relations of the form $y = Ax^n$ and $z = mx + c$ apply and find the values of A, n, m and c. Use your results to find two values of x for which $y = z$.

EXERCISES 15 (f)

1. Find the equation to a straight line which passes through the point $(3, 3)$ and forms with the coordinate axes a triangle in the first quadrant of area 18 units.

2. $A(4, 7)$ and $B(-2, -1)$ are two vertices of an isosceles triangle ABC having a right angle at B. Find the length and equation of AB: hence write down the equation to BC and find the coordinates of the two positions of C.

(L.U.)

3. Find the equation to the line joining the two points $(1, -5/3)$, $(5/4, -2)$

and the equations to the two lines parallel to it and distant 2 units from it.

(L.U.)

4. Find the equations to the two straight lines which are parallel to the line $4x + 3y + 1 = 0$ and at a distance 2 from it, and also the areas of the triangles which these two lines respectively make with the coordinate axes. (Q.E.)

5. The vertices of a triangle are $A(3, 3)$, $B(-1, -1)$ and $C(-1, 5)$. Find the angle made by the median from B with the side AC. (L.U.)

6. Find the equation to a line at right angles to the line $(x/a) - (y/b) = 1$ through the point where it meets the x-axis.

7. Find the coordinates of the foot of the perpendicular from the point $(1, 0)$ upon the line $y = mx + (1/m)$.

8. Write down the equation to the straight line passing through the two points $(3, 2)$, $(6, 6)$.

Find where this line is cut by each of the two straight lines drawn through the point $(1, 4)$ with slopes of 0.5 and -1. What is the area of the triangle enclosed by these three straight lines? (Q.E.)

9. ABC is an isosceles triangle in which $AB = AC$; the equations to AB and BC are respectively $2x - y = 1$ and $x - 2y + 1 = 0$. Prove that AC is parallel to the line $2x + 11y = 0$. (O.C.)

10. A and B are the points $(2, 0)$ and $(4, 0)$ respectively: P is a variable point whose ordinate is always positive, such that the angle APB is $45°$. Show that P lies on the curve $x^2 + y^2 - 6x - 2y + 8 = 0$. (O.C.)

11. Find the equations to two straight lines which make angles of $45°$ with the straight line $4x + 3y = 21$ and which pass through the point $(1, -3)$.

(L.U.)

12. Find the coordinates of the foot of the perpendicular from the point $(5, 7)$ on the line which joins the points $(6, -1)$, $(1, 6)$. (O.C.)

13. Find the equations to the lines through the point $(2, 0)$ perpendicular to the lines $x - 3y + 8 = 0$, $x - 4y + 15 = 0$. Find also the equation to the line joining the feet of these perpendiculars. (O.C.)

14. Find the equations to the perpendiculars from the origin to the lines $x + y = 4$, $y - 2x = 5$. Find also the coordinates of the corners of the quadrilateral so formed and its area. (O.C.)

15. The line through $P(1, 11)$ perpendicular to the line joining $A(2, 3)$ and $B(-4, 6)$ meets AB at Q. Find the equations to AB and PQ. PQ is produced to R so that $PQ = QR$. Find the coordinates of R and prove that the area of the quadrilateral $APBR$ is 45. (L.U.)

16. A triangle is formed by the three straight lines

$$y = m_1 x + \frac{a}{m_1}, \quad y = m_2 x + \frac{a}{m_2}, \quad y = m_3 x + \frac{a}{m_3}.$$

Prove that its orthocentre always lies on the line $x + a = 0$. (L.U.)

17. Find the equation to the straight line which is such that the x-axis bisects the angle between it and the straight line $2x + 5y = 18$. (O.C.)

18. Prove that all points on the line $11x - 3y + 11 = 0$ are equidistant from the lines $12x + 5y + 12 = 0$ and $3x - 4y + 3 = 0$.   (O.C.)

19. Prove that the straight lines which join the point $(-2, 3)$ to the points $(6, 7)$ and $(0, -1)$ are perpendicular. Calculate the coordinates of the fourth vertex of the rectangle which has these points as three of its vertices.

20. Two perpendicular lines are drawn through the origin so as to form an isosceles right-angled triangle with the line $lx + my + n = 0$. Show that their equations are $(l - m)x + (l + m)y = 0$ and $(l + m)x + (m - l)y = 0$.
                                                                       (O.C.)

21. Find the equation to a line perpendicular to the line $3x + 4y + 5 = 0$ which passes through the point of intersection of the two lines $3x - y = 1$ and $x + y = 3$.

22. Find the equations to the diagonals of a parallelogram whose sides are the lines $3x - 2y = 1$, $4x - 5y = 6$, $3x - 2y = 2$, $4x - 5y = 3$.

23. Show that the angle between the lines $6x^2 - xy - y^2 = 0$ is $45°$.

24. Find the equation to the bisectors of the angles between the lines $3x^2 + 4xy - 5y^2 = 0$.

25. Express in a single equation the pair of perpendicular straight lines through the origin, one of which is the line $ax + by = 0$.

26. The income tax code numbers for allowances between £150 and £315 are shown in the following table:—

Code number (C)	60	70	80	90	100	110
Allowance (£A)	160	190	220	250	280	310

By using squared paper, verify that the law connecting C and A is
$$\frac{A - 160}{C - 60} = 3.$$
Express A in terms of C and calculate A when $C = 97$. Also express C in terms of A and calculate C when $A = 226$.   (L.U.)

27. The resistance R kg wt to the motion of a train is given in the table for various values of the speed V metres per second:—

R	66·8	70·6	72·4	74·3	78·5
V	20	25	27	29	33

By graphing R against V^2, show that these values are consistent with the formula $R = A + BV^2$ and find the values of A and B.   (L.U.)

28. Show that the values

v	1·42	1·51	1·63	1·78	1·90	2·00
P	4·28	4·70	5·26	6·01	6·62	7·15 (5)

are approximately related by the law $P^2 = Av^n$ and find values for the constants A and n.

29. Quantities x and y are connected by the relation $y = \log(a + bx)$ where a and b are constants. Plot 10^y against x from the table of values:—

x	1	2	3	4	5	6
y	0·857	0·924	0·982	1·033	1·079	1·121

and hence find estimates for a and b. (L.U.)

30. When a quantity of air is compressed to pressure p, the corresponding temperature t (degrees Centigrade) is given by:—

p	10	20	40	70	80
t	146	243	362	478	509

Verify that the law $T = kp^n$, where T is the absolute temperature of the air, is satisfied by these data and determine the values of the constants k and n.

CHAPTER 16

THE COORDINATE GEOMETRY OF THE CIRCLE

16.1. The equation to a circle with given centre and radius

Suppose that the radius of the circle is R and that its centre is the point $C(a, b)$. The circle is the locus of a point which moves so that its distance from the point (a, b) is always equal to R. Hence if P is any point (x, y) on the circle,

$$(x - a)^2 + (y - b)^2 = R^2, \qquad (16.1)$$

the left-hand side of this equation being the square of the distance between the points (x, y) and (a, b).

Equation (16.1) therefore represents a circle of radius R and centre the point (a, b). By writing $a = b = 0$, we find that the equation to a circle of radius R and centre the origin is

$$x^2 + y^2 = R^2. \qquad (16.2)$$

Example 1. *Write down the equation to the circle (i) with centre the origin and radius 2 units, (ii) with centre the point $(3, -2)$ and radius 3 units.*

(i) The required equation is $x^2 + y^2 = 2^2$, or $x^2 + y^2 = 4$.

(ii) From (16.1), the equation is $(x - 3)^2 + (y + 2)^2 = 3^2$, which can be written in the form $x^2 + y^2 - 6x + 4y + 4 = 0$.

16.2. The general equation to a circle

The general equation of the second degree in x and y is

$$ax^2 + 2hxy + by^2 + 2gx + 2fy + c = 0, \qquad (16.3)$$

where a, b, c, f, g, h are constants. From the way in which the equation (16.1) to a circle was formed it should be clear that *the coefficients of x^2 and y^2 in its equation must be equal and there must be no term containing the product xy.* In other words, equation (16.3) will represent a circle if $a = b$ and $h = 0$. There is no loss of generality in taking a to be unity and the general equation to a circle can be written

$$x^2 + y^2 + 2gx + 2fy + c = 0. \qquad (16.4)$$

The radius and coordinates of the centre of the circle given by the general equation (16.4) can be found as follows. Writing it in the form

$$(x + g)^2 + (y + f)^2 = g^2 + f^2 - c,$$

and comparing it with equation (16.1), which represents a circle of radius R and centre the point (a, b), we see that the radius of the circle given by equation (16.4) is

$$\sqrt{(g^2 + f^2 - c)}, \qquad (16.5)$$

and that its centre is the point

$$(-g, -f). \tag{16.6}$$

Example 2. *Find the radius and the coordinates of the centre of the circle*
$$x^2 + y^2 + 5x - 6y = 5.$$

Working from first principles, the equation can be written in the form

$$(x + \tfrac{5}{2})^2 + (y - 3)^2 = 5 + \tfrac{25}{4} + 9$$
$$= \tfrac{81}{4} = (\tfrac{9}{2})^2,$$

showing that the point (x, y) is always at a distance of 9/2 units from the point $(-5/2, 3)$. Hence the radius is 9/2 and the centre is the point $(-5/2, 3)$.

Alternatively, in the given equation $g = 5/2$, $f = -3$, $c = -5$ and (16.5), (16.6) give

$$\text{radius} = \sqrt{\{(5/2)^2 + (-3)^2 - (-5)\}} = 9/2,$$
$$\text{centre, } (-5/2, -(-3)) \quad \text{or} \quad (-5/2, 3).$$

16.3. The equation to the circle whose diameter is the join of the points (x_1, y_1) and (x_2, y_2)

In Fig. 115, A and B are the points (x_1, y_1), (x_2, y_2) respectively and P is any point (x, y) on the circle. Since AB is a diameter, the angle

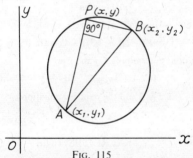

FIG. 115

APB is a right angle. The slopes of AP, PB are respectively

$$\frac{y - y_1}{x - x_1} \quad \text{and} \quad \frac{y - y_2}{x - x_2}.$$

Since AP, PB are perpendicular, the product of their slopes is -1, so that

$$\left(\frac{y - y_1}{x - x_1}\right)\left(\frac{y - y_2}{x - x_2}\right) = -1,$$

or, $$(x - x_1)(x - x_2) + (y - y_1)(y - y_2) = 0. \tag{16.7}$$

This relation is satisfied by the coordinates (x, y) of *any* point P on the circle and is therefore the required equation to the circle whose diameter is the join of the points (x_1, y_1), (x_2, y_2).

16.4. The equation to a circle through three given points

Suppose that the three given points are (x_1, y_1), (x_2, y_2), (x_3, y_3). Let the equation to the circle through these points be

$$x^2 + y^2 + 2gx + 2fy + c = 0.$$

Then if the point (x_1, y_1) lies on the circle, these coordinates must satisfy the equation and hence

$$x_1{}^2 + y_1{}^2 + 2gx_1 + 2fy_1 + c = 0.$$

Similarly $\quad x_2{}^2 + y_2{}^2 + 2gx_2 + 2fy_2 + c = 0,$

and $\quad\quad x_3{}^2 + y_3{}^2 + 2gx_3 + 2fy_3 + c = 0.$

These three equations are sufficient to enable the constants f, g and c, and hence the equation to the circle, to be determined.

Example 3. *Find the equation to the circle which passes through the origin and the points* $(2, 0)$, $(3, -1)$.   (O.C.)

Let the required equation be

$$x^2 + y^2 + 2gx + 2fy + c = 0.$$

If the origin $(0, 0)$ lies on the circle, $c = 0$.

If the point $(2, 0)$ is on the circle

$$(2)^2 + 2(2)g + c = 0,$$

and if the point $(3, -1)$ is also on the circle,

$$(3)^2 + (-1)^2 + 2(3)g + 2(-1)f + c = 0.$$

Hence the equations determining f, g and c are

$$c = 0, \quad 4 + 4g = 0, \quad 10 + 6g - 2f = 0.$$

Thus $c = 0$, $g = -1$, $f = 2$ and the required equation is

$$x^2 + y^2 - 2x + 4y = 0.$$

EXERCISES 16 (a)

1. Find the coordinates of the centre and radius of the circle
$$x^2 + y^2 - 10x + 12y = 0.$$ (O.C.)

2. Write down the equation to the circle which:—

    (i) has its centre at the point $(-5, -6)$ and whose radius is 10,
    (ii) has its centre at the point $(a, -b)$ and whose radius is $\sqrt{(a^2 - b^2)}$.

3. Find the equation to the circle which has the points $(0, -1)$ and $(2, 3)$ as ends of a diameter.   (L.U.)

4. Find the equation to the circle which passes through the points $(0, 0)$, $(3, 1)$ and $(3, 9)$.   (O.C.)

5. Find the equation to the circle which passes through the points $(5, 0)$, $(6, 0)$ and $(8, 6)$.   (O.C.)

6. Find the equation to the diameter of the circle $x^2 + y^2 - 6x + 2y = 15$, which, when produced, passes through the point $(8, -2)$.   (O.C.)

7. Find the equation to the circle whose centre lies on the line $y = 3x - 7$ and which passes through the points $(1, 1)$ and $(2, -1)$. (L.U.)

8. If O is the origin and P, Q are the intersections of the circle $x^2 + y^2 + 4x + 2y - 20 = 0$ and the straight line $x - 7y + 20 = 0$, show that OP and OQ are perpendicular. Find the equation to the circle through O, P and Q. (L.U.)

16.5. The equation to the tangent to a circle at a given point

Suppose that we require the equation to the tangent at the point (x_1, y_1) to the circle

$$x^2 + y^2 + 2gx + 2fy + c = 0. \qquad (16.8)$$

Differentiating the equation with respect to x

$$2x + 2y\frac{dy}{dx} + 2g + 2f\frac{dy}{dx} = 0,$$

so that the gradient of the circle at the point (x_1, y_1) is given by

$$\left(\frac{dy}{dx}\right)_{x=x_1} = -\frac{x_1 + g}{y_1 + f}.$$

The tangent is the line through the point (x_1, y_1) with slope equal to the gradient of the curve; its equation is

$$y - y_1 = -\left(\frac{x_1 + g}{y_1 + f}\right)(x - x_1).$$

This can be written

$$xx_1 + yy_1 + g(x - x_1) + f(y - y_1) = x_1^2 + y_1^2. \qquad (16.9)$$

Since the point (x_1, y_1) lies on the circle

$$x_1^2 + y_1^2 + 2gx_1 + 2fy_1 + c = 0,$$

and we can replace $x_1^2 + y_1^2$ by $-(2gx_1 + 2fy_1 + c)$. Hence equation (16.9) can be written

$$xx_1 + yy_1 + g(x + x_1) + f(y + y_1) + c = 0. \qquad (16.10)$$

This is the required equation to the tangent at the point (x_1, y_1). *It can be obtained from the equation* (16.8) *to the circle by replacing* x^2 *by* xx_1, y^2 *by* yy_1, $2x$ *by* $(x + x_1)$ *and* $2y$ *by* $(y + y_1)$. This is a particular case of a general rule which enables the equation to the tangent at (x_1, y_1) to be written down at sight for any of the curves in this and the next chapter.

In particular, the equation to the tangent at (x_1, y_1) to the circle, centre the origin and radius R, i.e. the circle $x^2 + y^2 = R^2$, is

$$xx_1 + yy_1 = R^2. \qquad (16.11)$$

Example 4. *Write down the equation to the tangent* (i) *at the point* $(1, -7)$ *to the circle* $x^2 + y^2 = 50$, (ii) *at the point* $(2, 1)$ *to the circle*

$$4x^2 + 4y^2 - x + 5y = 23.$$

(i) The required equation is

$$x(1) + y(-7) = 50, \quad \text{or} \quad x - 7y = 50.$$

(ii) Here the required equation is

$$4x(2) + 4y(1) - \tfrac{1}{2}(x + 2) + \tfrac{5}{2}(y + 1) = 23,$$

or, $15x + 13y = 43$.

16.6. The points of intersection of a straight line and circle

Here we consider the points of intersection of the straight line $y = mx + c$ and the circle $x^2 + y^2 = R^2$. The coordinates of the points of intersection of the line and circle will satisfy the simultaneous equations

$$y = mx + c,$$
$$x^2 + y^2 = R^2.$$

Hence writing $y = mx + c$ in the equation to the circle, the abscissae of the points of intersection will be given by

$$x^2 + (mx + c)^2 = R^2,$$

or, $$(1 + m^2)x^2 + 2mcx + c^2 - R^2 = 0. \qquad (16.12)$$

This quadratic equation has real, equal or imaginary roots according as

$$(2mc)^2 - 4(1 + m^2)(c^2 - R^2),$$

is positive, zero or negative; i.e., according as c^2 is less than, equal to, or greater than $R^2(1 + m^2)$.

The three possibilities are shown in Fig. 116. The three lines (a), (b) and (c) all have slope m and make successively greater intercepts on the y-axis. The line (a) corresponds to a value of c^2 which is less

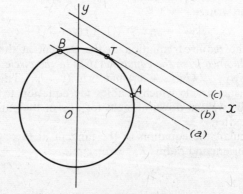

Fig. 116

than $R^2(1 + m^2)$ and it meets the circle in two real points A, B. The line (c), corresponding to a value of c^2 which is greater than $R^2(1 + m^2)$, does not meet the circle at all, or rather, it meets the circle in two imaginary points. The line (b), corresponding to a value of c^2 of $R^2(1 + m^2)$, meets the circle in two coincident points T; it is the tangent to the circle at T.

The equation to the line (b) for which $c^2 = R^2(1 + m^2)$ can be written

$$y = mx \pm R\sqrt{(1 + m^2)}. \qquad (16.13)$$

These lines always touch the circle $x^2 + y^2 = R^2$. One of them is shown as line (b) of Fig. 116: the other is the tangent to the circle whose point of contact is diametrically opposite the point T.

The length of the chord intercepted by the circle on the line (a) of Fig. 116 can be found as follows. The abscissae of the end-points of the chord are given by the roots of equation (16.12). If the roots of this equation are denoted by x_1 and x_2,

$$x_1 + x_2 = \frac{-2mc}{1 + m^2}, \quad x_1 x_2 = \frac{c^2 - R^2}{1 + m^2}.$$

Hence

$$(x_1 - x_2)^2 = (x_1 + x_2)^2 - 4x_1 x_2$$
$$= \frac{4}{(1 + m^2)^2}\{m^2 c^2 - (c^2 - R^2)(1 + m^2)\}$$
$$= \frac{4}{(1 + m^2)^2}\{R^2(1 + m^2) - c^2\}.$$

If y_1, y_2 are the ordinates of the end-points we have, since the points lie on the line $y = mx + c$,

$$y_1 - y_2 = mx_1 + c - (mx_2 + c) = m(x_1 - x_2).$$

The square of the length of the chord

$$= (x_1 - x_2)^2 + (y_1 - y_2)^2$$
$$= (1 + m^2)(x_1 - x_2)^2$$
$$= \frac{4}{1 + m^2}\{R^2(1 + m^2) - c^2\}.$$

Example 5. *A circle, the coordinates of whose centre are both positive, touches both axes of coordinates. If it also touches the line $3x - 4y + 6 = 0$, find its equation and the coordinates of its point of contact with this line.* (L.U.)

Let the required equation to the circle be $x^2 + y^2 + 2gx + 2fy + c = 0$. The circle meets the x-axis (the line $y = 0$), where

$$x^2 + 2gx + c = 0,$$

and touches it if this equation has equal roots. This requires that $g^2 = c$. Similarly the circle touches the y-axis if $f^2 = c$. Hence $c = g^2 = f^2$, and the equation to the circle can be written

$$x^2 + y^2 + 2gx + 2gy + g^2 = 0.$$

To find where the line $3x - 4y + 6 = 0$ meets this circle, we write $y = (3x + 6)/4$ in its equation, giving

$$x^2 + \frac{(3x + 6)^2}{16} + 2gx + \frac{2g(3x + 6)}{4} + g^2 = 0$$

or, $25x^2 + 4(9 + 14g)x + 4(9 + 12g + 4g^2) = 0.$ (16.14)

The line touches the circle if this equation has equal roots. The condition for this is

$$16(9 + 14g)^2 = 4.(25).(4).(9 + 12g + 4g^2).$$

This reduces to $2g^2 - g - 3 = 0$ giving $g = -1$ or $3/2$. The positive value is excluded because g must be negative if the centre $(-g, -f)$ is to lie in the first quadrant. Hence the required equation to the circle is obtained by using $g = -1$ and hence is

$$x^2 + y^2 - 2x - 2y + 1 = 0.$$

With $g = -1$, the equation (16.14) giving the abscissae of the points of intersection of line and circle reduces to $25x^2 - 20x + 4 = 0$. This has two roots each equal to $2/5$ (we should expect equal roots for the line is a tangent). Hence the abscissa of the point of contact is $2/5$ and, since $y = (3x + 6)/4$, the ordinate is $9/5$.

Example 6. *A circle, which passes through the origin, cuts off intercepts of lengths 4 and 6 units on the positive x- and y-axes respectively. Find the equation to the circle, and the equations to the tangents to the circle at the points (other than the origin) where it cuts the axes.* (L.U.)

Let the equation to the circle be $x^2 + y^2 + 2gx + 2fy + c = 0$. Since the circle passes through the origin, its equation is satisfied by $x = 0$, $y = 0$ and hence $c = 0$. The x-axis (the line $y = 0$) meets the curve where

$$x^2 + 2gx = 0,$$

i.e., where $x = 0$ (the origin) and where $x = -2g$. The second point of intersection is at 4 units from the origin, so that $-2g = 4$ giving $g = -2$. Similarly the y-axis (the line $x = 0$) meets the circle where

$$y^2 + 2fy = 0,$$

and this equation must have roots zero (the origin) and 6. Hence $2f = -6$ giving $f = -3$. Hence the required equation to the circle is $x^2 + y^2 - 4x - 6y = 0$.

The equation to the tangent at the point $(4, 0)$ is

$$x(4) + y(0) - 2(x + 4) - 3(y + 0) = 0,$$

or, $2x - 3y = 8$: the tangent at the point $(0, 6)$ is

$$x(0) + y(6) - 2(x + 0) - 3(y + 6) = 0,$$

or, $2x - 3y + 18 = 0$.

EXERCISES 16 (b)

1. Write down the equation to the tangent to the circle
$$x^2 + y^2 - 4x - 6y + 3 = 0$$
at the point $(5, 4)$ and that to the tangent to
$$x^2 + y^2 - 2x - 3y + 3 = 0$$
at the point $(1, 2)$.

2. (i) Write down the equation to the tangent at the origin to the circle
    $x^2 + y^2 + 2gx + 2fy = 0$.

    (ii) Find the equation to the tangent at the point $(3, 2)$ to the circle
    $(x - 1)^2 + (y + 2)^2 = 20$.

3. Find the equations to those tangents to the circle $x^2 + y^2 = 25$ which are
    parallel to the line $4x - 3y = 0$.

4. Show that the line $2x - 3y + 26 = 0$ is a tangent to the circle
    $x^2 + y^2 - 4x + 6y - 104 = 0$ and find the equation to the diameter
    through the point of contact.   (O.C.)

5. Find the equation to the circle which has its centre at the point $(2, -1)$
    and touches the line $3x + y = 0$.   (O.C.)

6. Tangents are drawn to the circle $x^2 + y^2 - 6x - 4y + 9 = 0$ from the
    origin. If θ is the angle between them, find the value of $\tan \theta$.   (L.U.)

7. Show that $y = mx$ is a tangent to the circle
$$x^2 + y^2 + 2gx + 2fy + c = 0,$$
    if $(g + mf)^2 = c(1 + m^2)$.
        Find the equations to the tangents from the origin to the circle
    $x^2 + y^2 - 6x - 3y + 9 = 0$ and the coordinates of their points of contact.
                                                                    (L.U.)

8. Find the length of the chord joining the points in which the straight line
    $(x/a) + (y/b) = 1$ meets the circle $x^2 + y^2 = R^2$.

9. A circle touches the x-axis and cuts off a constant length $2a$ from the y-axis.
    Show that the equation to the locus of its centre is the curve $y^2 - x^2 = a^2$.

10. The point (a, b) is the middle point of a chord of the circle $x^2 + y^2 = R^2$.
    Show that the equation to the chord is
$$ax + by = a^2 + b^2.$$

16.7. The length of the tangent from a given external point to a circle

In Fig. 117, T is a given point with coordinates (a, b) and TQ is one
of the tangents from T to the circle $x^2 + y^2 + 2gx + 2fy + c = 0$,

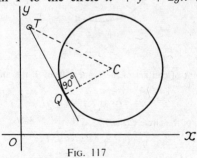

FIG. 117

its point of contact being Q. If C is the centre of the circle, TQ is
perpendicular to the radius CQ and hence
$$TQ^2 = TC^2 - CQ^2. \tag{16.15}$$

Now T is the point (a, b), and from (16.6), C is the point $(-g, -f)$, so that

$$TC^2 = (a + g)^2 + (b + f)^2.$$

Also, from (16.5), the radius CQ of the circle is given by

$$CQ^2 = g^2 + f^2 - c.$$

Substitution in (16.15) then gives for the square of the length of the tangent

$$TQ^2 = (a + g)^2 + (b + f)^2 - g^2 - f^2 + c$$
$$= a^2 + b^2 + 2ga + 2fb + c. \qquad (16.16)$$

Thus *the square of the length of the tangent drawn to the circle from the point (a, b) is obtained by writing a for x and b for y in the left-hand side of the equation to the circle.*

Example 7. *Find the length of the tangent from the point $(5, -1)$ to the circle $(x - \frac{1}{2})^2 + y^2 = \frac{25}{4}$.*

The given circle is one of radius $5/2$ with centre C at the point $(\frac{1}{2}, 0)$. If T is the point $(5, -1)$, see Fig. 118,

$$CT^2 = (5 - \frac{1}{2})^2 + (-1 - 0)^2 = \frac{81}{4} + 1 = \frac{85}{4}.$$

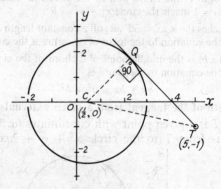

Fig. 118

If Q is the point of contact of the tangent from T, the angle CQT is $90°$ and hence

$$TQ^2 = CT^2 - CQ^2.$$

But the radius of the circle CQ is $5/2$, so that

$$TQ^2 = \frac{85}{4} - \frac{25}{4} = 15,$$

giving $TQ = \sqrt{15}$ units.

Alternatively, the equation to the circle written in standard form is $x^2 + y^2 - x - 6 = 0$. Hence (16.16) gives

$$TQ^2 = (5)^2 + (-1)^2 - (5) - 6 = 15.$$

16.8. Orthogonal circles

Two circles are said to be *orthogonal* when the tangents at their points of intersection are at right angles.

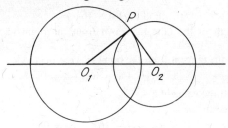

FIG. 119

In Fig. 119 one of the points of intersection of two orthogonal circles is P and O_1, O_2 are the centres of the circles. Since the tangents to the two circles at P are perpendicular, the radii O_1P, O_2P which are at right angles to the tangents, are also perpendicular. Hence the angle O_1PO_2 is a right angle and therefore

$$O_1O_2{}^2 = O_1P^2 + O_2P^2,$$

i.e., the square of the distance between the centres of two orthogonal circles is equal to the sum of the squares of the radii.

If the equations to the circles are

$$x^2 + y^2 + 2gx + 2fy + c = 0,$$

and

$$x^2 + y^2 + 2g'x + 2f'y + c' = 0,$$

$$\tag{16.17}$$

the centres of the circles are $(-g, -f)$ and $(-g', -f')$. The square of the distance between the centres is

$$(-g + g')^2 + (-f + f')^2,$$

and the squares of the radii of the circles are $g^2 + f^2 - c$ and $g'^2 + f'^2 - c'$. The circles therefore cut orthogonally if

$$(-g + g')^2 + (-f + f')^2 = g^2 + f^2 - c + g'^2 + f'^2 - c',$$

i.e., if, $2gg' + 2ff' = c + c'.$ (16.18)

Example 8. *Find the equation to the circle which passes through the origin and cuts both of the circles*

$$x^2 + y^2 - 6x + 8 = 0 \quad and \quad x^2 + y^2 - 2x - 2y = 7 \ orthogonally.$$

The equation to any circle which passes through the origin is

$$x^2 + y^2 + 2gx + 2fy = 0.$$

This circle cuts the first of the given circles orthogonally if

$$2g(-3) + 2f(0) = 8, \quad \text{i.e., if } -6g = 8 \quad \text{or} \quad g = -4/3.$$

It cuts the second of the given circles orthogonally if

$$2g(-1) + 2f(-1) = -7.$$

This gives $f = \frac{7}{2} - g = \frac{7}{2} + \frac{4}{3} = \frac{29}{6}$, and the required equation to the circle is

$$x^2 + y^2 - \tfrac{8}{3}x + \tfrac{29}{3}y = 0,$$

or, $\qquad\qquad 3x^2 + 3y^2 - 8x + 29y = 0.$

EXERCISES 16 (c)

1. Find the length of the tangent drawn from the point $(2, 5)$ to the circle $x^2 + y^2 - 2x - 3y = 1$.

2. The length of the tangent from the point $(1, 1)$ to the circle

    $$x^2 + y^2 - 4x - 6y + k = 0$$

    is 2 units. Find the value of k. (L.U.)

3. Given the three circles,

    $$x^2 + y^2 - 16x + 60 = 0,$$
    $$x^2 + y^2 - 12x + 20 = 0,$$
    $$x^2 + y^2 - 16x - 12y + 84 = 0,$$

    find (i) the coordinates of the point such that the lengths of the tangents from it to each of the three circles are equal, (ii) the length of each tangent. (L.U.)

4. Show that the circles $x^2 + y^2 - 2ax + c^2 = 0, x^2 + y^2 - 2by - c^2 = 0$ are orthogonal.

5. A circle, centre C, cuts the circle $x^2 + y^2 = 4$ at right angles and passes through the point $(1, 3)$. Find the equation to the locus of C. (L.U.)

6. The line $x + y = 3$ meets the circle $x^2 + y^2 + x - 5y + 4 = 0$ at points P, Q. Find the equations to the tangents at P, Q to the circle and the coordinates of their point of intersection R. Find also the equation to the circle centre R which cuts the given circle orthogonally. (L.U.)

16.9. The radical axis of two circles

The radical axis of two circles is the locus of a point which moves so that the lengths of the tangents drawn from it to the two circles are equal.

Suppose the equations to the two circles are

$x^2 + y^2 + 2gx + 2fy + c = 0$ and $x^2 + y^2 + 2g'x + 2f'y + c' = 0.$

The square of the length of the tangent from a point (x, y) to the first circle is, by (16.16), $x^2 + y^2 + 2gx + 2fy + c$, while that of the tangent from the same point to the second circle is $x^2 + y^2 + 2g'x + 2f'y + c'$. If the point (x, y) lies on the radical axis, these quantities are equal so that

$$x^2 + y^2 + 2gx + 2fy + c = x^2 + y^2 + 2g'x + 2f'y + c' \quad (16.19)$$

or, $\qquad 2(g - g')x + 2(f - f')y + c - c' = 0. \qquad (16.20)$

This is the equation to the radical axis of the two circles, and since it is of the first degree in x and y it is a straight line.

The equation (16.19) to the radical axis can be written in the form

$$x^2 + y^2 + 2gx + 2fy + c - (x^2 + y^2 + 2g'x + 2f'y + c') = 0.$$

In this form it is apparent that any point which lies on both the given circles lies also on the radical axis. Hence the radical axis corresponds, in the case of two circles which intersect in real points, to the line through the common points.

Example 9. *Find the equation to the common chord of the two circles*

$$x^2 + y^2 + 10x + 8y + 32 = 0 \quad \text{and} \quad x^2 + y^2 - 4x - 6y + 12 = 0$$

and show that this line is perpendicular to the line of centres of the circles.

The equation to the common chord (or radical axis) is obtained by equating to zero the expression corresponding to the difference of the left-hand sides of the equations to the circles. In the case of the circles given here it is

$$14x + 14y + 20 = 0, \quad \text{or} \quad 7x + 7y + 10 = 0.$$

The coordinates of the centres of the two given circles are respectively $(-5, -4)$ and $(2, 3)$. The slope of the line of centres is therefore

$$\frac{3 - (-4)}{2 - (-5)} \quad \text{or} \quad 1$$

while the slope of the common chord is -1. The product of these two slopes being -1, the common chord is perpendicular to the line of centres.

16.10. The circle through the points of intersection of two given circles

Consider the equation

$$x^2 + y^2 + 2gx + 2fy + c + \lambda(x^2 + y^2 + 2g'x + 2f'y + c') = 0, \quad (16.21)$$

where λ is any constant. Since this is an equation of the second degree in which the coefficients of x^2 and y^2 are equal (they are both $1 + \lambda$) and there is no term in xy, it is the equation to a circle. Further, the equation is satisfied for points whose coordinates (x, y) simultaneously satisfy the equations

$$\left.\begin{array}{l} x^2 + y^2 + 2gx + 2fy + c = 0, \\ x^2 + y^2 + 2g'x + 2f'y + c' = 0. \end{array}\right\} \quad (16.22)$$

Such points are those which are common to the curves represented by equations (16.22). Hence equation (16.21) represents a circle passing through the points of intersection of the circles given by equations (16.22).

The foregoing is another example of the device mentioned in § 15.8. Equation (16.21) represents a family of circles, all of which pass through the common points of the circles given by (16.22). Individual circles of this family correspond to different values of λ. Generally it will be required to find the equation to a circle which, besides passing through the intersection of two given circles, fulfils some additional condition such as passing through another given point or touching a given line. A further condition of this sort enables the value of λ to be fixed for the particular circle required.

Example 10. *Find the equation to the circle through the origin and through the points of intersection of the circles*

$$x^2 + y^2 - 2x - 4y - 4 = 0 \quad \text{and} \quad x^2 + y^2 + 8x - 4y + 6 = 0.$$

L.U.)

Any circle through the common points of the given circles is

$$x^2 + y^2 - 2x - 4y - 4 + \lambda(x^2 + y^2 + 8x - 4y + 6) = 0,$$

or, $(1 + \lambda)(x^2 + y^2) + (8\lambda - 2)x - (4\lambda + 4)y + 6\lambda - 4 = 0.$

If this circle passes through the origin, its equation must be satisfied by $x = 0$, $y = 0$, so that

$$6\lambda - 4 = 0$$

giving $\lambda = 2/3$. With this value of λ the equation is

$$(1 + \tfrac{2}{3})(x^2 + y^2) + (\tfrac{16}{3} - 2)x - (\tfrac{8}{3} + 4)y = 0,$$

i.e., $x^2 + y^2 + 2x - 4y = 0$.

EXERCISES 16 (d)

1. Find the radical axis of the circles $x^2 + y^2 + 4x + 5y = 6$ and $x^2 + y^2 + 5x + 4y = 9$.

2. Find the equations to the common chord and the line of centres of the two circles

$$x^2 + y^2 + 6x - 3y + 4 = 0$$
and $$2x^2 + 2y^2 - 3x - 9y + 2 = 0.$$

3. Show that the length of the common chord of the circles

$$x^2 + y^2 + ax + by + c = 0, \quad x^2 + y^2 + bx + ay + c = 0$$

is $\{\tfrac{1}{2}(a + b)^2 - 4c\}^{\frac{1}{2}}.$

4. Show that the circles $x^2 + y^2 + 4x + y = 3$, $x^2 + y^2 - x - y = 1$ and $x^2 + y^2 + 14x + 5y = 7$ are co-axial (i.e., they have the same radical axis).

5. Find the equation to the circle which passes through the point $(1, 2)$ and through the points of intersection of the circles $x^2 + y^2 + 2x + 3y = 7$ and $x^2 + y^2 + 3x - 2y = 1$.

6. Find the equations to the circles passing through the points of intersection of the circles

$$x^2 + y^2 - 18x - 2y + 8 = 0,$$
$$x^2 + y^2 - 26x + 6y = 24$$

and touching the straight line $y = 10$.   (L.U.)

EXERCISES 16 (e)

1. (i) Find the radius and coordinates of the centre of the circle $x^2 + y^2 - 2x - 6y + 6 = 0$.

    (ii) If the line $x = 2y$ meets the circle $x^2 + y^2 - 8x + 6y - 15 = 0$ at the points P, Q, find the coordinates of P and Q and the equation to the circle passing through P, Q and the point $(1, 1)$.   (L.U.)

2. Show that the locus of a point such that the square of its distance from the point $(3, 4)$ is proportional to its distance from the line $x + y = 0$, one point on the locus being the point $(1, 2)$, is a circle and find its centre and radius. (O.C.)

3. Find the equation to the circle of which the points $(8, -2)$ and $(-2, 6)$ are ends of a diameter. (O.C.)

4. Find the equation to the circle which passes through the points $(-2, 2)$, $(2, 4)$, $(5, -5)$. Show that this circle touches the circle
$$2x^2 + 2y^2 - 17x + 16y + 65 = 0$$
at the point $(5, -5)$. (L.U.)

5. Show that the circle $x^2 + y^2 - 2ax - 2ay + a^2 = 0$ touches the axes of x and y.
    Also find the equation to the circle which passes through the points $(2, 3)$, $(4, 5)$, $(6, 1)$. (Q.E.)

6. Find the equations to two circles which touch the x-axis at the origin and also touch the line $12x + 5y = 60$. (L.U.)

7. Find the coordinates of the centre and the radius of the circle $x^2 + y^2 - 2x - 8y + 1 = 0$. Show that this circle touches the x-axis and that the point (h, k) of contact of the other tangent from the point $(3, 0)$ must satisfy the condition $h - 2k = 1$. (L.U.)

8. A circle of radius 5 has its centre in the positive quadrant, touches the x-axis and intercepts a chord of length 6 on the y-axis. Show that its equation is $x^2 + y^2 - 8x - 10y + 16 = 0$. If $y = mx$ is a tangent from the origin apart from the x-axis, find m. (L.U.)

9. Prove that the points $(3/2, 6)$, $(-9/2, -2)$ lie on the circle $4x^2 + 4y^2 + 12x - 16y - 75 = 0$, and that the tangents at these points are parallel. (O.C.)

10. Find the equation to the tangent at the point $(3, -4)$ to the circle $x^2 + y^2 = 25$. What are the equations to the two tangents parallel to the y-axis? Show that the first tangent intersects these two in points which subtend a right angle at the origin. (O.C.)

11. Find the equations to the two circles each of which touch the three circles $x^2 + y^2 = 4a^2$, $x^2 + 2ax + y^2 = 0$, $x^2 - 2ax + y^2 = 0$. (Q.E.)

12. Find the equation to the circle touching the x-axis at the point $(5, 0)$ and passing through the point $(7, 4)$. What are the coordinates of the point on the circle other than $(5, 0)$, the tangent at which passes through the origin? (L.U.)

13. Prove that the equation to the circle whose centre lies in the first quadrant, which touches the x-axis and which passes through the points $A(0, 6)$, $B(0, 24)$ is $x^2 + y^2 - 24x - 30y + 144 = 0$. Find also the equation to the other chord through the origin whose length is equal to that of the chord AB. (L.U.)

14. Find the coordinates of the centre and the radius of the circle $2x^2 + 2y^2 + 4x - 12y + 15 = 0$. Find also the equation to the tangent to the circle which is furthest from the origin. Calculate the length of the chord intercepted by the circle on the line $x + y = 0$. (L.U.)

15. Find the equations of the circles that touch the circle $x^2 + y^2 = 4$ and the straight lines $y = 0$ and $x = 7$. (Q.E.)

16. Verify that the circle $x^2 + y^2 - 8x - 7y + 12 = 0$ passes through the point $(2, 0)$ and calculate the coordinates of the other points of intersection of the circle with the axes. Find the equation to the tangent to the circle at the other end of the diameter through $(2, 0)$. Calculate the length of a tangent to the circle from the point $(9, 2)$. (L.U.)

17. Show that the line $x + 3y = 1$ touches the circle
$$x^2 + y^2 - 3x - 3y + 2 = 0$$
and find the coordinates of the point of contact. Prove, by calculation, that the point $P(3, 2\cdot5)$ lies outside the circle and calculate also the length of the tangent drawn to the circle from P. (L.U.)

18. Find the length of the tangents from the point $(5, -4)$ to the circle $x^2 + y^2 + 8 = 2k(x + y - 1)$. If, in this equation k be given various values, show that all the circles so obtained have the same radical axis.

19. Find the equation to the circle which cuts orthogonally the circle $x^2 + y^2 - 4x + 6y - 7 = 0$, passes through the point $(0, 3)$ and touches the x-axis.

20. $A(3, 0)$ and $B(0, 2)$ are two fixed points. Find the equation to the locus of a point P such that $2PA = 3PB$. Show that the locus is a circle passing through the origin and find the equation to the tangent at the origin. Show also that this circle cuts at right angles the circle $x^2 + y^2 - 3x - 2y = 0$. (L.U.)

21. Show that the circles
$$x^2 + y^2 + 4x - 2y - 11 = 0 \quad \text{and} \quad x^2 + y^2 - 4x - 8y + 11 = 0$$
intersect at right angles and find the length of their common chord. (L.U.)

22. Prove that the circles
$$x^2 + y^2 + 2x - 8y + 8 = 0,$$
$$x^2 + y^2 + 10x - 2y + 22 = 0$$
touch one another. Find (i) the point of contact, (ii) the equation to the common tangent at this point, (iii) the area of the triangle enclosed by this common tangent, the line of centres and the y-axis. (L.U.)

23. Find the equations to two circles which pass through the point $(4, 1)$ and touch both the lines $x = 6$, $y = 5$. Prove that the equation to the common chord of the circles is $x + y = 5$.

24. Find the equation to the circle passing through the point $(-2, -6)$ and through the two points of intersection of the circles
$$x^2 + y^2 - 3x + 4y - 2 = 0 \quad \text{and} \quad x^2 + y^2 + 5x - 3y = 8.$$

25. Find the radius and the coordinates of the centre of a circle which passes through the points of intersection of the circles $x^2 + y^2 = 4$ and $x^2 + y^2 - 2x - 4y + 4 = 0$ and which touches the straight line $x + 2y = 0$.

CHAPTER 17

THE PARABOLA, ELLIPSE AND HYPERBOLA

17.1. Introduction

The equation of the second degree in x and y represents, in certain circumstances, a pair of straight lines or a circle. For example, if the equation contains only terms in x^2, y^2 and xy it represents a pair of lines through the origin (§ 15.9), while, if the coefficients of the terms in x^2 and y^2 are equal and there is no term in xy it represents a circle (§ 16.2). Other curves can be represented by an equation of the second degree and these are the subject of the present chapter.

Consider the locus of a point P which moves so that its distance from a fixed point (the *focus S*) is always in a constant ratio (the *eccentricity* ϵ) to its perpendicular distance from a fixed straight line (the *directrix AB*). We shall show in the sections which follow that the locus is given by an equation of the second degree in x and y. A plot of the locus reveals that its shape, which depends of course on the eccentricity ϵ, changes significantly when ϵ passes through unity. The locus is called a *parabola* when $\epsilon = 1$, an *ellipse* when $\epsilon < 1$ and a *hyperbola* when $\epsilon > 1$.

It can be shown, but we shall not attempt to do so here, that the parabola, ellipse and hyperbola are, like the pair of straight lines and circle, all sections of a right circular cone by a cutting plane. For this reason, the equation of the second degree is sometimes said to represent a *conic section*.

17.2. The equation to a parabola

The equation to a parabola takes its simplest form when we take the focus S to be the point $(a, 0)$ and the directrix AB as the line $x = -a$. If in Fig. 120, P is the point (x, y) and PM is drawn perpendicular to AB, the parabola is the curve on which the point P lies when it moves so that the distance PM is always equal to the distance PS.

If the line AB meets the x-axis at C, C is the point $(-a, 0)$ and PM is clearly the sum of the abscissa of P and the length CO. Hence $PM = x + a$, and since P and S are respectively the points (x, y) and $(a, 0)$,

$$PS^2 = (x - a)^2 + y^2.$$

Since $PS = PM$, we have

$$(x - a)^2 + y^2 = (x + a)^2,$$

or, $$y^2 = 4ax, \tag{17.1}$$

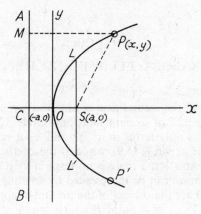

FIG. 120

and this is the required equation to the parabola with focus the point $(a, 0)$ and directrix the line $x + a = 0$.

To trace the curve, we first observe that y is imaginary when x is negative and there is thus no part of the curve to the left of the origin O. If y be zero, so also is x, so the x-axis meets the curve only at the origin O. This point is called the *vertex* of the parabola. If x be zero, $y^2 = 0$, showing that the y-axis meets the curve in two coincident points at the vertex; i.e., the y-axis is the tangent to the parabola at its vertex. For every positive value of x, equation (17.1) shows that y has two equal and opposite values, so that corresponding to any point P on the curve there is another point P', the image of P in the x-axis, also on the curve. In other words, the curve possesses symmetry about the x-axis, and this line is often referred to as the *axis* of the parabola. Finally, as x increases in magnitude so does y and the general shape of the curve is as shown in the diagram.

The double ordinate LSL' (Fig. 120) drawn through the focus S is known as the *latus-rectum* of the parabola. Since LS is the value of y when $x = a$, equation (17.1) gives at once $LS = 2a$ and hence the length of the latus-rectum of the parabola $= LSL' = 2LS = 4a$.

17.3. The tangent and normal to the parabola at a given point

Differentiating the equation $y^2 = 4ax$ with respect to x,

$$2y\frac{dy}{dx} = 4a,$$

so that the gradient of the parabola at the point (x_1, y_1) is given by

$$\left(\frac{dy}{dx}\right)_{x=x_1} = \frac{2a}{y_1}.$$

The tangent is the line through the point (x_1, y_1) with slope equal to the gradient of the curve; its equation is

$$y - y_1 = \frac{2a}{y_1}(x - x_1),$$

or, $$yy_1 - y_1^2 = 2ax - 2ax_1.$$

Since the point (x_1, y_1) lies on the parabola, y_1^2 can be replaced by $4ax_1$, and the equation to the tangent becomes

$$yy_1 = 2a(x + x_1). \tag{17.2}$$

It should be noted that equation (17.2) can be written down by observing the general rule that *the equation to the tangent at the point* (x_1, y_1) *is obtained from the equation to the curve by replacing* y^2 *by* yy_1 *and* $2x$ *by* $(x + x_1)$.

The *normal* to a curve at any point P is the straight line which passes through P and is perpendicular to the tangent at P. Since the slope of the tangent at the point (x_1, y_1) to the parabola is $2a/y_1$, that of the normal is $-y_1/2a$ and the equation to the normal is therefore

$$y - y_1 = -\frac{y_1}{2a}(x - x_1). \tag{17.3}$$

Example 1. *Find the equations to the tangents and normals to the parabola $y^2 = 16x$ at the points $(16, 16)$, $(1, -4)$. The tangents intersect at the point T and the normals intersect at R. Prove that the line TR is parallel to the axis of the parabola.* (O.C.)

Here $4a = 16$, so that $a = 4$. For the point $(16, 16)$, $x_1 = y_1 = 16$ and equations (17.2), (17.3) give for the tangent and normal

$$16y = 8(x + 16),$$

and $$y - 16 = -\tfrac{16}{8}(x - 16)$$

respectively. These equations can be written in the simpler forms $x - 2y + 16 = 0$ and $2x + y = 48$.

For the point $(1, -4)$, $x_1 = 1$, $y_1 = -4$ and equations (17.2), (17.3) give, after a little reduction, the equation to the tangent as

$$2x + y + 2 = 0$$

and that of the normal as $x - 2y = 9$.

The coordinates of T are given by solving the simultaneous equations $x - 2y + 16 = 0$, $2x + y + 2 = 0$, and are found to be $x = -4$, $y = 6$. The coordinates of R are similarly found, from the solution of the simultaneous equations $2x + y = 48$, $x - 2y = 9$, to be $x = 21$, $y = 6$. Thus both T and R are at height 6 above the x-axis and the line TR is therefore parallel to the x-axis which is also the axis of the parabola.

17.4. The points of intersection of a straight line and parabola

The coordinates of the points of intersection of the straight line

$y = mx + c$ and the parabola $y^2 = 4ax$ are the values of x and y which simultaneously satisfy both equations. Writing $y = mx + c$ in the equation to the parabola, the abscissae of the points of intersection are therefore given by

$$(mx + c)^2 = 4ax,$$

or, $$m^2x^2 + 2(mc - 2a)x + c^2 = 0. \qquad (17.4)$$

This quadratic equation has real, equal or imaginary roots according as

$$\{2(mc - 2a)\}^2 - 4m^2c^2$$

is positive, zero or negative; i.e., according as c is less than, equal to, or greater than a/m.

The three possibilities can be illustrated as was done for the circle in Fig. 116. When $c < a/m$, the line intersects the parabola in two real points. When $c > a/m$, it does not meet the parabola at all, or rather, it meets the curve in two imaginary points. If $c = a/m$, the line touches the parabola.

Substituting $c = a/m$ in the equation $y = mx + c$ to the line, we find that the line

$$y = mx + \frac{a}{m}, \qquad (17.5)$$

touches the parabola $y^2 = 4ax$ for all values of m.

Example 2. *Find the ordinates of the points in which the line $x + 2y = c$ meets the parabola $y^2 = 10x$, and find the value of c when this line is a tangent to the parabola.* (O.C.)

Since we here require the values of y at the points of intersection, we substitute $x = c - 2y$ from the equation to the line in the equation to the parabola. The ordinates of the points of intersection are then given by

$$y^2 = 10(c - 2y),$$

or, $$y^2 + 20y - 10c = 0,$$

giving $$y = \frac{-20 \pm \sqrt{(400 + 40c)}}{2} = -10 \pm \sqrt{\{10(10 + c)\}}.$$

The line touches the parabola if this quadratic in y has equal roots. This is so when $(20)^2 = 4(-10c)$, i.e., when $c = -10$.

Example 3. *Show that the point of intersection of two perpendicular tangents to a parabola lies on the directrix.*

By (17.5), the line $y = mx + (a/m)$ is a tangent to the parabola $y^2 = 4ax$. Writing $-1/m$ in place of m, the perpendicular line $y = -(x/m) - am$ is also a tangent to the parabola. By subtraction, the abscissa of the point of intersection of these two lines is given by

$$\left(m + \frac{1}{m}\right)x + \frac{a}{m} + am = 0,$$

i.e., by $x + a = 0$, and this is the equation to the directrix.

EXERCISES 17 (a)

1. Find the equation to the normal to the parabola $y^2 = 8x$ at the point $(4 \cdot 5, -6)$. (O.C.)

2. Find the equations to the tangents to the parabola $y^2 = 144x$ at the points $(144, 144)$, $(9, -36)$, and show that they are perpendicular. Find also the coordinates of their point of intersection. (O.C.)

3. Find the coordinates of the points in which the line $y = 8x - a$ meets the parabola $y^2 = 4ax$. Find the equations to the tangents to the parabola at these points and the coordinates of their point of intersection. (L.U.)

4. Find the equation to the normal at the point $P(1, 2)$ to the parabola $y^2 = 4x$. This normal meets the x-axis in G and M is the mid-point of PG. A line through M parallel to the y-axis meets the x-axis in N and the parabola in Q. Prove that $QN = PG$. (L.U.)

5. A point P moves so that its distance from the line $y = 2$ is equal to its distance from the point $x = 2$, $y = 4$. Find the equation to the locus of P and the equations to the tangents through the origin that touch the locus of P. (Q.E.)

6. Find the angles at which the parabolae $y^2 = ax$ and $x^2 = by$ intersect. (Q.E.)

7. If the tangent at the point P to the parabola $y^2 = 4ax$ meets the parabola $y^2 = 4a(x + b)$ at points Q and R, prove that P is the middle point of QR. (O.C.)

8. Prove that the normals to the parabola $y^2 = 4ax$ at its points of intersection with the straight line $2x - 3y + 4a = 0$ meet on the parabola. (O.C.)

17.5. The parametric equations to a parabola

The point (x_1, y_1) lies on the parabola $y^2 = 4ax$ only if the relation $y_1^2 = 4ax_1$ between its two coordinates is satisfied. It is often convenient to be able to write down the coordinates of a point which always lies on the parabola. Such a point is one with coordinates $(at^2, 2at)$ for it is clear that if

$$x = at^2, \quad y = 2at, \qquad (17.6)$$

then, $y^2 = (2at)^2 = 4a(at^2) = 4ax$ for all values of t. The equations (17.6) are called the *parametric equations* to the parabola $y^2 = 4ax$. They express the coordinates of a point on the curve in terms of the *parameter t*, and we shall, for brevity, refer to the point with coordinates $(at^2, 2at)$ as the point "t".

The equation to the chord joining the points $(at_1^2, 2at_1)$, $(at_2^2, 2at_2)$ on the parabola is, by equation (15.5),

$$\frac{x - at_1^2}{at_2^2 - at_1^2} = \frac{y - 2at_1}{2at_2 - 2at_1},$$

which reduces to

$$2x - (t_1 + t_2)y + 2at_1t_2 = 0. \qquad (17.7)$$

If we write $t_1 = t_2 = t$, so that the two points on the parabola coincide at the point $(at^2, 2at)$, the chord (17.7) becomes the tangent at this point, and its equation reduces to

$$x - ty + at^2 = 0. \qquad (17.8)$$

Alternatively, the equation to the tangent at the point "t" can be found as follows. The parabola can be written

$$x = at^2, \quad y = 2at,$$

and the gradient at any point is

$$\frac{dy}{dx} = \frac{(dy/dt)}{(dx/dt)} = \frac{2a}{2at} = \frac{1}{t}.$$

Hence the tangent is the line passing through the point $(at^2, 2at)$ with slope $1/t$ and its equation is

$$y - 2at = \frac{1}{t}(x - at^2),$$

or, $x - ty + at^2 = 0$ as in (17.8). The normal at the point "t" is the line through the point $(at^2, 2at)$ with slope $-t$ and its equation is therefore

$$y - 2at = -t(x - at^2),$$

or, $$y + tx = 2at + at^3. \qquad (17.9)$$

Many problems on the parabola are best solved by using the parametric equations. The above equations to the chord, tangent and normal are very useful in such work. The student should either remember them or (preferably) be able to derive them quickly.

Example 4. *Show that the tangents to a parabola at the extremities of a focal chord are at right angles to each other.*

If we take the parabola as $y^2 = 4ax$, and the extremities of the chord as the points "t_1", "t_2", the equation to the chord is, by (17.7),

$$2x - (t_1 + t_2)y + 2at_1t_2 = 0.$$

If the chord passes through the focus, the point $(a, 0)$, the above equation shows that $2a + 2at_1t_2 = 0$, or $t_1t_2 = -1$. But equation (17.8) shows that the slopes of the tangent at the extremities of the chord are $1/t_1$ and $1/t_2$. Since $t_1t_2 = -1$, the product of the slopes is -1 and the tangents are therefore at right angles to each other.

17.6. An important property of the parabola

In Fig. 121, P is the point "t" on the parabola $y^2 = 4ax$ whose focus is S. PT is the tangent at P meeting the x-axis at Q. Since the coordinates of P and S are respectively $(at^2, 2at)$ and $(a, 0)$,

$$PS^2 = (at^2 - a)^2 + (2at - 0)^2 = a^2(t^4 - 2t^2 + 1 + 4t^2)$$
$$= a^2(t^2 + 1)^2,$$

so that $PS = a(t^2 + 1)$. The equation to the tangent PT is, from equation (17.8), $x - ty + at^2 = 0$. This meets the x-axis where

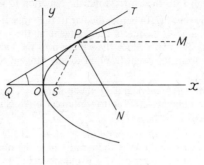

FIG. 121

$x + at^2 = 0$, so that $QO = at^2$. Since $OS = a$,

$$QS = QO + OS = at^2 + a = a(t^2 + 1).$$

Hence $QS = PS$ and the triangle QSP is isosceles. Hence

$$\text{angle } PQS = \text{angle } SPQ.$$

If PM is drawn through P parallel to the x-axis, it follows that

$$\text{angle } TPM = \text{angle } PQS = \text{angle } SPQ,$$

so that the lines PS and PM are equally inclined to the tangent at P. If PN is the normal at P, it follows immediately that *the lines PS and PM are equally inclined to the normal at P.*

This is an important property of the parabola. It means that if a ray of light starting from the focus S of a parabolic mirror strikes the mirror at P, the reflected ray, which makes an equal angle with the normal, will be parallel to the axis of the mirror. Since P is any point on the parabola, all incident rays from a source at the focus will be reflected as rays which are all parallel to the mirror's axis. Conversely, parallel rays from a distant source striking such a mirror will all be reflected so as to pass through the focus. Motor-car headlamps, searchlights and some electric fires are often designed so as to make use of this property.

EXERCISES 17 (b)

1. If the chord joining the points $(at_1{}^2, 2at_1)$, $(at_2{}^2, 2at_2)$ on the parabola $y^2 = 4ax$ passes through the focus $(a, 0)$, find t_2 in terms of t_1. PQ is a focal chord and PL, QM are perpendicular to the axis of the parabola. Prove that $PL.QM$ is constant.   (L.U.)

2. If PQ is one of a series of parallel chords of a parabola, prove that the midpoint of PQ always lies on a line parallel to the axis of the parabola.   (L.U.)

3. P is the point $(at^2, 2at)$ on the parabola $y^2 = 4ax$. From a fixed point $Q(h, k)$ a line is drawn perpendicular to the tangent at P to meet, at R, the parallel through P to the x-axis. Find the equation to the locus of R. (L.U.)

4. A line from the point $(2, 0)$ perpendicular to the tangent at the point $(2t^2, 4t)$ to the parabola $y^2 = 8x$ meets that tangent at the point (h, k). Express h and k in terms of t and deduce the equation to the locus of the foot of the perpendicular from the point $(2, 0)$ on to any tangent to this parabola. (L.U.)

5. P is the point $(at_1^2, 2at_1)$ and Q the point $(at_2^2, 2at_2)$ on the parabola $y^2 = 4ax$. The tangents at P and Q intersect at R. Show that the area of the triangle PQR is $\frac{1}{2}a^2(t_1 - t_2)^3$. (L.U.)

6. P is the point $(at^2, 2at)$ on the parabola $y^2 = 4ax$. If PN is the perpendicular from P to the x-axis and M is the point where the normal at P meets the x-axis, prove that the distance MN is independent of t. (L.U.)

7. P is a point on a parabola whose focus is S. D is the foot of the perpendicular from P to the directrix. Show that the tangent to the parabola at P bisects the angle SPD. (L.U.)

8. The normal to the parabola $y^2 = 4ax$ at the point $P(at^2, 2at)$ meets the axis of the parabola at G and GP is produced, beyond P, to Q so that $GP = PQ$. Show that the equation to the locus of Q is $y^2 = 16a(x + 2a)$. (L.U.)

17.7. The equation to an ellipse

The equation to an ellipse of eccentricity ϵ (less than unity) takes its simplest form when we take the focus S as the point $(-a\epsilon, 0)$ and

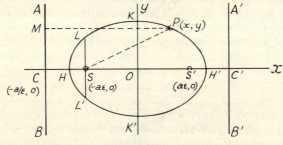

FIG. 122

the directrix AB as the line $x = -a/\epsilon$. In Fig. 122, P is the point (x, y) and PM is perpendicular to AB. The ellipse is the curve on which P lies when it moves so that the distance PS is ϵ times the distance PM.

If the directrix AB meets the x-axis at C, C is the point $(-a/\epsilon, 0)$ and PM is the sum of the abscissa of P and the length CO. Hence $PM = x + (a/\epsilon)$, and since P and S are respectively the points (x, y) and $(-a\epsilon, 0)$

$$PS^2 = (x + a\epsilon)^2 + y^2.$$

Since $PS = \epsilon . PM$,

$$(x + a\epsilon)^2 + y^2 = \epsilon^2 \left(x + \frac{a}{\epsilon} \right)^2,$$

or, $$(1 - \epsilon^2)x^2 + y^2 = a^2(1 - \epsilon^2).$$

This can be written

$$\frac{x^2}{a^2} + \frac{y^2}{a^2(1 - \epsilon^2)} = 1,$$

and writing, $$b^2 = a^2(1 - \epsilon^2), \tag{17.10}$$

the equation to the ellipse becomes

$$\frac{x^2}{a^2} + \frac{y^2}{b^2} = 1. \tag{17.11}$$

To trace the curve we observe that, since only even powers of x and y occur, the curve is symmetrical about both the coordinate axes. From this symmetry we can deduce the existence of a second focus S' at the point $(a\epsilon, 0)$ and a second directrix $A'B'$ along the line $x = a/\epsilon$. The curve cuts the x-axis where $(x^2/a^2) = 1$, i.e. in the points $(\pm a, 0)$ and it cuts the y-axis where $(y^2/b^2) = 1$, i.e., at the points $(0, \pm b)$. These points are denoted by H, H', K, K' in Fig. 122. By writing equation (17.11) in the form

$$\frac{y^2}{b^2} = 1 - \frac{x^2}{a^2},$$

it is clear that y^2 is negative, and therefore there is no part of the curve, for values of x greater than a or less than $-a$. Similarly, there are no parts of the curve for which y is greater than b or less than $-b$. The general shape of the curve is shown in the diagram.

The points H, H' are called the *vertices* of the ellipse. The lines HH' and KK' are called its *axes*; since $\epsilon < 1$, equation (17.10) shows that $b < a$, and the axes HH', KK' are referred to respectively as the *major* and *minor* axes. The origin O is called the *centre* of the ellipse, and a chord passing through the centre is called a *diameter*. The double ordinate LSL' through the focus S is the *latus-rectum* and there will be a second latus-rectum through the second focus S'. Since LS is the value of y when $x = -a\epsilon$ in equation (17.11),

$$LS = b \sqrt{1 \left(- \frac{a^2 \epsilon^2}{a^2} \right)} = b\sqrt{(1 - \epsilon^2)}$$

$$= b^2/a,$$

when use is made of equation (17.10). Thus the latus-rectum is of length $2b^2/a$.

To sum up, the curve

$$\frac{x^2}{a^2} + \frac{y^2}{b^2} = 1 \tag{17.11}$$

is an ellipse of eccentricity ϵ, less than unity, given by

$$\epsilon^2 = 1 - (b^2/a^2). \tag{17.12}$$

The foci are the points $(\pm a\epsilon, 0)$, the directrices the lines $x = \pm(a/\epsilon)$, the semi-axes a, b and the centre the origin of coordinates.

Example 5. *Show that the sum of the focal distances of any point on an ellipse is equal to the length of the major axis. Deduce a simple mechanical method for constructing the curve.*

Using Fig. 122 and the definition of the ellipse, if x is the abscissa of P,

$$PS = \epsilon.PM = \epsilon\left(\frac{a}{\epsilon} + x\right) = a + \epsilon x.$$

Similarly, if PM' is drawn perpendicular to the second directrix $A'B'$,

$$PS' = \epsilon.PM' = \epsilon\left(\frac{a}{\epsilon} - x\right) = a - \epsilon x.$$

The sum of the focal distances PS, PS' is therefore equal to $2a$, the length of the major axis.

By fixing two pins at S, S' and keeping stretched by a pencil point an endless piece of string passing round the two pins, the pencil will describe an ellipse with S, S' as foci.

17.8. The tangent and normal to the ellipse at a given point

Differentiating the equation (17.11) to the ellipse with respect to x,

$$\frac{2x}{a^2} + \frac{2y}{b^2}\frac{dy}{dx} = 0,$$

so that the gradient of the ellipse at a given point (x_1, y_1) is given by

$$\left(\frac{dy}{dx}\right)_{x=x_1} = -\frac{b^2 x_1}{a^2 y_1}.$$

The tangent at the point (x_1, y_1) is the line through this point with slope equal to the gradient of the curve; its equation is

$$y - y_1 = -\frac{b^2 x_1}{a^2 y_1}(x - x_1),$$

or, $$b^2 xx_1 + a^2 yy_1 = b^2 x_1^2 + a^2 y_1^2.$$

Dividing by $a^2 b^2$, and using the relation $(x_1^2/a^2) + (y_1^2/b^2) = 1$ which is the condition that the point (x_1, y_1) shall lie on the ellipse, the equation to the tangent at the point (x_1, y_1) can be written

$$\frac{xx_1}{a^2} + \frac{yy_1}{b^2} = 1. \tag{17.13}$$

Again it should be noted that *the equation to the tangent at the point (x_1, y_1) can be obtained from the equation to the curve by replacing x^2, y^2 by xx_1, yy_1 respectively.*

The normal at the point (x_1, y_1) is the line through this point at right angles to the tangent. Its slope is therefore $(a^2 y_1)/(b^2 x_1)$ and its equation is

$$y - y_1 = \frac{a^2 y_1}{b^2 x_1}(x - x_1).$$

This can be written in the more symmetrical form

$$\frac{x - x_1}{x_1/a^2} = \frac{y - y_1}{y_1/b^2}. \tag{17.14}$$

Example 6. *Find the equations to the tangent and normal to the ellipse* $5x^2 + 3y^2 = 137$ *at the point in the first quadrant whose ordinate is 2.*

The abscissae of points on the ellipse at which the ordinate is 2, are given by writing $y = 2$ in the equation to the ellipse. This gives $5x^2 = 137 - 3(2)^2$, or $x = \pm 5$. The point in the first quadrant with this ordinate is therefore the point $(5, 2)$. The equation to the ellipse can be written in the form

$$\frac{x^2}{(137/5)} + \frac{y^2}{(137/3)} = 1,$$

so that $a^2 = 137/5$ and $b^2 = 137/3$. Equation (17.13) then gives for the tangent at the point $(5, 2)$,

$$\frac{5x}{(137/5)} + \frac{2y}{(137/3)} = 1,$$

or, $25x + 6y = 137$. The normal is given by equation (17.14) as

$$\frac{x - 5}{5/(137/5)} = \frac{y - 2}{2/(137/3)},$$

i.e., $6x - 25y + 20 = 0$.

17.9. The points of intersection of a straight line and ellipse

The coordinates of the points of intersection of the straight line $y = mx + c$ and the ellipse $(x^2/a^2) + (y^2/b^2) = 1$ are the values of x and y which simultaneously satisfy both equations. Writing $y = mx + c$ in the equation to the ellipse, the abscissae of the points of intersection are therefore given by

$$\frac{x^2}{a^2} + \frac{(mx + c)^2}{b^2} = 1,$$

or, $$(a^2 m^2 + b^2)x^2 + 2a^2 mcx + a^2(c^2 - b^2) = 0. \tag{17.15}$$

This quadratic equation has real, equal or imaginary roots according as

$$(2a^2 mc)^2 - 4(a^2 m^2 + b^2)a^2(c^2 - b^2)$$

is positive, zero or negative; i.e., according as c^2 is less, equal to or greater than $a^2 m^2 + b^2$.

Again the three possibilities can be illustrated as was done for the circle in Fig. 116. When $c^2 < a^2 m^2 + b^2$, the line intersects the ellipse in two real points. When $c^2 > a^2m^2 + b^2$, the line intersects the ellipse only in imaginary points. If $c^2 = a^2m^2 + b^2$, the line is a tangent to the ellipse.

Writing $c = \sqrt{(a^2m^2 + b^2)}$ in the equation $y = mx + c$ to the line, we find that the line

$$y = mx + \sqrt{(a^2m^2 + b^2)}, \qquad (17.16)$$

always touches the ellipse. Further, since the radical sign on the right-hand side of equation (17.16) may have either positive or negative signs attached to it, we see that there are two tangents to the ellipse having the same m. In other words, there are two tangents parallel to any given direction.

Example 7. *Find the locus of the point of intersection of tangents to an ellipse which are at right angles to one another.*

Taking the ellipse in the usual form (17.11), the line

$$y = mx + \sqrt{(a^2m^2 + b^2)},$$

is always a tangent. A perpendicular tangent is obtained by replacing m by $-1/m$, and its equation is

$$y = -\frac{x}{m} + \sqrt{\left(\frac{a^2}{m^2} + b^2\right)}.$$

These equations can be written

$$y - mx = \sqrt{(a^2m^2 + b^2)},$$
$$my + x = \sqrt{(a^2 + b^2m^2)}.$$

The coordinates of the point of intersection of the tangents simultaneously satisfy these two equations. If therefore we eliminate m between the equations we shall obtain the locus of the point of intersection. Squaring and adding we find

$$(1 + m^2)(x^2 + y^2) = a^2m^2 + b^2 + a^2 + b^2m^2,$$

or,

$$x^2 + y^2 = a^2 + b^2.$$

The required locus is therefore a circle with centre coincident with the centre of the ellipse and with radius $\sqrt{(a^2 + b^2)}$. This circle is called the *director circle*.

17.10. The parametric equations to an ellipse. The eccentric angle

As with the parabola, it is often convenient to express the coordinates of any point on the ellipse in terms of one variable. It is easy to verify that equation (17.11) is satisfied by a point whose coordinates are given by

$$x = a \cos \phi, \quad y = b \sin \phi. \qquad (17.17)$$

These can therefore be regarded as the parametric equations to the ellipse. They express the coordinates of a point on the curve in terms of a parameter, the angle ϕ, and, for brevity, the point $(a \cos \phi, b \sin \phi)$ is often referred to as the point "ϕ".

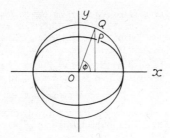

FIG. 123

In Fig. 123, Q is a point on the circle whose centre coincides with the centre O of the ellipse $(x^2/a^2) + (y^2/b^2) = 1$ and whose diameter is equal to the major axis $(2a)$ of the ellipse. Such a circle is called the *auxiliary circle* of the ellipse. The line QP is drawn perpendicular to the x-axis to meet the ellipse at the point P shown. If the angle $QOx = \phi$, it is clear that the coordinates of Q are $(a\cos\phi, a\sin\phi)$. The abscissa of P is equal to that of Q and so is also $a\cos\phi$. Substitution of this value of x in the equation to the ellipse gives the corresponding value of the ordinate to be $b\sin\phi$ and hence P is the point $(a\cos\phi, b\sin\phi)$. The angle ϕ, defined in this way, is known as the *eccentric angle* of the point P.

To obtain the equation to the tangent to the ellipse at a point whose eccentric angle is ϕ, we have

$$x = a\cos\phi, \quad y = b\sin\phi,$$

and the gradient of the ellipse at this point is given by

$$\frac{dy}{dx} = \frac{(dy/d\phi)}{(dx/d\phi)} = \frac{b\cos\phi}{-a\sin\phi} = -\frac{b}{a}\cot\phi.$$

The tangent is therefore the line through the point $(a\cos\phi, b\sin\phi)$ having a slope $-(b/a)\cot\phi$. Its equation is

$$y - b\sin\phi = -\frac{b}{a}\cot\phi(x - a\cos\phi),$$

which reduces to

$$\frac{x}{a}\cos\phi + \frac{y}{b}\sin\phi = 1. \qquad (17.18)$$

The normal at this point is the line through the point $(a\cos\phi, b\sin\phi)$ with slope $(a/b)\tan\phi$. Its equation will therefore be

$$y - b\sin\phi = \frac{a}{b}\tan\phi(x - a\cos\phi).$$

This reduces to

$$ax\sec\phi - by\csc\phi = a^2 - b^2. \qquad (17.19)$$

Example 8. Q *is the point* $(a \cos \phi, a \sin \phi)$ *on the auxiliary circle of the ellipse* $(x^2/a^2) + (y^2/b^2) = 1$. P *is the point on the ellipse with coordinates* $(a \cos \phi, b \sin \phi)$. *If* S *is a focus of the ellipse show that the length of the perpendicular from* S *on to the tangent at* Q *to the circle is equal to* SP. (L.U.)

The equation to the auxiliary circle is $x^2 + y^2 = a^2$ and the equation to the tangent at the point $(a \cos \phi, a \sin \phi)$ is $xa \cos \phi + ya \sin \phi = a^2$, or,

$$x \cos \phi + y \sin \phi = a.$$

If S is the focus $(a\epsilon, 0)$, where ϵ is the eccentricity of the ellipse, the length of the perpendicular from S on to this line is

$$\frac{a\epsilon \cos \phi - a}{\sqrt{(\cos^2 \phi + \sin^2 \phi)}} = a(\epsilon \cos \phi - 1).$$

The distance SP between the points $(a\epsilon, 0)$, $(a \cos \phi, b \sin \phi)$ is given by

$$SP^2 = (a\epsilon - a \cos \phi)^2 + (0 - b \sin \phi)^2$$
$$= a^2(\epsilon - \cos \phi)^2 + a^2(1 - \epsilon^2) \sin^2 \phi,$$

when use is made of the relation $b^2 = a^2(1 - \epsilon^2)$. This, by use of the identity $\cos^2 \phi + \sin^2 \phi = 1$, reduces to $SP = a(\epsilon \cos \phi - 1)$, showing that SP is equal to the length of the perpendicular from S on to the tangent at Q to the circle. A similar proof holds if we take S to be the second focus $(-a\epsilon, 0)$.

EXERCISES 17 (c)

1. Find the distance between the foci, the eccentricity and the length of the latus-rectum of the ellipse $3x^2 + 4y^2 = 12$.

2. Find the equation to an ellipse whose centre is the origin, whose latus-rectum is 10 and whose minor axis is equal to the distance between the foci. The axes of the ellipse lie along the coordinate axes.

3. Find the locus of a point which moves so that the sum of its distances from two fixed points 6 units apart is always 10 units. Your answer should be in the form of an equation referred to axes of symmetry. (O.C.)

4. Show that the intercepts made on the axes by the tangent at the point $(16/5, 9/5)$ to the ellipse $(x^2/16) + (y^2/9) = 1$ are equal. (O.C.)

5. Find the equations to those tangents to the ellipse $x^2 + 2y^2 = 8$ which are parallel to the line $y = 2x$. (O.C.)

6. Find the equations to the normals at the points $(6, 4)$ and $(8, 3)$ to the ellipse $x^2 + 4y^2 = 100$. Prove that the line joining the origin to the middle point of the chord joining these two points is perpendicular to the line joining the origin to the point of intersection of the normals. (O.C.)

7. Two diameters of an ellipse are said to be *conjugate* when each bisects all chords parallel to the other. Show that the diameters $y = mx$, $y = m'x$ of the ellipse $(x^2/a^2) + (y^2/b^2) = 1$ are conjugate if $mm' = -b^2/a^2$.

8. P is the point $(a \cos \phi, b \sin \phi)$ on the ellipse $(x^2/a^2) + (y^2/b^2) = 1$. The normal at P to the ellipse meets the x-axis at Q. Show that the locus of the mid-point of PQ is an ellipse whose semi-axes are $(2a^2 - b^2)/2a$ and $b/2$. (O.C.)

9. Show that tangents to the ellipse $(x^2/a^2) + (y^2/b^2) = 1$ at points whose eccentric angles differ by $90°$ meet on the ellipse $(x^2/a^2) + (y^2/b^2) = 2$.

10. Show that the equation to the chord joining the two points whose eccentric angles are ϕ, ϕ' on the ellipse $(x^2/a^2) + (y^2/b^2) = 1$ is

$$\frac{x}{a}\cos\tfrac{1}{2}(\phi + \phi') + \frac{y}{b}\sin\tfrac{1}{2}(\phi + \phi') = \cos\tfrac{1}{2}(\phi - \phi').$$

Deduce the equation to the tangent at the point "ϕ".

17.11. The equation to a hyperbola

The equation to a hyperbola can be derived by a similar method to that used in § 17.7 to obtain the equation to an ellipse. Again (Fig. 124) we take the focus S as the point $(-a\epsilon, 0)$ and the directrix AB

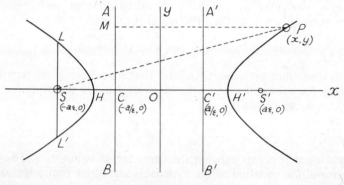

FIG. 124

as the line $x = -a/\epsilon$. Since the eccentricity ϵ is now greater than unity, the relative positions of the focus S and the point of intersection C of AB with the x-axis are interchanged—here C is nearer to the origin O than S.

If P is the point (x, y) on the curve, PM is again the sum of the abscissa of P and the length CO, i.e., $PM = x + (a/\epsilon)$. The length PS is again given by

$$PS^2 = (x + a\epsilon)^2 + y^2,$$

and since, by the definition of a hyperbola, P is a point such that $PS = \epsilon.PM$

$$(x + a\epsilon)^2 + y^2 = \epsilon^2\left(x + \frac{a}{\epsilon}\right)^2.$$

This can be written

$$(\epsilon^2 - 1)x^2 - y^2 = a^2(\epsilon^2 - 1),$$

or,

$$\frac{x^2}{a^2} - \frac{y^2}{a^2(\epsilon^2 - 1)} = 1.$$

Writing

$$b^2 = a^2(\epsilon^2 - 1), \qquad\qquad (17.20)$$

the equation to the curve becomes

$$\frac{x^2}{a^2} - \frac{y^2}{b^2} = 1. \tag{17.21}$$

In tracing the curve, we first observe that its equation contains only even powers of x and y and it is therefore symmetrical about both the coordinate axes. From this symmetry we can deduce the existence of a second focus S' at the point $(a\epsilon, 0)$ and a second directrix $A'B'$ along the line $x = a/\epsilon$. The curve cuts the x-axis where $(x^2/a^2) = 1$, i.e., in the points $(\pm a, 0)$ shown as H' and H in Fig. 124. By writing $x = 0$ in equation (17.21), the points in which the curve cuts the y-axis are given by $-(y^2/(b^2) = 1$, showing that such points are imaginary. By writing the equation in the form

$$\frac{y^2}{b^2} = \frac{x^2}{a^2} - 1,$$

it is clear that y^2 is negative, and therefore there is no part of the curve, for values of x which lie between $\pm a$. On the other hand, the equation can be written

$$\frac{x^2}{a^2} = 1 + \frac{y^2}{b^2},$$

showing that points exist on the curve for *all* values of y. The above forms of the equation to the hyperbola also show that y increases as x increases and vice versa. The curve consists of two portions, one of which extends in an infinite direction towards the positive direction of the x-axis, and the other in an infinite direction towards the negative side of the axis and is shown in the diagram.

The points H, H' are called the *vertices* and the line HH' the *transverse* axis of the hyperbola. The origin O is the *centre* and chords through the centre are called *diameters*. The double ordinate LSL' through the focus S is the *latus-rectum* and there will be a second latus-rectum through the second focus S'. From equation (17.21), since LS is the value of y when $x = -a\epsilon$,

$$LS = b \left(\frac{a^2\epsilon^2}{a^2} - 1\right) = b\sqrt{(\epsilon^2 - 1)} = \frac{b^2}{a}$$

when use is made of equation (17.20). The length of the latus-rectum is therefore $2b^2/a$.

To summarise, the curve

$$\frac{x^2}{a^2} - \frac{y^2}{b^2} = 1, \tag{17.21}$$

is a hyperbola of eccentricity ϵ, greater than unity, given by

$$\epsilon^2 = 1 + (b^2/a^2). \tag{17.22}$$

The foci are the points $(\pm a\epsilon, 0)$, the directrices the lines $x = \pm(a/\epsilon)$, the transverse axis is of length $2a$ and the centre is the origin of coordinates.

Example 9. *Show that the difference of the focal distances of any point on a hyperbola is equal to the length of the transverse axis.*

Using Fig. 124 and the definition of the hyperbola, if x is the abscissa of P,

$$PS = \epsilon.PM = \epsilon\left(x + \frac{a}{\epsilon}\right) = \epsilon x + a.$$

Similarly, if PM' is drawn perpendicular to the second directrix $A'B'$,

$$PS' = \epsilon.PM' = \epsilon\left(x - \frac{a}{\epsilon}\right) = \epsilon x - a.$$

The difference of the focal distances PS, PS' is therefore equal to $2a$, the length of the transverse axis.

17.12. The tangent and normal to the hyperbola at a given point

Differentiating the equation (17.21) to the hyperbola with respect to x,

$$\frac{2x}{a^2} - \frac{2y}{b^2}\frac{dy}{dx} = 0,$$

so that the gradient of the hyperbola at a given point (x_1, y_1) is given by

$$\left(\frac{dy}{dx}\right)_{x=x_1} = \frac{b^2 x_1}{a^2 y_1}.$$

The tangent at the point (x_1, y_1) is the line through this point with slope equal to the gradient of the curve; its equation is

$$y - y_1 = \frac{b^2 x_1}{a^2 y_1}(x - x_1),$$

or, $$b^2 x x_1 - a^2 y y_1 = b^2 x_1{}^2 - a^2 y_1{}^2.$$

Dividing by $a^2 b^2$, and using the relation $(x_1{}^2/a^2) - (y_1{}^2/b^2) = 1$ which is the condition that the point (x_1, y_1) shall lie on the hyperbola, the equation to the tangent at the point (x_1, y_1) can be written

$$\frac{x x_1}{a^2} - \frac{y y_1}{b^2} = 1. \tag{17.23}$$

Once again *the equation to the tangent at the point (x_1, y_1) can be obtained from the equation to the curve by replacing x^2, y^2 by $x x_1$, $y y_1$ respectively.*

The normal at the point (x_1, y_1) is the line through this point at right angles to the tangent. Its slope is therefore $-(a^2 y_1)/(b^2 x_1)$ and its equation is

$$y - y_1 = -\frac{a^2 y_1}{b^2 x_1}(x - x_1).$$

This can be written in the more easily memorised form

$$\frac{x - x_1}{x_1/a^2} = \frac{y - y_1}{y_1/(-b^2)}. \tag{17.24}$$

Example 10. *Write down the equations to the tangent and normal to the hyperbola* $9x^2 - 4y^2 = 36$ *at the point* $(4, 3\sqrt{3})$.

The equation to the hyperbola can be written as

$$\frac{x^2}{4} - \frac{y^2}{9} = 1,$$

so that $a^2 = 4$, $b^2 = 9$. From (17.23) the equation to the tangent at the point $(4, 3\sqrt{3})$ is

$$\frac{4x}{4} - \frac{3\sqrt{3}y}{9} = 1,$$

or, $3x - \sqrt{3}y = 3$. Equation (17.24) gives the normal at this point as the line

$$\frac{x - 4}{4/4} = \frac{y - 3\sqrt{3}}{3\sqrt{3}/(-9)}.$$

or, $x + \sqrt{3}y = 13$.

17.13. The points of intersection of a straight line and hyperbola

Since the equation to the hyperbola only differs from that to the ellipse in having $-b^2$ in place of b^2, many of the results derived for the ellipse can be used for the hyperbola if the sign of b^2 is changed. For example, equations (17.23), (17.24) for the tangent and normal to the hyperbola at the point (x_1, y_1) can be obtained from the corresponding results (17.13), (17.14) for the ellipse by replacing b^2 by $-b^2$.

In the same way, the analysis given in § 17.9 for the points of intersection of the line $y = mx + c$ and the ellipse will apply if the sign of b^2 is changed throughout. We shall find that the line meets the hyperbola in real, coincident or imaginary points according as c^2 is greater than, equal to or less than $a^2m^2 - b^2$. We shall also find that the line

$$y = mx + \sqrt{(a^2m^2 - b^2)},$$

always touches the hyperbola.

Working as in Example 7, we shall find that the locus of the point of intersection of perpendicular tangents to the hyperbola is the circle $x^2 + y^2 = a^2 - b^2$. This circle is again called the *director circle*. It should be noticed that whereas for the ellipse $b < a$, there is (see equation (17.20)) no corresponding limitation in the case of the hyperbola. The radius of the director circle being $\sqrt{(a^2 - b^2)}$, the circle is real whenever $b^2 < a^2$. If $b^2 = a^2$, the radius of the circle is zero and it reduces to a point circle at the origin and in this case the centre of the hyperbola is the only point from which perpendicular tangents can be drawn to the curve. If $b^2 > a^2$, the radius of the director circle is imaginary and no perpendicular tangents can be drawn to the hyperbola.

17.14. The parametric equations to a hyperbola

An ordinate of the hyperbola does not meet the auxiliary circle on HH' as diameter in real points (see Fig. 124). There is thus no real eccentric angle as in the case of the ellipse. It is, however, often useful to be able to express the coordinates of any point on the curve in terms of one variable. Since $\sec^2 \phi = 1 + \tan^2 \phi$, the equations

$$x = a \sec \phi, \quad y = b \tan \phi, \tag{17.25}$$

may be used, for these expressions clearly satisfy the equation $(x^2/a^2) - (y^2/b^2) = 1$ to the hyperbola.

It is possible to give a geometrical definition of the angle ϕ used above but it is not very important to do so. The parametric equations (17.25) are, however, very useful in solving some problems and we shall derive the equations to the tangent and normal to the hyperbola at the point "ϕ", i.e., the point whose coordinates are given by equations (17.25).

From (17.25),

$$\frac{dx}{d\phi} = a \sec \phi \tan \phi, \quad \frac{dy}{d\phi} = b \sec^2 \phi,$$

so that the gradient of the hyperbola at the point $(a \sec \phi, b \tan \phi)$ is given by

$$\frac{dy}{dx} = \frac{(dy/d\phi)}{(dx/d\phi)} = \frac{b \sec^2 \phi}{a \sec \phi \tan \phi} = \frac{b \sec \phi}{a \tan \phi}.$$

The tangent is the line through the point $(a \sec \phi, b \tan \phi)$ with slope $(b \sec \phi)/(a \tan \phi)$. Its equation is

$$y - b \tan \phi = \frac{b \sec \phi}{a \tan \phi}(x - a \sec \phi),$$

which reduces to

$$\frac{x}{a} \sec \phi - \frac{y}{b} \tan \phi = 1. \tag{17.26}$$

The normal is the line through the point $(a \sec \phi, b \tan \phi)$ with slope $-(a \tan \phi)/(b \sec \phi)$. Its equation will therefore be

$$y - b \tan \phi = -\frac{a \tan \phi}{b \sec \phi}(x - a \sec \phi).$$

This reduces to

$$ax \sin \phi + by = (a^2 + b^2) \tan \phi. \tag{17.27}$$

Example 11. *If S, S' are the foci and P is any point on a hyperbola, show that SP, S'P are equally inclined to the tangent at P.* (L.U.)

In Fig. 125, S is the focus $(-a\epsilon, 0)$ and S' the focus $(a\epsilon, 0)$ of the hyperbola.

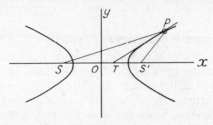

FIG. 125

P is the point $(a \sec \phi, b \tan \phi)$ and the tangent at P cuts the x-axis at the point T. The equation to the tangent at P is

$$(x \sec \phi)/a - (y \tan \phi)/b = 1,$$

and the abscissa of T is obtained by writing $y = 0$ in this equation. The abscissa of T is therefore given by $x = a/\sec \phi = a \cos \phi$. Hence $ST = a\epsilon + \cos \phi$, and $TS' = a\epsilon - a \cos \phi$. Hence

$$\frac{ST}{TS'} = \frac{a\epsilon + a\cos \phi}{a\epsilon - a\cos \phi} = \frac{\epsilon + \cos \phi}{\epsilon - \cos \phi}.$$

But we have shown in Example 9 that, if x is the abscissa of P, $PS = \epsilon x + a$, $PS' = \epsilon x - a$. Here $x = a \sec \phi$, so that

$$\frac{PS}{PS'} = \frac{a\epsilon \sec \phi + a}{a\epsilon \sec \phi - a} = \frac{\epsilon + \cos \phi}{\epsilon - \cos \phi}.$$

Hence

$$\frac{ST}{TS'} = \frac{PS}{PS'},$$

and it follows from a well-known geometrical theorem that PT bisects the angle SPS'.

EXERCISES 17 (d)

1. Find the coordinates of the foci, the eccentricity and the length of the latus-rectum of the hyperbola $4x^2 - 9y^2 = 36$.

2. The centre of a hyperbola is at the origin and its transverse axis lies along the x-axis. Find the equation to the hyperbola if the distance between its foci is equal to $4a$, where $2a$ is the length of the transverse axis.

3. Find the equations to the tangent and normal at the point $(-3, -1)$ to the hyperbola $x^2 - 6y^2 = 3$.

4. Find the equations to those tangents to the hyperbola $4x^2 - 9y^2 = 1$ which are parallel to the line $4y = 5x + 9$.

5. Two diameters of a hyperbola are said to be *conjugate* when each bisects all chords parallel to the other. Show that the diameters $y = mx$, $y = m'x$ of the hyperbola $(x^2/a^2) - (y^2/b^2) = 1$ are conjugate if $mm' = b^2/a^2$.

6. P is a point on the hyperbola $(x^2/a^2) - (y^2/b^2) = 1$ and N is the foot of the perpendicular from P on the x-axis. The tangent to the hyperbola at the point P meets the x-axis at T. Show that, if O is the origin, $OT.ON = a^2$.

7. Show that the coordinates of any point on the hyperbola

$$(x^2/a^2) - (y^2/b^2) = 1$$

can be represented by $2x = a(t + t^{-1})$, $2y = b(t - t^{-1})$ and find the equation to the tangent at such a point.

8. Show that the equation to the chord joining the points $(a \sec \phi, b \tan \phi)$, $(a \sec \phi', b \tan \phi')$ on the hyperbola $(x^2/a^2) - (y^2/b^2) = 1$ is

$$\frac{x}{a} \cos \tfrac{1}{2}(\phi - \phi') - \frac{y}{b} \sin \tfrac{1}{2}(\phi + \phi') = \cos \tfrac{1}{2}(\phi + \phi').$$

Deduce the equation to the tangent at the point "ϕ".

17.15. The asymptotes of a hyperbola

The abscissae of the points of intersection of the straight line $y = mx + c$ and the hyperbola $(x^2/a^2) - (y^2/b^2) = 1$ are given by writing $y = mx + c$ in the latter equation. This gives

$$\frac{x^2}{a^2} - \frac{(mx + c)^2}{b^2} = 1,$$

or, arranged as a quadratic equation in $1/x$,

$$\frac{a^2(c^2 + b^2)}{x^2} + \frac{2a^2 mc}{x} + a^2 m^2 - b^2 = 0.$$

If $a^2 m^2 - b^2$ and $a^2 mc$ are both zero, this equation in $1/x$ has two zero roots. In other words, if

$$m = \pm b/a \quad \text{and} \quad c = 0,$$

the line $y = mx + c$ meets the hyperbola in two points at each of which $1/x$ is zero, i.e., the line meets the hyperbola in two points situated at an infinite distance from the centre of the curve.

Lines which meet a hyperbola in two points both of which are situated at an infinite distance, but which are not themselves altogether at infinity are called *asymptotes*. The asymptotes of the hyperbola $(x^2/a^2) - (y^2/b^2) = 1$ are therefore the lines

$$y = \pm \frac{b}{a} x. \tag{17.28}$$

These two lines pass through the centre of the hyperbola (the origin) and are equally inclined to the x-axis, the inclinations being $\pm \tan^{-1}(b/a)$. They are shown as the lines LOL', MOM' in Fig. 126. Written as a single equation, the asymptotes (17.28) would be given by

$$\left(y - \frac{b}{a} x \right) \left(y + \frac{b}{a} x \right) = 0,$$

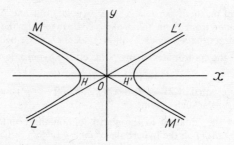

Fig. 126

or, after division by b^2 and slight rearrangement,

$$\frac{x^2}{a^2} - \frac{y^2}{b^2} = 0. \tag{17.29}$$

Example 12. *P is a point on the hyperbola $(x^2/a^2) - (y^2/b^2) = 1$ and the tangent at P meets the asymptotes at Q and Q'. Show that P is the mid-point of QQ'.*

If P is the point $(a \sec \phi, b \tan \phi)$, the tangent at P is, from equation (17.26),

$$\frac{x}{a} \sec \phi - \frac{y}{b} \tan \phi = 1.$$

The asymptotes of the hyperbola are

$$\frac{x^2}{a^2} - \frac{y^2}{b^2} = 0,$$

and the abscissae of the points Q, Q' are given by eliminating y from these two equations. Writing

$$y = \frac{b}{\tan \phi}\left(\frac{x}{a} \sec \phi - 1\right)$$

from the first equation in the second, we have

$$\frac{x^2}{a^2} - \cot^2 \phi\left(\frac{x}{a} \sec \phi - 1\right)^2 = 0.$$

This, using $\operatorname{cosec}^2 \phi = 1 + \cot^2 \phi$ and division by $\cot^2 \phi$, reduces to

$$\frac{x^2}{a^2} - \frac{2x}{a} \sec \phi + 1 = 0.$$

If the roots of this quadratic equation are x_1, x_2,

$$x_1 + x_2 = 2a \sec \phi,$$

and $\qquad \frac{1}{2}(x_1 + x_2) = a \sec \phi =$ abscissa of point P.

Similarly, by eliminating x from the equation to the tangent and that to the asymptotes, we shall find that half the sum of the ordinates of the points Q, Q' is equal to the ordinate of the point P. Hence P is the mid-point of QQ'.

17.16. The rectangular hyperbola

If in the hyperbola $(x^2/a^2) - (y^2/b^2) = 1$, the quantities a and b are equal, the equations to the asymptotes are, by (17.28), $y = \pm x$.

The asymptotes are therefore inclined at angles of $\pm 45°$ with the x-axis and are perpendicular to each other. The equation to the hyperbola can be written

$$x^2 - y^2 = a^2, \tag{17.30}$$

and, because of the perpendicularity of its asymptotes, it is often called a *rectangular* hyperbola. Because of the equality of a and b, this hyperbola is also sometimes said to be *equilateral*.

17.17. The equation to a rectangular hyperbola referred to its asymptotes

The equation to a rectangular hyperbola takes a very simple form when the axes of coordinates coincide with the asymptotes. If in

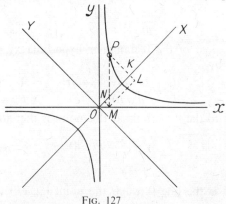

FIG. 127

Fig. 127, P is a point on a rectangular hyperbola and the coordinates (X, Y) of P are measured from lines OX, OY bisecting the angles between the two asymptotes Ox, Oy, X and Y are related by

$$X^2 - Y^2 = a^2. \tag{17.31}$$

PK, PM are drawn perpendicular to OX, Ox respectively and ML, MN are drawn perpendicular to PK, OX as shown. Since the angle XOx is $45°$,

$$OK = NK + ON = ML + ON$$
$$= PM \cos 45° + OM \cos 45° = (PM + OM)/\sqrt{2},$$

and $$PK = PL - KL = PL - MN$$
$$= PM \sin 45° - OM \sin 45° = (PM - OM)/\sqrt{2}.$$

If Ox, Oy are taken as coordinate axes, $OM = x$, $PM = y$, and since $OK = X$, $PK = Y$, the above results give

$$X = (y + x)/\sqrt{2}, \quad Y = (y - x)/\sqrt{2}.$$

Substitution in (17.31) yields

$$\frac{(y + x)^2}{2} - \frac{(y - x)^2}{2} = a^2,$$

reducing to $\qquad\qquad xy = a^2/2.$

Writing $2c^2 = a^2$, the equation takes the very simple form

$$xy = c^2. \qquad (17.32)$$

When a rectangular hyperbola is referred to its asymptotes as axes, a point whose coordinates are given by

$$x = ct, \quad y = c/t, \qquad (17.33)$$

always lies on it, for these coordinates satisfy the equation $xy = c^2$. The equations (17.33) give a parametric representation to a rectangular hyperbola and the point given by them may be called the point "t".

If we differentiate equations (17.33) with respect to t

$$\frac{dx}{dt} = c, \quad \frac{dy}{dt} = -\frac{c}{t^2},$$

and the gradient of the rectangular hyperbola $xy = c^2$ at the point "t" is given by

$$\frac{dy}{dx} = \frac{(dy/dt)}{(dx/dt)} = \frac{-c/t^2}{c} = -\frac{1}{t^2}.$$

The tangent to the hyperbola at this point is therefore the line

$$y - \frac{c}{t} = -\frac{1}{t^2}(x - ct),$$

or, $\qquad\qquad x + t^2 y = 2ct. \qquad (17.34)$

The normal is the line through the point $(ct, c/t)$ with slope t^2 and its equation is

$$y - \frac{c}{t} = t^2(x - ct),$$

or, $\qquad\qquad t^2 x - y = c(t^3 - 1/t). \qquad (17.35)$

Example 13. *Find the equation to the normal at the point* (3, 4) *to the rectangular hyperbola* $xy = 12$, *and the coordinates of its second point of intersection with the curve.* (O.C.)

Here $c^2 = 12$ and $c = 2\sqrt{3}$. The parameter "t" of the point (3, 4) is given by $3 = 2\sqrt{3}t$, so that $t = \frac{1}{2}\sqrt{3}$. From (17.35) the equation to the normal is

$$\left(\frac{\sqrt{3}}{2}\right)^2 x - y = 2\sqrt{3}\left\{\left(\frac{\sqrt{3}}{2}\right)^3 - \frac{2}{\sqrt{3}}\right\},$$

or, $3x - 4y + 7 = 0$.

Let the second point of intersection of the normal be the point $(2\sqrt{3}t, 2\sqrt{3}/t)$. Since this point lies on the line $3x - 4y + 7 = 0$

$$6\sqrt{3}t - \frac{8\sqrt{3}}{t} + 7 = 0,$$

giving $6\sqrt{(3t^2} + 7t - 8\sqrt{3} = 0$. This can be written in the form

$$(2t - \sqrt{3})(3\sqrt{3}t + 8) = 0,$$

so that $t = \frac{1}{2}\sqrt{3}$ or $-8/(3\sqrt{3})$. The first value of t corresponds to the point $(3, 4)$, so that the coordinates of the second point of intersection are found by using $t = -8/(3\sqrt{3})$. With this value of t, the point $(2\sqrt{3}t, 2\sqrt{3}/t)$ is the point $(-16/3, -9/4)$.

EXERCISES 17 (e)

1. A tangent to the hyperbola $(x^2/a^2) - (y^2/b^2) = 1$ meets the asymptotes in Q and Q'. If O is the centre of the hyperbola, show that the area of the triangle OQQ' is ab.

2. S is the focus on the positive x-axis of the hyperbola $(x^2/9) - (y^2/4) = 1$ and Y is the foot of the perpendicular from S on that asymptote which lies in the first and third quadrants. Prove that Y lies on the directrix and also on the circle whose centre is the origin and whose radius is the semi-transverse axis.

3. Find the eccentricity of a rectangular hyperbola.

4. Show that the equation to the line joining the points $(ct, c/t)$, $(ct', c/t')$ on the rectangular hyperbola $xy = c^2$ is $x + tt'y = c(t + t')$. Deduce the equation to the tangent at the point "t".

5. The normal to the rectangular hyperbola $xy = 8$ at the point $(4, 2)$ meets the asymptotes at M and N. Find the length of MN.

6. PN is the perpendicular to an asymptote from a point on a rectangular hyperbola. Prove that the locus of the mid-point of PN is a rectangular hyperbola with the same axes.   (O.C.)

7. TP, TQ are the tangents at two points P, Q on the rectangular hyperbola $xy = c^2$. Show that the line joining the centre of the hyperbola to the point T bisects the chord PQ.

8. Find the locus of the mid-point of a straight line which moves so that it always cuts off a constant area k^2 from the corner of a square.

EXERCISES 17 (f)

1. Write down the coordinates of the focus S and the equation to the directrix of the parabola $y^2 = 4x$.
   PQ is a focal chord of this parabola and PR is a chord perpendicular to the axis. The tangents at Q and R meet at T. Prove that ST is parallel to the directrix.   (L.U.)

2. The tangent to the parabola $x^2 = 4ay$ at the point for which $y = at^2$ meets the axes of x and y at the points P and Q. Obtain an expression for the area of the triangle POQ in terms of t, O being the origin.   (L.U.)

3. Find the coordinates of the point of intersection R of the tangents to the parabola $y^2 = 4ax$ at the points $P(at_1^2, 2at_1)$, $Q(at_2^2, 2at_2)$. If the tangents at P, Q are inclined to one another at an angle of $45°$, show that the locus of R is the curve $y^2 = x^2 + 6ax + a^2$.   (L.U.)

4. P and Q are two points on the parabola $y^2 = 4ax$ whose coordinates are $(at_1^2, 2at_1)$ and $(at_2^2, 2at_2)$. O is the origin of coordinates and OP is perpendicular to OQ. Show that $t_1t_2 + 4 = 0$ and that the tangents to the curve at P and Q meet on the line $x + 4a = 0$.   (L.U.)

5. P is the point $(at^2, 2at)$ on the parabola $y^2 = 4ax$. N is the foot of the perpendicular drawn from the origin to the tangent at P. Show that, as P varies, the locus of N is the curve $x(x^2 + y^2) + ay^2 = 0$. (L.U.)

6. The normal at the point $P(at^2, 2at)$ to the parabola $y^2 = 4ax$ meets the parabola again at the point $R(aT^2, 2aT)$. Prove that $T = -t - 2/t$. Prove also that, if the normal at $Q(at_1{}^2, 2at_1)$ passes through R, then $t_1 = 2/t$. (O.C.)

7. The normal at the point $P(at^2, 2at)$ to the parabola $y^2 = 4ax$ meets the curve again at the point $Q(at'^2, 2at')$. Find t' in terms of t and hence, or otherwise, prove that the lines joining the origin to P and Q are at right angles if $t^2 = 2$.

8. From a point P on a parabola with vertex A and focus S, the line PN is drawn perpendicular to the axis AS. The tangents at A and P intersect at Q. Prove that $PQ^2 = AN \cdot SP$.

9. A variable chord through the focus of the parabola $y^2 = 4ax$ cuts the curve at P and Q. The straight line joining P to the point $(0, 0)$ cuts the line joining Q to the point $(-a, 0)$ at R. Show that the equation to the locus of the point R is $y^2 + 8x^2 + 4ax = 0$.

10. Find the equations to the tangents to the parabola $y^2 = 9x$ which pass through the point $(4, 10)$.

11. Find the equations to the tangents to the curves

$$x^2 + y^2 = 9, \quad 4x^2 + y^2 = 16$$

at one of the points where the curves intersect, indicating on a sketch which point of intersection you have chosen.

Calculate the angle at which the curves intersect. (Q.E.)

12. AB and BC are two rods each of length a jointed at B. A is pivoted to a fixed point and C can move in a straight slot, which passes through A. P is a point on BC such that $BP = b$. Find the coordinates of P referred to AC as axis of x and a perpendicular through A to AC as axis of y, when each of the angles BAC, BCA is θ, and show that, as the rods move, P traces out an ellipse whose semi-axes are $(a + b)$ and $(a - b)$. (O.C.)

13. Show that the tangents to the ellipse $x^2 + 2y^2 = 18$ at the points $(0, -3)$, $(-72/17, -3/17)$ intersect on the normal at the point $(4, 1)$. (O.C.)

14. The equation to a chord of the ellipse $x^2 + 4y^2 = 260$ is $x + 6y = 50$. Find the coordinates of its middle point. (O.C.)

15. The tangent at the point $P(a \cos \phi, b \sin \phi)$ to an ellipse centre C and semi-axes a, b meets the major axis at T. N is the foot of the perpendicular from P to the major axis. Show that $CN \cdot CT = a^2$. (O.C.)

16. S, S' are the foci of an ellipse of semi-axes a and b. The normal at a point P on the ellipse meets the minor axis at G. Show that the square of the distance of G from either focus is $(a^2 - b^2)SP \cdot S'P/b^2$. (O.C.)

17. Find the ratio of a to b for which the ellipse $(x^2/a^2) + (y^2/b^2) = 1$ and the parabola $y^2 = 4ax$ cut at right angles. (Q.E.)

18. The centre of a hyperbola is the origin and its transverse axis lies along

the axis of x. If the distance between the foci is 16 and the eccentricity is $\sqrt{2}$, write down the equation to the hyperbola.

19. If h and k are the intercepts on the coordinate axes of any tangent to the hyperbola $(x^2/a^2) - (y^2/b^2) = 1$, show that

$$(a^2/h^2) - (b^2/k^2) = 1.$$

20. The normal at any point P on the hyperbola $(x^2/a^2) - (y^2/b^2) = 1$ meets the x-axis at G. Q is a point on either asymptote such that PQ is parallel to the y-axis. Show that GQ is perpendicular to the asymptote on which Q lies.

21. P is the point $(5/4, 3/4)$ on the rectangular hyperbola $x^2 - y^2 = 1$. The normal at P cuts the axes of x and y at G and g respectively. Prove that, if O is the origin, $PG = Pg = PO$.

22. Show that the equation to the chord joining two points (x_1, y_1), (x_2, y_2) on the rectangular hyperbola $xy = c^2$ is

$$\frac{x}{x_1 + x_2} + \frac{y}{y_1 + y_2} = 1.$$

23. PP' is a diameter of the rectangular hyperbola $xy = c^2$. The tangent at P meets lines through P' parallel to the asymptotes in Q and Q'. Prove that P is the middle point of QQ' and that the equation to the locus of Q is $xy + 3c^2 = 0$.

24. The perpendicular from the origin to the tangent at a point P on the rectangular hyperbola $xy = c^2$ meets the curve at Q and R. The chords PQ and PR meet the x-axis at U and V. Prove that the mid-point of UV is the foot of the perpendicular from P to the x-axis. (O.C.)

25. The tangent at P to the rectangular hyperbola $xy = c^2$ meets the lines $x - y = 0$ and $x + y = 0$ at A and B, and Δ denotes the area of the triangle OAB where O is the origin. The normal at P meets the x-axis at C and the y-axis at D. If Δ_1 denotes the area of the triangle ODC show that $\Delta^2\Delta_1 = 8c^6$. (O.C.)

CHAPTER 18

SOME THEOREMS IN PURE GEOMETRY

18.1. Introduction

The student is assumed to be familiar with the usual theorems on angles at a point, parallel straight lines, the congruence of triangles, areas of triangles and simple rectilinear figures and Pythagoras' theorem. Such theorems form one part of what may be called "school" geometry and, except in so far that they may be appealed to in the solution of some examples and exercises, will not be further considered here.

A knowledge of the usual theorems on the circle and similar triangles will also be assumed. This part of the subject is also usually studied at the school stage but a statement of the more important theorems and a few revision examples may not be out of place here. For proofs of these theorems of elementary geometry the reader is referred to one of the usual texts.

The remainder of this chapter is devoted to theorems on similar rectilinear figures, further properties of the triangle, concurrency of lines and collinearity of points. It is assumed that the reader has not previously studied such theorems, proofs are given and rather more examples and exercises are provided.

18.2. Statements of some theorems on the circle

The more important theorems on the circle can be stated as follows:—

(a) A straight line drawn from the centre of a circle to bisect a chord, which is not a diameter, is at right angles to the chord.
   Conversely, the perpendicular to a chord from the centre bisects the chord.

(b) There is one, and only one, circle which passes through three given points not in a straight line.

(c) Equal chords of a circle are equidistant from the centre, and the converse.

(d) The tangent to a circle and the radius through the point of contact are perpendicular to each other.

(e) The angle which an arc of a circle subtends at the centre is double that which it subtends at any point on the remaining part of the circumference.

(f) Angles in the same segment of a circle are equal, and the converse.

(g) The angle in a semicircle is a right angle.

344

(h) The opposite angles of any quadrilateral inscribed in a circle are supplementary, and the converse.

(i) If a straight line touch a circle, and from the point of contact a chord be drawn, the angles which this chord makes with the tangent are equal to the angles in the alternate segments, and the converse.

(j) If two chords of a circle intersect either inside or outside a circle, the rectangle contained by the parts of the one is equal to the rectangle contained by the parts of the other, and the converse.

(k) If, from any point outside a circle, a secant and a tangent are drawn, the rectangle contained by the whole secant and the part of it outside the circle is equal to the square on the tangent, and the converse.

18.3. Some revision examples on the geometry of the circle

In this section two examples on the theorems of § 18.2 are given. Further problems will be found in Exercises 18 (a).

Example 1. *A line DE parallel to the base BC of a triangle ABC cuts AB, AC in D and E respectively. The circle which passes through D, and touches AC at E, meets AB at F. Prove that F, E, C, B lie on a circle.* (L.U.)

Join *EF* (Fig. 128). By theorem (i) above,

$$\text{angle } CED = \text{angle } EFD.$$

Since *DE* is parallel to *BC*, the angles *CED*, *BCA* are supplementary.

Hence the angles *BCA*, *EFD* are supplementary, and, by the converse of theorem (h) above, the points *F, E, C, B* lie on a circle.

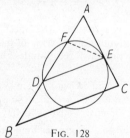

FIG. 128

Example 2. *In a triangle ABC, the side AB is greater than the side AC, and D is a point in AB such that AD is equal to AC; the internal bisectors of the angles B and C meet in I. Show that the four points B, D, I, C lie on a circle.* (O.C.)

Since *AD = AC*, the triangle *ADC* is isosceles and (Fig. 129),

$$\text{angle } ADC = \text{angle } DCA.$$

Since the sum of the angles of the triangle *ADC* is 180°,

$$2 \text{ angle } ADC + \text{angle } CAD = 180°,$$

giving   angle $ADC = 90° - \frac{1}{2}A,$

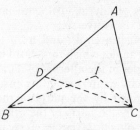

FIG. 129

where A = angle CAD. The angle CDB is the supplement of the angle ADC so that angle $CDB = 180° - (90° - \frac{1}{2}A) = 90° + \frac{1}{2}A$.

Since angle $IBC = \frac{1}{2}B$, angle $BCI = \frac{1}{2}C$, the triangle IBC gives

$$\text{angle } CIB = 180° - \tfrac{1}{2}B - \tfrac{1}{2}C$$
$$= 180° - \tfrac{1}{2}(180° - A) = 90° + \tfrac{1}{2}A,$$

since $A + B + C = 180°$.

Hence, angle CDB = angle CIB and the converse of theorem (f) shows that the points B, D, I, C lie on a circle.

EXERCISES 18 (a)

1. O is a fixed point and P is a variable point on a fixed line l. Q is a point on the line OP such that $OP \cdot OQ$ is constant. Show that the locus of Q is a circle passing through O.   (L.U.)

2. The points A, B, C, D lie on a circle in that order. From B perpendiculars BX, BY are drawn to AD and CD respectively. Prove that the angle BYX is equal to the angle BDA.   (L.U.)

3. The altitudes AD, BE of a triangle ABC meet at the point H. If AD is produced to meet the circumscribing circle of the triangle at K, prove that $HK = 2HD$. Prove also that the three altitudes all pass through the point H.   (L.U.)

4. With a point A on the circumference of a circle S_1 as centre, a circle S_2 is described, cutting S_1 at B and C. A straight line through A meets S_1 at P, S_2 at Q, and the chord BC at R. Prove that $AQ^2 = AP \cdot AR$.   (L.U.)

5. In a triangle ABC the angle C is greater than the angle B. The bisector of the angle A meets the circumcircle of the triangle at E, and F, G are the feet of the perpendiculars from E to AB and AC respectively. Prove that $AF = AG = \frac{1}{2}(AB + AC)$ and that $BF = CG = \frac{1}{2}(AB - AC)$.   (L.U.)

6. X, Y are the points of contact of tangents drawn to a circle from an external point P. A straight line through P cuts the circle in the points Q and R. RY (produced where necessary) meets the circle through P, Q and Y, in the point K. Prove that $PK = PY$ and that the angle $KPX = 2$ angle XYR.   (L.U.)

7. AB is a diameter of a circle whose centre is C, and X is any point on the circumference. P is the foot of the perpendicular drawn from X to AB. H

and K are points on the circumference such that $XH = XK = XP$. HK meets XP in M and XC in N. Prove that the points M, N, C and P lie on the same circle.   (L.U.)

8. A chord AB of a circle, whose centre is C, is produced to a point T, and a point X is taken on CT such that $TC.TX = TA.TB$. Prove that the angles AXC, BXT are equal.

18.4. Statements of some theorems on proportion and similar triangles

The more important theorems on proportion and similar triangles can be stated as follows:—

(*l*) If a straight line is drawn parallel to one side of a triangle, the other two sides are divided proportionally, and the converse.

(*m*) If two triangles are equiangular their corresponding sides are proportional, and the converse.

(*n*) If two triangles have one angle of the one equal to one angle of the other and the sides about these equal angles proportional, the triangles are similar.

(*o*) If a perpendicular is drawn from the right angle of a right-angled triangle to the hypotenuse, the triangles on each side of the perpendicular are similar to the whole triangle and to one another.

(*p*) The internal bisector of an angle of a triangle divides the opposite side in the ratio of the sides containing the angle, and likewise the external bisector externally. The converse is also true.

18.5. Some revision examples on proportion and similar triangles

Here a few worked examples on the theorems of § 18.4 are given; some problems of this type are given in Exercises 18 (*b*).

Example 3. *Any line parallel to the base BC of a triangle ABC cuts AB, AC in H and K respectively. P is any point on a line through A parallel to BC. If PH, PK produced cut BC at Q and R respectively, prove that BQ = CR.*   (L.U.)

Since (Fig. 130) AP, BC are parallel, the angles HAP, HBQ are equal and so also are the angles HPA, BQH. Hence the triangles APH, HBQ are equiangular. By theorem (*m*) above,

$$\frac{BQ}{AP} = \frac{BH}{AH}.$$

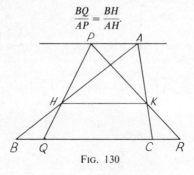

FIG. 130

Similarly the triangles APK, KCR are equiangular and

$$\frac{CR}{AP} = \frac{CK}{AK}.$$

Since HK is parallel to BC, theorem (l) above shows that

$$\frac{BH}{AH} = \frac{CK}{AK},$$

and hence

$$\frac{BQ}{AP} = \frac{CR}{AP}$$

leading to $BQ = CR$.

Example 4. *In a triangle ABC, the angle BAC is a right angle and AB = 2AC. AD is the perpendicular from A on to BC. Show that BD = 4DC.*

By theorem (o) above, the triangles (Fig. 131) ABD, ABC are similar and therefore

$$\frac{BD}{AD} = \frac{AB}{AC} = 2,$$

giving $BD = 2AD$.

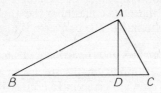

FIG. 131

Similarly, the triangles ADC, ABC are similar, so that

$$\frac{DC}{AD} = \frac{AC}{AB} = \frac{1}{2},$$

giving $AD = 2DC.$

Hence $BD = 2AD = 2(2DC) = 4DC.$

Example 5. *ABC is a triangle, right-angled at A. AN is perpendicular to BC, BK bisects the angle B and meets AC at K and AN at L. Prove that*
$$AL:LN = CK:KA.$$ (L.U.)

Applying theorem (p) above to the triangle ABN (Fig. 132),

$$\frac{AL}{LN} = \frac{BA}{BN}.$$

Similarly, from the triangle ABC,

$$\frac{CK}{KA} = \frac{BC}{BA}.$$

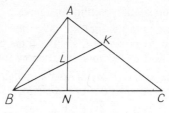

FIG. 132

But theorem (o) shows that the triangles ABN, ABC are similar and therefore

$$\frac{BA}{BN} = \frac{BC}{BA}.$$

Hence $AL:LN = CK:KA$.

EXERCISES 18 (b)

1. Two circles ABP, PDC intersect at P and APD, BPC are straight lines. Prove that, if the radius of the circle APB is twice the radius of the circle PDC, then the chord AB is twice the chord CD. (L.U.)

2. In a right-angled triangle ABC prove that the perpendicular AD from the right angle A to the hypotenuse BC is a mean proportional between the segments BD, DC of the hypotenuse.

   A variable tangent to a given circle meets two fixed parallel tangents at P, Q and touches the circle at R. Prove that the rectangle $PR.RQ$ is constant. (L.U.)

3. Two straight lines OAB, OCD cut a circle at A, B, C, D. Through O a line is drawn parallel to BC to meet AD (produced) in X. Prove that $OX^2 = AX.DX$. (L.U.)

4. Two circles, centres A and B, touch externally at a point C. A common tangent touches the circles at P and Q respectively and meets AB produced at S. If T is the point in PQ such that $PT:TQ = PS:QS$, prove that (i) the triangles PAT, QBT are similar, (ii) the internal bisector of the angle ATB passes through C. (L.U.)

5. Tangents are drawn to a circle at the points A and B, and P is any other point on the circle. Prove that the product of the perpendiculars from P to the tangents is equal to the square of the perpendicular from P to AB. (L.U.)

6. A triangle ABC is inscribed in a circle. Lines drawn through A parallel to the tangents at B and C meet BC in D and E respectively. Prove that (i) $AD = AE$, (ii) $BD/CE = AB^2/AC^2$. (L.U.)

7. Two equal circles intersect at A and B. A line through A cuts one circle at C and the other at D, and the circle BCD cuts BA, produced if necessary, at T. Prove that $TC/TD = AC/AD$. (L.U.)

8. X is the mid-point of the base BC of a triangle ABC. The bisectors of the angles AXB, AXC meet AB, AC at H and K. Prove that HK is parallel to BC.

18.6. Two theorems on similar rectilinear figures

Polygons which are equiangular and have their corresponding sides proportional are said to be *similar*. If also their corresponding sides are parallel they are said to be *similarly situated* (or *homothetic*).

Theorem 1. *The ratio of the areas of similar triangles (or polygons) is equal to the ratio of the squares on corresponding sides.*

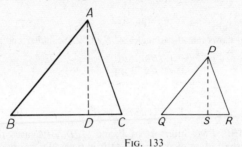

FIG. 133

In Fig. 133, ABC, PQR are similar triangles and AD, PS are their altitudes. Since the angle ABD equals the angle PQS, and the angle BDA equals the angle QSP, both the latter being right angles, the triangles ABD, PQS are equiangular. Hence

$$\frac{AD}{PS} = \frac{AB}{PQ} = \frac{BC}{QR},$$

the last equality following from the fact that the triangles ABC, PQR are similar. Thus

$$\frac{\triangle ABC}{\triangle PQR} = \frac{\frac{1}{2}AD \cdot BC}{\frac{1}{2}PS \cdot QR} = \frac{BC^2}{QR^2}.$$

If two polygons are similar, they can be divided up into the same number of similar triangles and it follows that the ratio of the areas of similar polygons is equal to the ratio of the squares on corresponding sides.

Theorem 2. *If O is any fixed point and $ABCD \ldots P$ is any polygon, and if points A', B', C', $\ldots$, P' are taken on OA, OB, OC, $\ldots$, OP (or these lines produced either way), such that*

$$OA'/OA = OB'/OB = \ldots = OP'/OP = \lambda,$$

then the polygons $ABCD \ldots P$, $A'B'C'D' \ldots P'$ are similar and similarly situated.

Since $OA'/OA = OB'/OB$, AB is parallel to $A'B'$ and the triangles OAB, $OA'B'$ are similar. Hence

$$\frac{A'B'}{AB} = \frac{OA'}{OA} = \lambda.$$

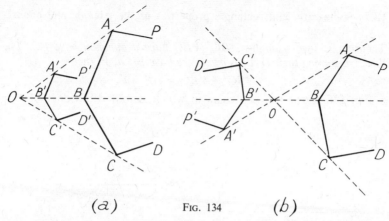

(a)   Fig. 134   (b)

Corresponding sides of the two polygons are therefore proportional and parallel and the two polygons are therefore similar and similarly situated.

In Fig. 134, O is said to be the *centre of similitude* of the two polygons. If corresponding points of the two polygons lie on the same side of O, the polygons are said to be *directly homothetic* with respect to O and O is said to be the *external* centre of similitude (Fig. (a)). If corresponding points lie on opposite sides of O, the polygons are said to be *inversely homothetic* with respect to O and O is in this case called the *internal* centre of similitude (Fig. (b)).

Example 6. *PQR is an acute-angled triangle. Show how to construct a square with two vertices on QR, one vertex on PQ and one vertex on PR.* (L.U.)

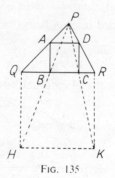

Fig. 135

On QR describe a square $QHKR$ externally to the triangle (Fig. 135). Join PH, PK and let these lines meet QR at B and C respectively. Draw BA, CD perpendicular to QR to meet PQ, PR at A and D respectively. Then $ABCD$ is the required square, for regarding P as a centre of similitude, $ABCD$ is similar to $QHKR$ and is therefore a square.

18.7. Some ratio and rectangle properties of the triangle and quadrilateral

Theorem 3. *Two triangles ABC, A'BC have a common base BC. The line AA' joining their vertices meets the base BC at P; then*

$$\triangle ABC/\triangle A'BC = AP/A'P.$$

Fig. 136 shows the two cases in which A and A' lie on the same

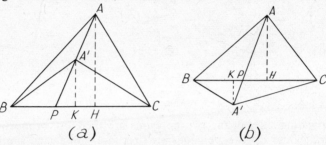

(a) (b)

FIG. 136

and opposite sides of the base BC. AH, $A'K$ are drawn perpendicular to BC. Then

$$\frac{\triangle ABC}{\triangle A'BC} = \frac{\frac{1}{2}AH.BC}{\frac{1}{2}A'K.BC} = \frac{AH}{A'K}$$

$$= \frac{AP}{A'P},$$

for the triangles APH, $A'PK$ are similar.

Theorem 4. *If A, B are two fixed points and P is a moving point such that the ratio PA/PB is constant, the locus of P is a circle.*

In Fig. 137, AB is divided internally at C and externally at D in the given ratio PA/PB. Since

$$\frac{PA}{PB} = \frac{AC}{CB} = \frac{AD}{DB},$$

PC and PD are the internal and external bisectors of the angle APB.

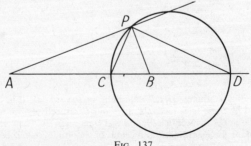

FIG. 137

Hence the angle DPC is a right angle and P therefore lies on the circle whose diameter is CD. This circle is called the circle of *Apollonius*.

Theorem 5 (Ptolemy's theorem). *If ABCD is a cyclic quadrilateral, then* $AB.CD + BC.DA = AC.BD.$

In Fig. 138, AX is drawn to meet BD at X so that the angle XAD

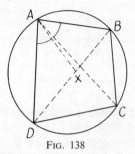

FIG. 138

is equal to the angle BAC. Since these angles and the angles ADX, ACB are equal (angles in the same segment), the triangles ADX, CBA are similar and

$$\frac{DA}{DX} = \frac{AC}{BC}, \quad \text{giving } BC.DA = AC.DX.$$

Since the angle $CAD =$ angle $CAX +$ angle $XAD =$ angle CAX $+$ angle $BAC =$ angle BAX, and the angles in the same segment DCA, XBA are also equal, the triangles ADC, AXB are similar so that

$$\frac{CD}{AC} = \frac{XB}{AB}, \quad \text{giving } AB.CD = AC.XB.$$

By addition $AB.CD + BC.DA = AC(DX + XB) = AC.BD.$

Example 7. *If P is a point on the smaller arc BC of the circumscribed circle of an equilateral triangle ABC, show that PB + PC = PA.*

Applying Ptolemy's theorem to the cyclic quadrilateral $ABPC$ (Fig. 139),

$$PB.CA + PC.AB = AP.BC.$$

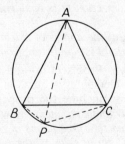

FIG. 139

Since the triangle ABC is equilateral,

$$CA = AB = BC,$$

and the result follows immediately.

EXERCISES 18 (c)

1. Give the steps of a purely geometrical construction for an equilateral triangle whose area is twice the area of a given quadrilateral. (L.U.)

2. Prove that the areas of the rectangles formed by the external and internal bisectors, respectively, of the angles of a parallelogram are in the ratio $(a + b)^2 : (a + b)^2$, where a, b are the lengths of the sides of the parallelogram.

3. Show how to trisect the area of a triangle by means of two lines drawn parallel to one of the sides. (O.C.)

4. A chord AB of a circle (when produced) meets the tangent at a point P to the circle at the point T. Show that $(PA/PB)^2 = TA/TB$.

5. The base BC of a triangle ABC is fixed and X, Y are the mid-points of the sides AB, AC respectively. If $BY = 2CX$ and BY, CX intersect at P, show that, as the point A moves, the point P always lies on a circle.

6. A and B are two fixed points and P is a point such that $PA/PB = x/y$, where the ratio x/y is constant. The circle on which P lies meets AB in X, AB produced in Y and O is the mid-point of AB. If $AB = 2(x + y)$ verify that (i) $2/AB = (1/AX) + (1/AY)$ and (ii) $OX . OY = OB^2$.

7. $ABCD$ is a cyclic quadrilateral. The sides AB, CD are 0·06 m and 0·12 m respectively and are parallel. If the length of the diagonal BD is 0·11 m, show that $AC = BD$ and that $AD = BC = 0·07$ m.

8. AC is a diameter of a circle and B, D are two points on the circle, one on either side of AC. If the angles CAD, BAC are denoted by α, β respectively, use Ptolemy's theorem to show that

$$\sin(\alpha + \beta) = \cos \alpha \sin \beta + \sin \alpha \cos \beta.$$

18.8. Some further properties of a triangle

The two elementary properties:—

(a) *the perpendicular bisectors of the sides of a triangle are concurrent and their point of intersection O is equidistant from the vertices o the triangle,*

(b) *the internal bisectors of the angles of a triangle are concurrent and their point of intersection I is equidistant from the sides of the triangle,*

should be already known. Proofs of (a) and (b) are omitted here as they form part of every elementary course in geometry. Property (b) above can be extended to include the property (c) that *the internal bisector of one angle of a triangle is concurrent with the external bisectors of the other two angles and the three points of concurrence I_1, I_2, I_3 are each equidistant from the three sides of the triangle.* This too should be already known and again a proof is omitted here. The points O, I, I_1, I_2 and I_3 are known respectively as the *circumcentre*, *in-centre* and *ex-centres* of the triangle. Other properties of a similar character are considered below.

Property (d). *The altitudes of a triangle are concurrent.*

In Fig. 140, AD, BE, CF are the altitudes of the triangle ABC. YZ, ZX and XY are lines through A, B, C parallel to BC, CA and AB respectively. Since $ZBCA$ is a parallelogram, $ZA = BC$ and since

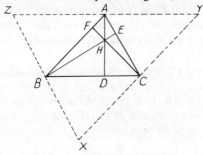

Fig. 140

$ABCY$ is a parallelogram, $AY = BC$. Hence $ZA = AY$ and, since YZ, BC are parallel, AD is perpendicular to YZ. Thus AD is the perpendicular bisector of YZ. Similarly BE, CF are the perpendicular bisectors of ZX and XY and hence, by property (a), the lines AD, BE, CF are concurrent.

The point H of concurrence of the altitudes AD, BE, CF is known as the *orthocentre* of the triangle ABC, and the triangle DEF is called the *pedal triangle* of the triangle ABC.

Property (e). *The lines joining the vertices of a triangle to the mid-points of the opposite sides are concurrent.*

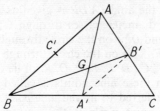

Fig. 141

In Fig. 141, A', B', C' are the mid-points of the sides BC, CA, AB of the triangle ABC. The lines AA', BB' meet at the point G. Since

$CA' = \frac{1}{2}CB$ and $CB' = \frac{1}{2}CA$, $A'B'$ is parallel to AB and $A'B' = \frac{1}{2}AB$. Since $A'B'$ is parallel to AB, the triangles $GA'B'$, GAB are similar and

$$\frac{A'G}{AG} = \frac{A'B'}{AB} = \frac{1}{2}.$$

Thus, $A'G = \frac{1}{2}GA$, leading to $A'G = \frac{1}{3}AA'$. Hence BB' meets AA' at the point of trisection of AA' nearest to A'. Similarly it can be shown that CC' meets AA' at the same point.

The three lines AA', BB', CC' are known as the *medians* of the triangle, and the point G of concurrence of the medians is called the *centroid* of the triangle ABC. It can be shown that the point G coincides with the centre of mass of a triangular lamina ABC of uniform surface density.

Property (f). *A', B', C' are the mid-points of the sides BC, CA, AB of a triangle ABC and D, E, F are the feet of the perpendiculars from A, B, C on to BC, CA, AB. H is the orthocentre of the triangle ABC and P, Q, R are the mid-points of HA, HB and HC. The nine points A', B', C', D, E, F, P, Q, R lie on a circle.*

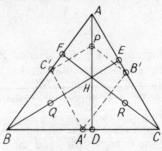

Fig. 142

Since in Fig. 142, $AP = PH$ and $AC' = C'B$, PC' is parallel to BH. Since A', C' are the mid-points of BC, AB respectively $A'C'$ is parallel to AC. Hence the angle $PC'A'$ is a right angle. Similarly, the angle $PB'A'$ is a right angle, and the angle $A'DP$ is given as a right angle. Hence the circle on $A'P$ as diameter passes through C', B' and D. In other words, the points P and D lie on the circle through A', B', C'. Similarly, the points E, Q and F, R lie on the circle through A', B', C', and hence the nine points A', B', C', D, E, F, P, Q, R all lie on the same circle.

The circle through these nine points is called the *nine-point* circle and its centre N is referred to as the *nine-point centre*. It should be noted that the nine-point circle is the circumscribed circle of the triangle $A'B'C'$. Since the sides of this triangle are each half those of the corresponding sides of the triangle ABC, it follows that the radius of the nine-point circle of the triangle ABC is $\frac{1}{2}R$ where R is the radius of the circumscribed circle of the triangle.

Property (g). *The circumcentre, orthocentre, centroid and nine-point centre of a triangle all lie on a single straight line (the Euler line).*

In Fig. 143, O and G are respectively the circumcentre and centroid of the triangle ABC. A' and C' are the mid-points of BC, BA, and OG is produced to H so that $OG = \frac{1}{2}GH$. The first stage of the proof consists of showing that H is the orthocentre of the triangle ABC.

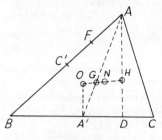

Fig. 143

Let AH (produced) meet BC at D. Since G lies at the point of tri-section of AA' nearest to A', $A'G = \frac{1}{2}GA$ and the triangles $OA'G$, AGH contain equal angles $A'GO$, AGH and the sides about these angles are proportional, the triangles are similar and therefore AH is parallel to OA'. But OA' is perpendicular to BC and hence AH is perpendicular to BC. It can be shown similarly that CH is perpendicular to AB and hence that H is the orthocentre.

Since $A'O$, DH are both perpendicular to BC, the perpendicular bisector of $A'D$ bisects OH. Similarly, the perpendicular bisector of $C'F$ bisects OH. But both $A'D$ and $C'F$ are chords of the nine-point circle and therefore the mid-point of OH is the nine-point centre N.

Property (h). *If P is a point on the circumcircle of a triangle ABC and L, M, N are the feet of the perpendiculars from P to BC, CA, AB, then L, M, N lie on a straight line (the Simson or pedal line of P with respect to the triangle ABC).*

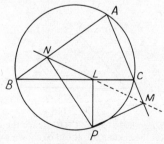

Fig. 144

Since (Fig. 144), the angles *PMC*, *CLP* are right angles, the points *P, M, C, L* are concyclic. Hence the angle *MLP* equals the angle *MCP*. Since the points *P, C, A, B* are concyclic, the angle *MCP* equals the angle *ABP*. But the angles *PNB*, *PLB* are both right angles so that the points *P, L, N, B* are concyclic and the angles *ABP, PLN* are supplementary. Hence

$$\text{angle } MLP = \text{angle } MCP = \text{angle } ABP$$
$$= 180° - \text{angle } PLN,$$

so that the sum of the angles *MLP, PLN* is two right angles. Thus *MLN* is a straight line.

Example 8. *I is the in-centre, O the circumcentre and H the orthocentre of a triangle ABC. Prove that AI bisects the angle HAO.*

In Fig. 145, *AI* produced meets the circumscribed circle at *X*. Since the angles *XAB, CAX* are equal, so are the arcs *BX, XC*. Hence *OX* is perpendicular to *BC* and is therefore parallel to *AH*.

Hence the angle *HAI* is equal to the angle *OXI* and the latter angle is equal to the angle *IAO* as the triangle *XAO* is isosceles.

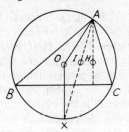

FIG. 145

Example 9. *H is the orthocentre and DEF the pedal triangle of a triangle ABC. Show that H is the in-centre of the triangle DEF.*

The points *B, D, H, F* (Fig. 146) are concyclic since *BDH, HFB* are both right angles. Therefore

$$\text{angle } FDH = \text{angle } FBH$$
$$= \text{angle } ABE = 90° - \text{angle } CAB.$$

Similarly, angle *HDE* = 90° − angle *CAB*. Hence the angles *FDH, HDE* are equal and *HD* bisects the angle *FDE*. It can be shown similarly that *HE* bisects the angle *DEF* and it follows that *H* is the in-centre of the triangle *DEF*.

EXERCISES 18 (*d*)

1. The vertices B, C of a triangle ABC are fixed and the vertex A moves so that the angle BAC is constant. Show that the in-centre I of the triangle ABC lies on a circle passing through B and C. (L.U.)

2. In a triangle ABC, AD is perpendicular to BC and H is the orthocentre of the triangle. AD meets the circumcircle of the triangle ABC at X. Prove that $HD = DX$ and that $AD \cdot HD = BD \cdot DC$. (L.U.)

3. O is the circumcentre and H the orthocentre of a triangle ABC. The line AO (produced) meets the circumcircle at L. Prove that $BHCL$ is a parallelogram.

4. E, F are the feet of the perpendiculars from B, C on to the opposite sides of a triangle ABC. X is the mid-point of EF and AA' is a median of the triangle ABC. Show that the angle $XAB =$ angle CAA'.

5. If H is the orthocentre of a triangle ABC, show that the triangle HBC has the same nine-point circle as the triangle ABC.

6. If I_1, I_2, I_3 are the centres of the three escribed circles of a triangle ABC, prove that the nine-point circle of the triangle $I_1 I_2 I_3$ is the circumcircle of the triangle ABC.

7. O, G, N, H are respectively the circumcentre, centroid, nine-point centre and orthocentre of a triangle. Show that
$$OG:GN:NH = 2:1:3.$$

8. ABC is a triangle in which $AB = AC$, and P, Q are the mid-points of the arcs AB, AC respectively of the circumcircle. Prove that the Simson lines of P and Q intersect at an angle of $90° - \frac{1}{2}A$.

18.9. The theorems of Ceva and Menelaus

Suppose that A, B and P, Q are pairs of points on the same or parallel straight lines. Then the segments AB, PQ are said to have the same or opposite *senses* according as the displacements from A to B and P to Q are in the same or opposite directions. It is convenient to take account of the sense of a line as in coordinate geometry. Thus if the points A, B lie on a line parallel to the x-axis, the length AB is said to be positive or negative according as the point B lies to the right or left of A.

With this convention, $AB = -BA$, or $AB + BA = 0$, and if A, X, B are three points in any order on a line (Fig. 147), then $AB = XB - XA$ in all cases.

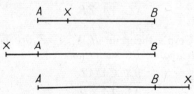

Fig. 147

Two useful theorems on the concurrency of lines and collinearity of points are those due to Ceva and Menelaus. These and their converses are given below.

Ceva's theorem. *If three straight lines through the vertices A, B, C of a triangle ABC are concurrent at P and meet the opposite sides of the triangle at X, Y, Z, then*

$$\frac{BX}{XC} \cdot \frac{CY}{YA} \cdot \frac{AZ}{ZB} = 1.$$

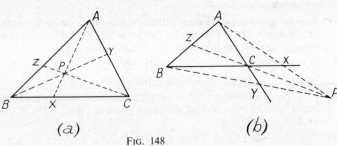

(a)   (b)

FIG. 148

In Fig. 148 (a), the point P lies inside the triangle ABC and in diagram (b), P lies outside the triangle. In diagram (a), each of the ratios BX/XC, CY/YA, AZ/ZB is positive, while in diagram (b), the ratios BX/XC, CY/YA are negative and AZ/ZB is positive. In either case, the product of the three ratios is positive.

By theorem 3, § 18.7,

in diagram (a),

$$\frac{BX}{XC} = \frac{\triangle BPA}{\triangle CPA}, \quad \frac{CY}{YA} = \frac{\triangle BPC}{\triangle BPA}, \quad \frac{AZ}{ZB} = \frac{\triangle CPA}{\triangle BPC},$$

in diagram (b),

$$\frac{BX}{XC} = -\frac{\triangle BPA}{\triangle CPA}, \quad \frac{CY}{YA} = -\frac{\triangle BPC}{\triangle BPA}, \quad \frac{AZ}{ZB} = \frac{\triangle CPA}{\triangle PBC}.$$

In either case, by multiplication,

$$\frac{BX}{XC} \cdot \frac{CY}{YA} \cdot \frac{AZ}{ZB} = 1.$$

The converse of Ceva's theorem. *If X, Y, Z are points on the sides BC, CA, AB of a triangle ABC such that*

$$\frac{BX}{XC} \cdot \frac{CY}{YA} \cdot \frac{AZ}{ZB} = 1,$$

then the three lines AX, BY, CZ are concurrent.

If AX, BY, CZ are not concurrent, let BY, CZ meet at P and let AP (produced if necessary) meet BC at X' (Fig. 149).

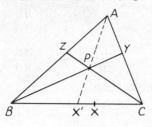

FIG. 149

Then, by Ceva's theorem,

$$\frac{BX'}{X'C} \cdot \frac{CY}{YA} \cdot \frac{AZ}{ZB} = 1;$$

but,

$$\frac{BX}{XC} \cdot \frac{CY}{YA} \cdot \frac{AZ}{ZB} = 1,$$

so that X' coincides with X and AX, BY, CZ, are concurrent.

Menelaus' theorem. *If a straight line meets the sides BC, CA, AB of a triangle ABC at P, Q, R respectively, then*

$$\frac{BP}{PC} \cdot \frac{CQ}{QA} \cdot \frac{AR}{RB} = -1.$$

The straight line cuts either one side (Fig. 150 (a)) or three sides (Fig. 150 (b)) of the triangle ABC externally. In diagram (a) the ratio

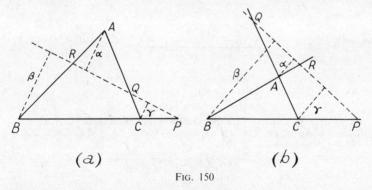

(a) (b)

FIG. 150

BP/PC is negative and the ratios CQ/QA, AR/RB are both positive and in diagram (b) all three of these ratios are negative. The product of the three ratios is therefore negative in each case.

Let α, β, γ be the lengths of the perpendiculars from A, B, C respectively on the line PQR. Then from similar triangles, in diagram (a),

$$\frac{BP}{PC} = -\frac{\beta}{\gamma}, \quad \frac{CQ}{QA} = \frac{\gamma}{\alpha}, \quad \frac{AR}{RB} = \frac{\alpha}{\beta}$$

in diagram (b),

$$\frac{BP}{PC} = -\frac{\beta}{\gamma}, \quad \frac{CQ}{QA} = -\frac{\gamma}{\alpha}, \quad \frac{AR}{RB} = -\frac{\alpha}{\beta}.$$

In either case, by multiplication,

$$\frac{BP}{PC} \cdot \frac{CQ}{QA} \cdot \frac{AR}{RB} = -1.$$

The converse of Menelaus' theorem. *If points P, Q, R are taken on the sides BC, CA, AB of a triangle ABC such that*

$$\frac{BP}{PC} \cdot \frac{CQ}{QA} \cdot \frac{AR}{RB} = -1,$$

then the points P, Q, R are collinear.

This can be proved by a "reductio ad absurdum" method similar to that used in proving the converse of Ceva's theorem. The proof is left as an exercise for the reader.

Example 10. *Use the converse of Ceva's theorem to prove that the medians of a triangle are concurrent.*

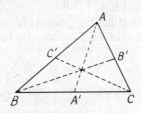

FIG. 151

If (Fig. 151) AA', BB', CC' are the medians of a triangle ABC, A', B', C' are the mid-points of BC, CA, AB respectively. Hence $BA' = A'C$, $CB' = B'A$, $AC' = C'B$ and therefore

$$\frac{BA'}{A'C} \cdot \frac{CB'}{B'A} \cdot \frac{AC'}{C'B} = 1.$$

The converse of Ceva's theorem then shows that AA', BB' and CC' are concurrent.

Example 11. *The tangents to the circumcircle of a triangle ABC at the vertices A, B, C meet the opposite sides at P, Q, R. Show that P, Q, R are collinear.*

Since the angle PAC (Fig. 152) equals the angle ABC in the alternate segment

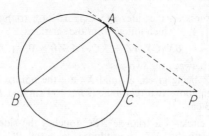

FIG. 152

and the angle BPA is common, the triangles PAC, ABP are similar. Hence

$$\frac{BP}{AP} = \frac{AP}{CP} = \frac{AB}{AC},$$

giving

$$\frac{BP}{AP} \cdot \frac{AP}{CP} = \frac{AB^2}{AC^2}$$

or,

$$\frac{BP}{CP} = \frac{AB^2}{AC^2}.$$

Thus

$$\frac{BP}{PC} = -\frac{AB^2}{AC^2},$$

and similarly,

$$\frac{CQ}{QA} = -\frac{BC^2}{BA^2}, \quad \frac{AR}{RB} = -\frac{CA^2}{CB^2}.$$

By multiplication,

$$\frac{BP}{PC} \cdot \frac{CQ}{QA} \cdot \frac{AR}{RB} = -1,$$

and the converse of Menelaus' theorem shows that P, Q, R are collinear.

EXERCISES 18 (e)

1. Use the converse of Ceva's theorem to show that the bisectors of the angles of a triangle are concurrent.

2. X, Y, Z are the points of contact of the inscribed circle of a triangle ABC with the sides BC, CA, AB respectively. Prove that AX, BY and CZ are concurrent. (L.U.)

3. X, Y, Z are points in the sides BC, CA, AB respectively of a triangle ABC such that the lines AX, BY, CZ are concurrent at an internal point O. Show that

$$\frac{OY}{OZ} = \frac{AZ}{AY} \cdot \frac{BY}{BZ} \cdot \frac{CY}{CZ}.$$

(L.U.)

4. AX, BY, CZ are concurrent straight lines meeting the sides BC, CA, AB of a triangle ABC at X, Y, Z. The circumcircle of the triangle XYZ meets the sides BC, CA, AB again at X', Y', Z' respectively. Show that AX', BY' and CZ' are concurrent.

5. Use the converse of Menelaus' theorem to show that the points at which the external bisectors of the angles of a triangle meet the opposite sides are collinear.

6. X, Y, Z are points on the sides BC, CA, AB of a triangle ABC. State the properties of X, Y, Z which follow from the relations

$$BX.CY.AZ \pm XC.YA.ZB = 0,$$

distinguishing between the two cases.

A, B, C, Y, Z being given, X_1 and X_2 are the points determined (as the point X) by the above two relations. If $BC = 4$ cm and $BX_1 = 3$ cm show that $X_1X_2 = 3$ cm.

7. The inscribed circle of a triangle ABC touches the sides BC, CA, AB at X, Y, Z respectively: YZ is produced to meet BC in P. Prove that $BP:CP = BX:XC$.

8. AA', BB', CC' are the medians of a triangle ABC. AA' meets $B'C'$ in P and CP meets AB in Q. Show that $AB = 3AQ$.

EXERCISES 18 (f)

1. Points P, Q, R are taken on the sides BC, CA, AB, respectively of a triangle ABC. If the circles AQR, BPR meet again at O, prove that the points C, P, O, Q are concyclic.   (L.U.)

2. Two circles intersect in A and A' and the common diameter meets the circles in B, B' and C, C' respectively. Prove that the circles circumscribing the triangles ABC and $AB'C'$ touch at A.   (L.U.)

3. AB is a fixed diameter of a circle and PQ is a variable chord parallel to AB. R is the mid-point of PQ and X is the foot of the perpendicular from P on to AB. Show that $AP^2 + PR^2 - AX^2$ is a constant.

4. AB is a chord of a circle and N is the middle point of AB. The diameter of the circle through N is PQ. Prove that AQ is a tangent to the circle through A, P and N.   (L.U.)

5. Two chords AC, BD of a circle intersect in E within the circle and AB, DC intersect in F. Prove that EF is the common chord of the circles circumscribing the triangles ABE and DCE.   (L.U.)

6. ABC is a triangle and the bisector of the angle BAC meets the circumcircle of the triangle again at X and BC at D. Prove that the triangles ABD and AXC are similar and deduce that

$$AB.AC = AD.AX = AD^2 + BD.DC.$$ (L.U.)

7. AX and BY are parallel lines and AY, BX intersect in P. A line through P parallel to XA meets AB in Q. Prove that

(i) $AX.BQ = BY.AQ,$

(ii) $\dfrac{1}{AX} + \dfrac{1}{BY} = \dfrac{1}{PQ}.$ (L.U.)

8. T is a point $0\cdot0625$ m from the centre of a circle of radius $0\cdot05$ m. The tangents from T touch the circle at A and B. Show that the length of AB is $0\cdot06$ m.
                                                                      (L.U.)

9. The triangle ABC is right-angled at C and O is the middle point of AB. If the internal and external bisectors of the angle ACB meet AB and AB produced in X and Y, prove that OC is a tangent to the circle XCY.

10. C is the centre of a circle, X is any point within the circle and CX is produced beyond X to Y so that $CX.CY = CA^2$, where A is the point of intersection of CX and the circle. If P is any point on the circle, prove that $PX:PY = CX:CA$.

11. AB is a diameter of a circle and C is any point on the circumference. P, Q are the feet of the perpendiculars from A, B respectively on to the tangent at C to the circle. Show that

$$\triangle ABC = \triangle ACP + \triangle BQC. \qquad \text{(L.U.)}$$

12. Points D, E, F are taken in the sides BC, CA, AB respectively of a triangle ABC such that $BD/DC = CE/EA = AF/FB = n/m$. AD, BE meet in P; BE, CF meet in Q and CF, AD meet in R. Prove that the areas of the triangles BQC, CRA, APB are each equal to

$$\frac{nm\Delta}{n^2 + nm + m^2},$$

where Δ is the area of the triangle ABC. (L.U.)

13. A and B are two fixed points and P is a point which moves so that $PA/PB = \lambda$ (greater than unity). Show that the radius of the circle on which P moves is

$$\frac{\lambda.AB}{\lambda^2 - 1}.$$

14. If $ABCDE$ is a regular pentagon inscribed in a circle and P is a point on the minor arc of AB, show that

$$\frac{PA + PD}{PB + PD} = \frac{PE}{PC}.$$

15. $ABCD$ is a cyclic quadrilateral and the bisector of the angle ABD meets AD in E. Show that the angles ACE, DCE are unequal. (O.C.)

16. If H is the orthocentre of the triangle ABC and L, M, N are the points of the circumcircle diametrically opposite to A, B, C, prove that HL, HM, HN bisect BC, CA, AB respectively. (O.C.)

17. Given two vertices and the circumcircle of a triangle, prove that the locus of the orthocentre is a circle.

18. If I is the in-centre and I_1 the centre of the escribed circle touching AB, AC externally for a triangle ABC and if the line AII_1 meets the circumcircle of the triangle ABC at P, show that

$$PI = PB = PC = PI_1.$$

19. O, G, H are respectively the circumcentre, centroid and orthocentre of a triangle ABC and A' is the mid-point of BC. Show that $AH = 2OA'$.

20. P is a point on the circumcircle of a triangle ABC. PR is perpendicular to BC and cuts the circumcircle again at R. Prove that the Simson line of P with respect to the triangle ABC is parallel to AR.

21. P is a point in the plane of a triangle ABC and the joins of P to the vertices A, B, C cut the opposite sides in D, E, F respectively. Either or both the joins and the sides may have to be produced to give section. Show that,

wherever P may be taken,

$$\frac{AP}{AD} + \frac{BP}{BE} + \frac{CP}{CF} = 2,$$

so long as a negative value is given to any of these fractions if the directions of the segments represented by the sequence of letters in the numerator and denominator are opposite. Hence, or otherwise, show that, if $AD = DP$, then $BP/CP = EB/CF$. (L.U.)

22. On the sides BC, CA, AB of a triangle ABC lie the points P, Q, R, one on each side respectively. O is any point inside the triangle, and AP', BQ', CR' are drawn parallel to OP, OQ, OR respectively to meet BC, CA, AB in P', Q', R' respectively. Prove that

$$\frac{OP}{AP'} + \frac{OQ}{BQ'} + \frac{OR}{CR'} = 1.$$ (L.U.)

23. P is a point in the side BC of a triangle ABC such that $PC = 2BP$ and Q is a point in the side CA such that $2QA = 3CQ$. AP and BQ intersect in H and CH meets AB in R. Show that $AR = 3RB$ and that $2AH = 9HP$. (L.U.)

24. ABC, $A'B'C'$ are two triangles such that AA', BB', CC' meet at a point O. If BC, $B'C'$ meet at L, CA, $C'A'$ meet at M and AB, $A'B'$ meet at N, show that L, M, N are collinear. (*Desargues' theorem*.)

25. If a straight line cuts the sides AB, BC, CD, DA of a quadrilateral at X, Y, Z, W show that $AX \cdot BY \cdot CZ \cdot DW = XB \cdot YC \cdot ZD \cdot WA$.

CHAPTER 19

ELEMENTARY GEOMETRY OF THE PLANE AND SPHERE

19.1. Introduction and definitions

This chapter is concerned with the elementary geometry of the plane and sphere and, in particular, with the angles made by planes and straight lines with one another. Considerations of space necessitate a brief treatment and certain results which are usually given as "theorems" are relegated to the exercises.

The following, with most of which the reader may be already familiar, may be regarded as a set of preliminary definitions.

(i) A surface such that straight lines through every pair of points in it lie wholly in the surface is a *plane*.

(ii) If two planes, a straight line and plane, or two straight lines in the same plane have no point in common, they are said to be *parallel*.

(iii) A solid figure bounded only by plane polygons is called a *polyhedron*. The bounding polygons are *faces*, consecutive faces intersect in *edges* and consecutive edges meet in *vertices* of the polyhedron. Some particular cases are:—

  (a) the *prism*, in which two faces are congruent polygons and the remaining faces parallelograms; the *parallelepiped*, in which all the faces are *parallelograms*; if every face of a parallelepiped is a rectangle, the figure is a rectangular parallelepiped or *cuboid* and if every face is a square it is a *cube*;

  (b) the *pyramid*, in which one face is a polygon and all the remaining faces are triangles with a common vertex; a special case is the *tetrahedron*, in which all the faces are triangles.

  Some typical examples are shown in Fig. 153 below.

Opposite edges are those which do not meet, for example, the edges OA, BC of the tetrahedron of Fig. 153 and the *diagonals* of a parallelepiped are the lines joining opposite vertices, i.e., the lines, AC', CA', BD', $B'D$ of the parallelepiped of the diagram. When the faces of a polyhedron are all congruent regular polygons and the numbers of edges which meet at each vertex are equal, the polyhedron is said to be *regular*.

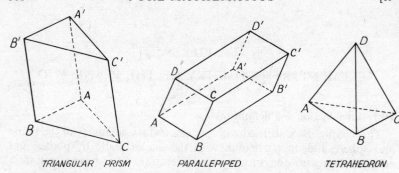

TRIANGULAR PRISM   PARALLELEPIPED   TETRAHEDRON

Fig. 153

(iv) A straight line which intersects a fixed line (curved, straight, or made up of segments of curves and straight lines) and remains parallel to another fixed line generates a *cylindrical* surface and is called a *generator* of the surface. The space bounded by a cylindrical surface and two parallel planes is a *cylinder*, one of the parallel planes being the *base*.

(v) A straight line which intersects a fixed line (curved, straight or made up of segments of curves and straight lines) and passes through a fixed point generates a *conical* surface and is called a *generator* of the surface. The space bounded by a conical surface and a plane is a *cone*, the plane being the *base*.

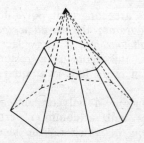 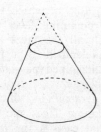

TRUNCATED PYRAMID   FRUSTUM OF CONE

Fig. 154

(vi) The part of a prism, pyramid, cylinder or cone intercepted between the base and any other plane is known as a *truncated* prism, pyramid, cylinder or cone. The part of a cone or pyramid intercepted between the base and a parallel plane is called a *frustum*.

(vii) A point which moves so that its distance from a fixed point (the centre) is constant lies on a *spherical* surface; the space inside a spherical surface is called a *sphere*.

19.2. Some axioms and further definitions

The following axioms are usually taken as the basis on which elementary solid geometry is built.

Axiom 1. There is one straight line, and only one, passing through two given points.

Axiom 2. There is one plane, and only one, passing through three given points which are not in the same straight line.

Axiom 3. If two planes have a common point, they have also a common straight line.

Axiom 4. Through any point in space there is one, and only one, straight line parallel to a given straight line. (*Playfair's axiom.*)

It is convenient here to list some further definitions. Lines lying in the same plane are said to be *coplanar* and lines not lying in the same plane are called *skew* lines. The *angle between two skew straight lines* is defined as the angle between two coplanar lines to which they are respectively parallel.

Example 1. *If three planes intersect, two by two, show that their lines of intersection are either concurrent or parallel.*

A convenient notation for working this type of problem is to denote the three planes by α, β, γ and to use $\alpha\beta$, $\beta\gamma$, $\gamma\alpha$ to denote the lines of intersection of the planes α and β, β and γ, γ and α respectively.

Since the straight lines $\gamma\alpha$ and $\alpha\beta$ lie in the same plane α, they either intersect or are parallel. If they intersect, their common point lies on all three planes (it can be denoted by the point $\alpha\beta\gamma$) and it lies on the line $\beta\gamma$. Hence the lines $\alpha\beta$, $\beta\gamma$, $\gamma\alpha$ are concurrent. If the lines $\gamma\alpha$ and $\alpha\beta$ are parallel, there is no point $\alpha\beta\gamma$ and neither $\gamma\alpha$ nor $\alpha\beta$ cuts $\beta\gamma$. But $\alpha\beta$, $\beta\gamma$ both lie in the plane β and therefore they are parallel Similarly $\gamma\alpha$ and $\beta\gamma$ can be shown to be parallel.

19.3. Some theorems on parallels

Theorem 1. *If a straight line is parallel to one straight line in a plane it is parallel to the plane.*

Suppose (Fig. 155) AB is parallel to CD in the plane $EDCF$. Then the lines AB, CD are coplanar. The planes $ABCD$, $EDCF$ meet in the straight line CD and in no other point (unless they coincide). But as AB, CD are parallel, they do not meet and hence AB does not meet the plane $EDCF$ unless it lies entirely in it. Thus AB is parallel to the plane $EDCF$.

It is left as an exercise for the reader to prove the converse that if a straight line in one plane is parallel to another plane, it is parallel to their line of intersection.

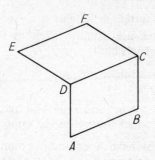

FIG. 155

Theorem 2. *If two straight lines are each parallel to a third, they are parallel to one another.*

Suppose (Fig. 156) the lines *AB*, *EF* are both parallel to the line *CD*. Let the plane *BAE* cut the plane *CEFD* in the line *EG*. Since *AB* is parallel to *CD*, theorem 1 shows that the plane *BAE* is parallel to the line *CD* and that the plane *DCE* is parallel to the line *AB*. Since

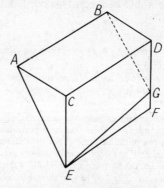

FIG. 156

EG is in the plane *DCE*, the line *EG* is parallel to *CD*. And since *EG* is in the plane *BAE*, *EG* is parallel to *AB*. Hence, by Axiom 4, *EG* coincides with *EF* and is parallel to *AB*.

Theorem 3. *If a plane cuts two parallel planes, the lines of intersection are parallel.*

In Fig. 157, the planes α and γ are parallel and are met by the plane β in lines $\alpha\beta$, $\beta\gamma$ respectively. As the planes α, γ are parallel, no line in α meets a line in γ. Hence the lines $\alpha\beta$, $\beta\gamma$ do not meet. But the lines $\alpha\beta$, $\beta\gamma$ are coplanar for they both lie in the plane β. Thus $\alpha\beta$, $\beta\gamma$ are coplanar lines which do not meet and they are therefore parallel to each other.

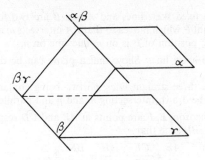

FIG. 157

Theorem 4. *If two straight lines are cut by three parallel planes, their intercepts are proportional.*

In Fig. 158, α, β, γ are three parallel planes, intersecting the lines *ABC, DEF* at *A, B, C* and *D, E, F*. The line *AF* meets the plane β at *P*. The plane *ACF* cuts the parallel planes β, γ in the parallel lines *BP, CF* respectively, so that

$$AB:BC = AP:PF.$$

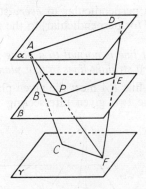

FIG. 158

Similarly the plane *AFD* cuts the parallel planes α, β in the parallel lines *AD, PE*, so that $AP:PF = DE:EF$.

Hence $AB:BC = DE:EF$.

EXERCISES 19 (a)

1. *ABCD* is a face of a cube. A plane passes through the diagonal *AG* of the cube and bisects the edge *BC* at *L*. Show that $AL = LG$.
2. Show that the lines joining the vertices of a tetrahedron to the centroids of the opposite faces are concurrent.

3. m and n are two fixed skew lines and A and B are two fixed points. Find the locus of a point P which moves so that PA intersects m and PB intersects n. Show that one position of P is on m and one on n. (L.U.)

4. n and m are two skew lines. Show that a plane can be drawn through m parallel to n.

n, m and p are three straight lines, no two being coplanar. Show that a straight line can be drawn intersecting m and p and parallel to n. (L.U.)

5. $ABCD$ is a tetrahedron. E, F are points in AB and CD respectively. If H, K are points in AC, BD such that

$$\frac{AE}{EB} = \frac{CF}{FD} = \frac{AH}{HC} = \frac{BK}{KD} = \lambda,$$

prove that EF and HK lie in the same plane. If EF and HK intersect in G, prove that $HG/KG = EG/GF = \lambda$. (L.U.)

6. l, m, n are three non-intersecting straight lines in space, no two of which are parallel. Prove that through any point on l just one straight line can be drawn intersecting both m and n. If l, m, n are all parallel to a given plane and PQR, $P'Q'R'$ are two straight lines meeting l in P and P', m in Q and Q', n in R and R', prove that $PQ/P'Q' = QR/Q'R'$. (L.U.)

19.4. Normals

If a straight line is perpendicular to *every* straight line in a plane, it is said to be *normal* (or perpendicular) to the plane.

Theorem 5. *A straight line is normal to a plane if it is perpendicular to each of two intersecting straight lines in that plane.*

In Fig. 159, let the line PA meet the given plane α at A and let AB, AC be parallel to the two given intersecting straight lines. Let AD

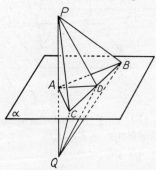

Fig. 159

be parallel to any other straight line in α and draw some line in α through D to cut AB, AC at B, C respectively. Produce PA to Q so that $PA = AQ$. Since PA is perpendicular to AB, the triangles PAB, QBA are congruent for the angles PAB, BAQ are right angles,

$PA = AQ$ and AB is common. Hence $PB = BQ$ and similarly it can be shown that $PC = CQ$. Because of these equalities and since BC is common, the triangles PCB, BQC are congruent and hence $PD = DQ$. The triangles PAD, QDA are congruent because $PA = AQ$, $PD = DQ$ and AD is common, so that the angles PAD, DAQ are equal. Since the sum of the angles PAD, DAQ is two right angles, the angle PAD is therefore one right angle and we have shown that PA is perpendicular to *any* straight line in the plane α and hence is a normal to the plane.

Theorem 6. *Planes which are normal to the same straight line are parallel to one another.*

In Fig. 160, the line CD is normal to the planes α and β. If a point common to the two planes exists, let it be P. Then in the triangle

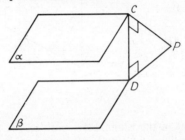

Fig. 160

PCD, both the angles PCD, CDP would be right angles. As this is impossible, the planes α, β have no common point and they are therefore parallel.

It follows as a corollary that through any point there is one and only one plane which is normal to a given straight line.

Example 2. *Show that straight lines which are normal to the same plane are parallel to one another.*

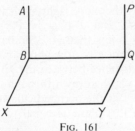

Fig. 161

Let (Fig. 161) AB, PQ be perpendicular to the plane $BXYQ$. BX is drawn in the plane $BXYQ$ perpendicular to BQ. Then AB, BQ, PQ are all perpendicular to BX and hence are coplanar. But AB, PQ are both perpendicular to BQ and hence they are parallel.

Theorem 7. *There is one, and only one, straight line cutting at right angles each of two given skew straight lines and its length is the shortest distance between them.*

In Fig. 162, let *AB*, *CD* be the given skew lines and let *DE* be a line through some point *D* on *CD* parallel to *AB*. Then *AB* is parallel to the plane *EDC*. Through any point *A* on *AB* draw *AR* normal to the plane *EDC* and let the plane *BAR* cut *CD* in *Q*. Since *AB* is parallel to the plane *EDC* and *AB*, *RQ* are coplanar, *AB* is parallel

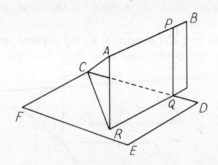

Fig. 162

to *RQ*. Draw *PQ* in the plane *BAR* parallel to *AR* to meet *AB* at *P*. Since *PQ* is parallel to *AR* and *AR* is normal to the plane *EDC*, *PQ* is normal to the plane *EDC*. Hence *PQ* is perpendicular to *CD* and *DE*. But *DE* is parallel to *AB*, so that *PQ* is perpendicular to *CD* and *AB*.

To show that the common perpendicular *PQ* is unique, let *P'Q'* be another common perpendicular. Then *P'Q'* would be perpendicular to the plane *EDC* and therefore *PQ*, *P'Q'* would be parallel. This is impossible as *AB*, *CD* are not coplanar and hence there is one, and only one, common perpendicular.

As *AR* is perpendicular to the plane *EDC*, *AR* is perpendicular to *CR*. The triangle *ACR* is therefore right-angled at *R* and *AR* is less than the hypotenuse *AC*. But *PARQ* is a parallelogram so that *PQ* = *AR* and hence *PQ* is less than *AC*. Similarly we can show that *PQ* is less than the length of any other line joining *AB* and *CD*. *PQ* is therefore the shortest distance between the given skew lines.

It follows that a plane can be drawn through one of two skew lines parallel to the other and that the perpendicular distance of any point on the second line from this plane is equal to the shortest distance between the skew lines.

Example 3. *One end of a rectangular box of length 2a is a square ABCD of edge a. If AP is a diagonal of the box, find the length of the shortest distance between AP and BC.* (L.U.)

From Fig. 163, it is clear that the edge BC is parallel to the plane ADP containing the diagonal AP. The shortest distance between AP and BC is therefore

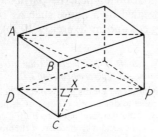

FIG. 163

equal to the distance of C from the plane ADP and this is equal to the length of the perpendicular CX from the right angle of the triangle DCP on to DP. Now

$$DP^2 = DC^2 + CP^2 = a^2 + (2a)^2 = 5a^2,$$

and the triangles DCX, DCP are similar for they both contain a right angle and the angle XDC is common. Hence

$$\frac{CX}{CD} = \frac{CP}{DP}, \quad \text{giving} \quad \frac{CX}{a} = \frac{2a}{\sqrt{5a}},$$

and the required shortest distance $= CX = 2a/\sqrt{5}$.

It follows also that parallel planes can be drawn through two skew

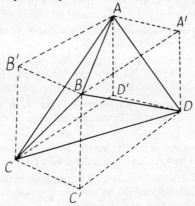

FIG. 164

lines. Use can be made of this fact to show that a tetrahedron can be inscribed in a parallelepiped with one edge of the tetrahedron lying on each face of the parallelepiped. Thus in Fig. 164, the edges of the tetrahedron $ABCD$ are the diagonals AB, BC, CD, DA, BD and CA of the faces $AB'BA'$, $BC'CB'$, $CC'DD'$, $AA'DD'$, $BA'DC'$ and $CD'AB'$ of the parallelepiped.

In working problems on the tetrahedron, it is often useful to sketch in the circumscribing parallelepiped. In particular cases, concealed properties are often revealed by a figure such as 164 and instances of its use occur in Example 4 and in Exercises 19 (b), No. 6, 19 (e), Nos. 1, 5 and 12.

Example 4. *In a tetrahedron ABCD, BC is perpendicular to AD and CA is perpendicular to BD. Prove that AB is perpendicular to CD.*

Referring to Fig. 164, it is clear that the diagonals $B'C'$, AD of the parallelograms $B'BC'C$, $AA'DD'$ are parallel to one another. Since BC is perpendicular to AD, BC is perpendicular to $B'C'$ and the parallelogram $B'BC'C$ has diagonals which are perpendicular to each other and it is therefore a rhombus.

Similarly, since CA is perpendicular to BD, it can be shown that the parallelogram $BC'DA'$ is also a rhombus. Hence

$$BB' = BC' = BA',$$

and the parallelogram $B'BA'A$ is also a rhombus. It follows that its diagonals BA, $B'A'$ are perpendicular and, since $B'A'$ is parallel to CD, BA is perpendicular to CD.

EXERCISES 19 (b)

1. The triangle ABC is right-angled at A. A point P not lying in the plane of ABC is equidistant from A, B and C. Prove that the line PN joining P to the middle point N of BC is perpendicular to the plane ABC. (L.U.)

2. The triangle ABC is right-angled at A. Any point P is taken on the perpendicular through A to the plane ABC, and a point Q is taken on CP (or CP produced) such that $BQ = BC$. Show that $AQ = AC$. (N.U.)

3. AB, CD are two skew lines equal in length and P, Q are the mid-points of AC, BD. Prove that PQ is less than AB.

4. OA, OB, OC are three concurrent straight lines each of which is perpendicular to the other two. The foot of the perpendicular from O to the plane ABC is H. Prove that H is the orthocentre of the triangle ABC. (L.U.)

5. $ABCD$ is the square base of a pyramid, vertex V, in which $AB = 6a$ and $VA = VB = VC = VD = 5a$. Prove that the shortest distance between AB and VC is $(3\sqrt{7a})/2$. (L.U.)

6. If the opposite edges of a tetrahedron are equal in pairs, show that the shortest distance between any pair is the join of the two mid-points. (L.U.)

7. Two equal skew straight lines AB, CD are inclined to each other at an angle of $60°$. The shortest distance BC between the two lines is also equal in length to AB and CD. Show that $AD^2 = 2AB^2$.

8. A square lamina $ABCD$, of side a, is rotated about AB through a right angle to take up the position $ABC'D'$. Show that the shortest distance between AC' and BD is $a/\sqrt{3}$. (L.U.)

19.5. Orthogonal projection and dihedral angles

The foot of the perpendicular from a point to a plane is called the *orthogonal projection* of the point on the plane. When a point traces out a given figure, its orthogonal projection on a given plane will trace out a figure in this plane. This second figure is called the orthogonal projection of the first figure on the given plane.

Theorem 8. *The orthogonal projection of a straight line on a plane is either a point or a straight line.*

If the line AB is normal to the plane α (Fig. 165) (a)), its orthogonal projection is clearly the point C in which the line meets the plane.

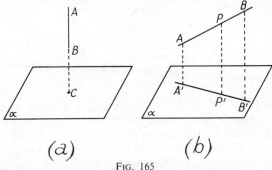

(a)   *(b)*

FIG. 165

If the line is not normal to the plane, let A, P, B (Fig. 165 (b)) be three points on the line AB and let A', P', B' be their orthogonal projections on the plane α. Since AA', PP', BB' are all normal to the plane α, they are parallel to one another and therefore coplanar. Hence P' lies on the line of intersection of the planes $A'B'BA$ and α. In other words, P' lies on the line $A'B'$.

It follows as a corollary that a straight line and its orthogonal projection are coplanar.

Theorem 9. *The angle between a straight line and its orthogonal projection on a plane is less than the angle between the line and any other line in the plane.*

In Fig. 166, $A'B$ is the projection of the line AB on the plane α. BC is drawn equal in length to BA' and parallel to any line m in the plane α.

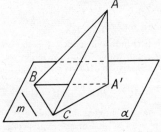

FIG. 166

Since AA' is normal to the plane α, the angle $CA'A$ is a right angle and is the largest angle of the triangle $CA'A$. Hence AA' is less than AC. In the triangles ABC, ABA', $BA' = BC$, BA is common and AA' is less than AC. Hence the angle ABA' is less than the angle ABC.

The angle between a straight line and its orthogonal projection on a plane is defined as the *angle between the line and plane*.

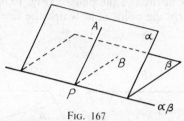

FIG. 167

Suppose (Fig. 167) that P is any point in the line of intersection $\alpha\beta$ of two planes α and β. PA, PB are lines, one in each plane which are perpendicular to the line of intersection $\alpha\beta$ of the two planes. Such lines are known as *lines of greatest slope*. The mutual inclination of the two planes α, β is called their *dihedral angle* and is measured by the angle APB, the angle between their lines of greatest slope.

Example 5. *An isosceles triangle ABC, in which $AB = AC = 2BC = 2a$, lying in a horizontal plane α, is rotated about the base BC until A is at a vertical height a above the plane α. Calculate the angle through which the triangle is rotated and the inclination of AC to the horizontal in its final position.* (L.U.)

XBC is the initial and ABC the final position of the triangle (Fig. 168). D is the mid-point of BC, and since the triangles ABC, XBC are both isosceles,

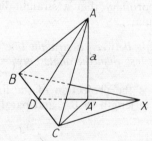

FIG. 168

AD and XD are both perpendicular to BC. A' is the orthogonal projection of A on the plane XBC and, from symmetry, it lies on DX. Since $AC = 2a$, $DC = \frac{1}{2}BC = \frac{1}{4}a$, and the triangle ADC is right-angled at D,

$$AD^2 = AC^2 - DC^2 = (2a)^2 - (\tfrac{1}{2}a)^2 = \frac{15a^2}{4}.$$

AD, DX are lines of greatest slope in the planes ABC, XBC and AA' $(=a)$ is perpendicular to the plane XBC and therefore perpendicular to $A'D$. Hence the angle through which the triangle is rotated, or the angle between the planes

ABC and XBC, is the angle ADA', and this is given from the right-angled triangle ADA' by

$$\sin(ADA') = \frac{AA'}{AD} = \frac{a}{(\sqrt{15}a/2)} = \frac{2}{\sqrt{15}} = 0{\cdot}5165,$$

so that the angle $ADA' = 31° \, 6'$.

The inclination of AC to the horizontal is the angle between AC and the plane XBC, and this is the angle between AC and its orthogonal projection $A'C$. Since AA' is normal to the plane XBC, the angle $CA'A$ is a right angle and the triangle $CA'A$ gives

$$\sin(ACA') = \frac{AA'}{AC} = \frac{a}{2a} = \frac{1}{2},$$

so that the angle ACA' is $30°$.

Theorem 10. *If the angle between a line AB and a plane α is θ, the length of the orthogonal projection of AB on α is $AB \cos \theta$.*

Let (Fig. 169) $A'B'$ be the orthogonal projection of AB on the plane

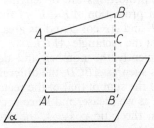

FIG. 169

α, so that AB and $A'B'$ are coplanar. Draw AC parallel to $A'B'$ to meet BB' in C.

Then $AA'B'C$ is a rectangle and $A'B' = AC$. Since AC is parallel to $A'B'$, the angle BAC is equal to θ, the angle between the line AB and the plane α, and the right-angled triangle BAC gives

$$A'B' = AC = AB \cos \theta.$$

It follows as a corollary that the ratio of lengths along the same or parallel lines is unaltered by orthogonal projection.

Theorem 11. *The area of the orthogonal projection of a plane figure of area A is $A \cos \phi$, where ϕ is the angle between the plane of the figure and the plane on to which it is projected.*

In Fig. 170, the plane figure has been divided into strips, of which $ABCD$ is typical, by lines AF, BC of greatest slope. Lines AE, CF are drawn parallel to the line of intersection XY of the plane of the figure and the plane α on to which it is projected.

Since AE, CF are parallel to one line (the line XY) of the plane α,

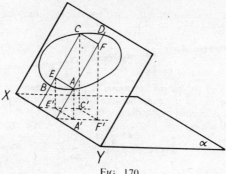

FIG. 170

they are (theorem 1) parallel to the plane α and therefore make zero angles with this plane. By theorem 10, their orthogonal projections $A'E'$, $C'F'$ are respectively equal to AE and CF. Since AF, CE are lines of greatest slope, they each make an angle ϕ with the plane α and their orthogonal projections are of lengths $AF \cos \phi$, $CE \cos \phi$. Hence the orthogonal projection $A'F'C'E'$ of the rectangle $AFCE$ is a rectangle of sides AE and $AF \cos \phi$ and its area is equal to $\delta A \cos \phi$, where δA is the area of the rectangle $AFCE$.

By increasing sufficiently the number and decreasing sufficiently the width of the strips such as $ABCD$, the difference between the area of the given figure and the sum of that of the rectangles such as $AFCE$ can be made as small as we please, and since each rectangle is diminished by projection in the ratio $\cos \phi : 1$, the projected area will, by a limiting process, be $A \cos \phi$.

Example 6. *XY is the line of intersection of two planes α, β mutually inclined at an angle θ. Z is a point in the plane α such that $XZ = 1$ and the angle $ZXY = \phi$. Prove that the length of the orthogonal projection of the line XZ on the plane β is $\sqrt{(\cos^2 \phi + \sin^2 \phi \cos^2 \theta)}$.*

Z' is the orthogonal projection of Z on the plane β (Fig. 171), and PZ, PZ' are lines of greatest slope in the two planes. From the right-angled triangle

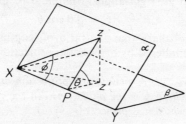

FIG. 171

ZXP, $XP = \cos \phi$, $ZP = \sin \phi$, since $XZ = 1$. The right-angled triangle $Z'ZP$ gives
$$Z'P = ZP \cos \theta = \sin \phi \cos \theta.$$
Finally the right-angled triangle $Z'XP$ gives
$$XZ' = \sqrt{(XP^2 + Z'P^2)} = \sqrt{(\cos^2 \phi + \sin^2 \phi \cos^2 \theta)}.$$

EXERCISES 19 (c)

1. $ABCD$ is the floor of a room, $A'B'C'D'$ is the ceiling, A' being above A and so on. If $AB = 5\cdot4$ m, $AD = 3\cdot6$ m, $AA' = 4\cdot2$ m, find the angle between the diagonal AC' and the floor of the room.

2. The three parallel edges AA', BB', CC' of a triangular prism are each of length 3 cm, and the ends ABC, $A'B'C'$ of the prism are equilateral triangles of sides 2 cm. Find the inclination of the plane $BA'C$ to the plane BAC.

3. One end of a rectangular box of length $2a$ is a square $ABCD$ of edge a. If AP is a diagonal of the box, calculate (i) the angle between AP and the plane $ABCD$, (ii) the angle between AP and BC. (L.U.)

4. The base ABC of a tetrahedron $OABC$ is a right-angled isosceles triangle and the length of the hypotenuse BC is 10 cm. $OA = 12$ cm, $OB = OC = 13$ cm. Find (i) the angle between the planes OBC and ABC, (ii) the angle between the planes OBA and ABC. (L.U.)

5. The edges VA, VB, VC of a tetrahedron are mutually perpendicular and $VA = a$, $VB = b$, $VC = c$. Prove that

   (i) $\cos BAC \cdot \cos CBA \cdot \cos ACB = \dfrac{a^2 b^2 c^2}{(a^2 + b^2)(b^2 + c^2)(c^2 + a^2)}$;

   (ii) the angle between the faces VBC and ABC is

   $$\cos^{-1}\left\{\frac{bc}{\sqrt{(b^2 c^2 + c^2 a^2 + a^2 b^2)}}\right\}.$$ (L.U.)

6. V is the apex of a pyramid on a square base $ABCD$.

   $$VA = VB = VC = VD = 0\cdot13 \text{ m}$$

   and the side of the base is $0\cdot10$ m. Find

       (i) the angle between a slant face and the base,
       (ii) the angle between adjacent slant faces,
       (iii) the angle between opposite slant faces. (L.U.)

7. If A is the area of the normal section of any prism, show that $A \sec \theta$ is the area of any section inclined at θ to the normal section. A cylinder whose normal section is a circle of radius a is cut obliquely by a plane. Prove that the area of the oblique section is πab where $2b$ denotes the longest diameter of the section. (L.U.)

8. The vertices A, B, C of a plane triangle ABC are at heights 5, 13 and 25 m above a horizontal plane. The orthogonal projection of the triangle ABC on to the horizontal plane is a triangle $A'B'C'$ (A' being the projection of A, etc.) and $B'C' = 16$ m, $C'A' = 21$ m, $A'B' = 15$ m. Find the lengths of the sides of the triangle ABC and the cosine of the angle which its plane makes with the horizontal.

19.6. Some geometrical properties of the sphere

Here we consider a few of the geometrical properties of the sphere. These are all of an elementary character and no attempt is made here to discuss what is usually called Spherical Geometry, the geometry of points and lines lying on the surface of a sphere.

Theorem 12. *Every plane section of a sphere is a circle.*

Let PQR be a section of a sphere by a plane and let O be the centre of the sphere (Fig. 172). Let N be the foot of the normal from O on

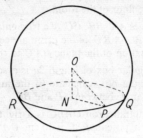

FIG. 172

to the plane PQR and join OP, NP. Then, since the angle ONP is a right angle,

$$NP^2 = OP^2 - ON^2.$$

Similarly, if Q is any other point on the boundary of the plane section, we can show that

$$NQ^2 = OQ^2 - ON^2,$$

and since $OP = OQ =$ radius of sphere, it follows that $NP = NQ$. Similarly all points on the intersection of the plane and sphere are equidistant from N and hence lie on a circle.

Theorem 13. *The curve of intersection of two spheres is a circle.*

Let O, O' be the centres of the two spheres and let P be any point on their curve of intersection (Fig. 173). Draw PN perpendicular to OO'. Let the spheres be cut by a plane through O, O' and P and let

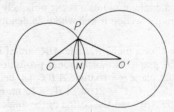

FIG. 173

the semi-circles bounded by the diameters along the line OO' revolve about OO'. These semi-circles will generate two spheres and their point of intersection P will generate a circle of radius NP and centre N. The curve of intersection of the spheres is therefore a circle whose plane is perpendicular to the line of centres of the spheres.

Theorem 14. *A sphere can be drawn through four points not in the same plane.*

In Fig. 174 let A, B, C, D be the four points, let Z be the middle point of BD and let X, Y be the centres of the circles through B, C, D and A, B, D respectively. In the plane XYZ, draw XW, YW perpendicular to XZ, YZ respectively to intersect at the point W.

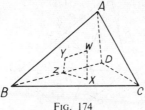

FIG. 174

Since BD is perpendicular to XZ and YZ, BD is perpendicular to the plane XYZ and therefore perpendicular to XW and YW. Hence XW is perpendicular to BD and XZ and therefore perpendicular to the plane BCD. Similarly YW is perpendicular to the plane ABD.

Since Y is the centre of the circle through A, B, D, $AY = BY = DY$, the angles AYW, BYW, DYW are all right angles, and WY is common to all three of the triangles AYW, BYW, DYW, so that these three triangles are congruent and therefore $AW = BW = DW$.

Similarly it can be shown that the triangles BXW, CXW, DXW are congruent and therefore $BW = CW = DW$. Hence

$$AW = BW = CW = DW,$$

and W is the centre of a sphere through the points A, B, C, D.

Example 7. *If O be a point outside a sphere and if two secants drawn from O cut the sphere in points A, B and C, D respectively, show that*

$$OA.OB = OC.OD.$$

Since the lines OAB, OCD intersect at O, they are coplanar and the section of the sphere by their plane is a circle (Fig. 175). Hence OAB, OCD are two secants

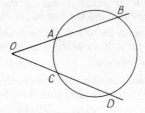

FIG. 175

drawn from O to meet a circle in A, B and C, D respectively and from theorem (*j*) of Chapter 18 it follows that

$$OA.OB = OC.OD.$$

Example 8. *If the middle points of the edges of a tetrahedron lie on a sphere, show that the opposite edges are at right angles to one another.*

In Fig. 176, let P, Q, R, X, Y, Z be the middle points of the edges BC, CA, AB, DA, DB, DC of the tetrahedron. Since X, Y are the middle points of DA, DB, XY is parallel to AB and equal to $\frac{1}{2}AB$. Similarly QP is parallel to and equal

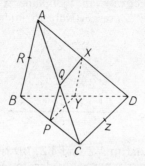

FIG. 176

to $\frac{1}{2}AB$. Hence XY and QP are parallel and equal and therefore $XYPQ$ is a parallelogram. The points X, Y, P and Q are therefore coplanar and, since they lie on a sphere, they lie also on a circle. Hence $XYPQ$ is a cyclic parallelogram and therefore it must be a rectangle. XY is therefore perpendicular to YP. But we have shown that XY is parallel to AB and we can show similarly that YP is parallel to DC, so that AB is perpendicular to DC. It can be shown in a similar way that BC is perpendicular to AD and that CA is perpendicular to BD.

EXERCISES 19 (d)

1. Two planes, inclined at 60°, intersect a sphere in equal circles of radius a. If the circles have two common points whose distance apart is a, show that the radius of the sphere is $(a\sqrt{5})/2$.   (L.U.)

2. If three points A, B, C are such that the angle ABC is a right angle and $AB = BC$, prove that the locus of points at which AB and BC both subtend a right angle is a circle of radius $AB/(2\sqrt{2})$.   (L.U.)

3. Prove that the points of contact of tangents from a point P to a sphere centre O lie on a circle. If OP equals d and r is the radius of the sphere, calculate the radius of the circle and the distance of its plane from P.   (L.U.)

4. The radii of two spheres are a and b, and the distance between their centres is $c < (a + b)$. Show that the radius r of the circle of intersection is given by
    $$2cr = \sqrt{\{(a + b + c)(b + c - a)(c + a - b)(a + b - c)\}}.$$

5. A sphere rests in a horizontal circular hole of diameter 0·02 m and the lowest point of the sphere is 0·005 m below the plane of the hole. Find the radius of the sphere.

6. Show that the locus of points in space whose distances from two given points are in a given ratio is a sphere.

7. A and B are points on a diameter of a sphere, of radius a, at equal distances b from the centre. A straight line through A perpendicular to AB meets the

sphere in P. A straight line through B perpendicular to AB and inclined to AP at an angle 2θ meets the sphere in Q. Prove that

$$PQ^2 = 4a^2 \sin^2 \theta + 4b^2 \cos^2 \theta. \qquad \text{(O.C.)}$$

8. Show that the radius of the sphere circumscribing a regular tetrahedron, each of whose edges is of length $2a$, is $\frac{1}{2}a\sqrt{6}$.

19.7 Rectangular Cartesian coordinates in three dimensions

The rectangular Cartesian coordinate system described in § 14.1 can easily be extended so that the position of a point in *space* can be specified. Thus, let $y'Oy$ and $z'Oz$ be two perpendicular straight lines intersecting at an origin O and let $x'Ox$ be a third straight line perpendicular to the plane containing the first two lines (Fig. 177). In interpreting the

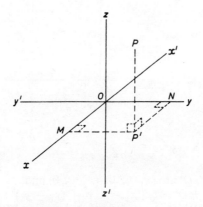

Fig. 177

diagram, the lines $y'Oy$, $z'Oz$ are assumed to lie in the plane of the paper and the line Ox to project towards the reader (x' being behind the paper). P' is the projection of any point P on the plane xOy and $P'M$, $P'N$ are drawn perpendicular respectively to $x'Ox$, $y'Oy$. If

$$OM = x, \quad ON = y, \quad PP' = z,$$

the position of P is specified when the three lengths x, y and z are known and the symbol (x, y, z) is used to denote the position of P.

The lines $x'Ox$, $y'Oy$, $z'Oz$ are said to form a set of rectangular Cartesian axes, the planes xOy, yOz, zOx are called the coordinate planes and the lengths x, y, z are the coordinates of the point P. The coordinate planes divide three-dimensional space into eight parts called octants and the signs of the coordinates of a point are determined by the octant in which it lies (as shown by the following table)

Octant / Coordinate	Oxyz	Oxyz'	Oxy'z'	Oxy'z	Ox'yz	Ox'yz'	Ox'y'z'	Ox'y'z
x	+	+	+	+	−	−	−	−
y	+	+	−	−	+	+	−	−
z	+	−	−	+	+	−	−	+

19.8. The distance between two points

Let P, Q be two points with coordinates (x_1, y_1, z_1), (x_2, y_2, z_2) and let P', Q' be their projections on the plane xOy (Fig. 178). The

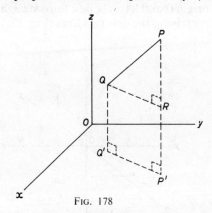

Fig. 178

coordinates of P', Q' are $(x_1, y_1, 0)$, $(x_2, y_2, 0)$ and, by § 14.3, the length $P'Q'$ is given by $P'Q' = \sqrt{\{(x_1 - x_2)^2 + (y_1 - y_2)^2\}}$. By drawing QR perpendicular to PP' it is apparent that the angle PRQ is a right angle and that

$$PQ^2 = QR^2 + PR^2 = (Q'P')^2 + PR^2$$
$$= (x_1 - x_2)^2 + (y_1 - y_2)^2 + (z_1 - z_2)^2, \qquad (19.1)$$

since $QR = Q'P'$ and $PR = PP' - RP' = PP' - QQ' = z_1 - z_2$. Formula (19.1) has been derived above only when the coordinates of both points are all positive but, when due consideration is given to the sign convention described in § 19.7, it will be found to be valid for all positions of the given points. In particular, the distance OP of the point $P(x_1, y_1, z_1)$ from the origin O is found by writing $x_2 = y_2 = z_2 = 0$ and we have

$$OP^2 = x_1^2 + y_1^2 + z_1^2. \qquad (19.2)$$

19.9. The direction-cosines of a straight line

In Fig. 179, PQ is any straight line and OA is a straight line of unit length drawn parallel to PQ and passing through the origin O. The direction of PQ is clearly determined by the coordinates of the point A and, since these coordinates are the orthogonal projections of the unit length OA on the coordinate axes, A is the point $(\cos \alpha, \cos \beta, \cos \gamma)$ where α, β, γ are respectively the angles xOA, yOA and zOA. Using formula (19.2) we have

$$\cos^2 \alpha + \cos^2 \beta + \cos^2 \gamma = OA^2 = 1. \qquad (19.3)$$

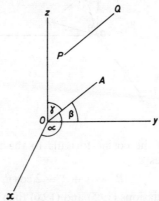

FIG. 179

The quantities $l = \cos \alpha$, $m = \cos \beta$, $n = \cos \gamma$ are said to be the *direction-cosines* of the line PQ and (19.3) can be written in the form

$$l^2 + m^2 + n^2 = 1 \qquad (19.4)$$

showing that *the sum of the squares of the direction-cosines of any straight line is equal to unity.* If the sense of the line PQ is reversed, OA is to be replaced by OA' where A' is the point with coordinates $(-\cos \alpha, -\cos \beta, -\cos \gamma)$ and the direction-cosines of QP are therefore equal in magnitude but opposite in sign to those of PQ. It follows that the direction-cosines of the coordinate axes, described in the positive sense, are respectively 1, 0, 0; 0, 1, 0 and 0, 0, 1 while those lines in the plane xOy which bisect the angle xOy have direction-cosines $1/\sqrt{2}, \pm 1/\sqrt{2}, 0$.

The angle θ between two lines with given direction-cosines can be found as follows. Let (Fig. 180) PQ and RS be the given lines with direction-cosines l_1, m_1, n_1 and l_2, m_2, n_2 respectively and let OA, OB be lines of unit length through the origin O parallel to PQ and RS. Then the coordinates of A and B are (l_1, m_1, n_1), (l_2, m_2, n_2) and, by equation (19.1), the length AB is given by

$$AB^2 = (l_1 - l_2)^2 + (m_1 - m_2)^2 + (n_1 - n_2)^2.$$

Since $l_1^2 + m_1^2 + n_1^2 = 1$ and $l_2^2 + m_2^2 + n_2^2 = 1$, this can be written

$$AB^2 = 2 - 2(l_1 l_2 + m_1 m_2 + n_1 n_2), \qquad (19.5)$$

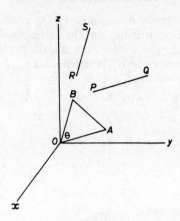

Fig. 180

and application of the cosine formula to the triangle OAB in which $OA = OB = 1$ gives

$$AB^2 = 1 + 1 - 2\cos\theta. \qquad (19.6)$$

Comparison of equations (19.5) and (19.6) then shows that the angle θ is given by

$$\cos\theta = l_1 l_2 + m_1 m_2 + n_1 n_2 \qquad (19.7)$$

and it should be noted that when the given lines are perpendicular to each other ($\theta = 90°$, $\cos\theta = 0$),

$$l_1 l_2 + m_1 m_2 + n_1 n_2 = 0. \qquad (19.8)$$

Example 9. *If P and Q are points with coordinates (x_1, y_1, z_1) and (x_2, y_2, z_2), show that the direction-cosines of the line PQ are*

$$\frac{x_2 - x_1}{r}, \quad \frac{y_2 - y_1}{r}, \quad \frac{z_2 - z_1}{r},$$

where $r^2 = (x_2 - x_1)^2 + (y_2 - y_1)^2 + (z_2 - z_1)^2$. Deduce that the angle between the diagonals of a cube is $\cos^{-1}(1/3)$.

The direction-cosines of the line PQ are the cosines of the angles made by PQ with the coordinate axes and these are obtained by dividing the lengths of the orthogonal projections of the line on the axes by the length of the line. The orthogonal projections of PQ being of lengths $x_2 - x_1, y_2 - y_1, z_2 - z_1$ and the length of the line PQ being given by $r = \sqrt{\{(x_2 - x_1)^2 + (y_2 - y_1)^2 + (z_2 - z_1)^2\}}$, the required results follow. In Fig. 181, $OABCDEFG$ is a cube of unit side. O and G are the points $(0, 0, 0)$ and $(1, 1, 1)$ so that the direction-cosines of the diagonal OG are

$$\frac{1 - 0}{r}, \quad \frac{1 - 0}{r}, \quad \frac{1 - 0}{r}$$

where $r^2 = (1 - 0)^2 + (1 - 0)^2 + (1 - 0)^2 = 3$, that is, they are $1/\sqrt{3}, 1/\sqrt{3}, 1/\sqrt{3}$. C and F are respectively the points $(0, 1, 0)$ and $(1, 0, 1)$ so that the

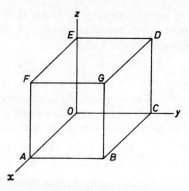

FIG. 181

direction-cosines of CF are

$$\frac{1-0}{r'}, \quad \frac{0-1}{r'}, \quad \frac{1-0}{r'}$$

where $(r')^2 = (1-0)^2 + (0-1)^2 + (1-0)^2 = 3$, that is, they are $1/\sqrt{3}, -1/\sqrt{3}, 1/\sqrt{3}$. Using equation (19.7), the angle θ between the diagonals OG and CF is given by

$$\cos\theta = \frac{1}{\sqrt{3}}\cdot\frac{1}{\sqrt{3}} + \frac{1}{\sqrt{3}}\left(-\frac{1}{\sqrt{3}}\right) + \frac{1}{\sqrt{3}}\cdot\frac{1}{\sqrt{3}} = \frac{1}{3}.$$

19.10. The equation of a plane

In Fig. 182, a plane meets the coordinate axes at points A, B, C and N is the foot of the perpendicular drawn from the origin O to the plane. If the length ON is p and if the direction-cosines of the

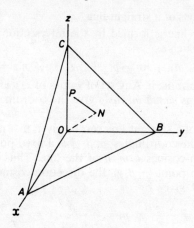

FIG. 182

line ON are l, m, n, the coordinates of N are (lp, mp, np) and, if P is any point in the plane with coordinates (x, y, z), the direction-cosines of the line PN are (by Example 9 above)

$$\frac{lp - x}{r}, \quad \frac{mp - y}{r}, \quad \frac{np - z}{r}$$

where $r^2 = (lp - x)^2 + (mp - y)^2 + (np - z)^2$. Since the lines ON, PN are at right angles to each other, equation (19.8) shows that the sum of the products of their direction-cosines is zero and hence that

$$\frac{l(lp - x)}{r} + \frac{m(mp - y)}{r} + \frac{n(np - z)}{r} = 0.$$

Since $l^2 + m^2 + n^2 = 1$, this can be written in the form

$$lx + my + nz = p \tag{19.9}$$

and this is the equation of a plane at distance p from the origin and whose normal has direction-cosines l, m, n.

Example 10. *Find the equation of the plane making intercepts of lengths 3, 5 and 6 on the coordinate axes.*

The equation of the plane ABC of Fig. 182 is $lx + my + nz = p$ and the lengths OA, OB, OC are 3, 5, 6. Hence the coordinates of A, B, C are respectively $(3, 0, 0)$, $(0, 5, 0)$, $(0, 0, 6)$ and these coordinates satisfy the equation of the plane. It follows that

$$3l = p, \quad 5m = p, \quad 6n = p$$

so that $l = p/3$, $m = p/5$, $n = p/6$ and the equation of the plane (after division by p) can be written

$$\frac{x}{3} + \frac{y}{5} + \frac{z}{6} = 1.$$

19.11. The equations of a straight line

Since a straight line is formed by the intersection of two planes, a pair of equations such as

$$lx + my + nz = p, \quad l'x + m'y + n'z = p' \tag{19.10}$$

is sufficient to determine it. Any set of values of x, y and z which satisfy these two equations *simultaneously* give the coordinates of a point on the line.

The equations of a straight line can be given in a more symmetrical form in terms of the coordinates (α, β, γ) of a fixed point on it together with the direction-cosines l, m, n of the line. Thus if P is the point (x, y, z) and Q the point (α, β, γ), the direction-cosines of the line QP are, by Example 9, given by

$$l = \frac{x - \alpha}{r}, \quad m = \frac{y - \beta}{r}, \quad n = \frac{z - \gamma}{r}$$

where $r = \sqrt{\{(x - \alpha)^2 + (y - \beta)^2 + (z - \gamma)^2\}}$ is the distance between the points P and Q. These equations can be written in the form

$$\frac{x - \alpha}{l} = \frac{y - \beta}{m} \quad \frac{z - \gamma}{n} = r \tag{19.11}$$

and this is a very useful form of the equations of a straight line.

Example 11. *Find the coordinates of the point in which the line of intersection of the planes $x - y + z = 2$, $2x - y - z = 4$ meets the plane $x = 0$. Find also the equations of the line of intersection in symmetrical form.*

The line meets the plane $x = 0$ where

$$-y + z = 2, \quad -y - z = 4$$

and these equations give $y = -3$, $z = -1$ so that the required point is $(0, -3, -1)$. The direction-cosines of the normals to the given planes are respectively proportional to $1, -1, 1$ and $2, -1, -1$. Each of these normals is perpendicular to the line of intersection of the planes so that, if l, m, n are proportional to the direction-cosines of the line

$$l - m + n = 0, \quad 2l - m - n = 0$$

and these equations are satisfied if $l:m:n = 2:3:1$. Hence the equations of the line can be written in the form

$$\frac{x}{2} = \frac{y + 3}{3} = \frac{z + 1}{1}.$$

EXERCISES 19 (e)

1. Find the distance between the points $A(-1, 2, 8)$ and $B(1, 5, 2)$. Find also the equation of the sphere with centre A and radius AB.

2. Find the equation of a right-circular cone of semi-vertical angle $45°$ whose axis lies along the coordinate axis Oy.

3. A straight line makes angles α, β, γ with the three coordinate axes. Show that $\sin^2 \alpha + \sin^2 \beta + \sin^2 \gamma = 2$.

4. Find the angle between straight lines whose direction-cosines are respectively proportional to $2, 3, 4$ and $1, 5, -2$.

5. If two straight lines have direction-cosines l_1, m_1, n_1 and l_2, m_2, n_2, show that the angle θ between them is given by

$$\sin^2 \theta = (m_1 n_2 - m_2 n_1)^2 + (n_1 l_2 - n_2 l_1)^2 + (l_1 m_2 - l_2 m_1)^2.$$

6. Find the direction-cosines of the normal to the plane $x + 2y - 2z = 9$ and find the lengths of the intercepts made by the plane on the coordinate axes.

7. If O is the origin of coordinates and P is the point $(2, 3, -1)$, find the equation of the plane through P at right angles to OP.

8. Find the equation of the plane through the point $(0, 1, 1)$ normal to the line joining the points $(1, 3, 4)$ and $(2, 4, 6)$. Find also the angle between this plane and the plane $2x - y + z = 6$.

9. Find the distance of the point $(-1, -5, -10)$ from the point of intersection of the line

$$\frac{x - 2}{3} = \frac{y + 1}{4} = \frac{z - 2}{12}$$

with the plane $x - y + z = 5$.

10. Find the conditions that the line

$$\frac{x - \alpha}{l} = \frac{y - \beta}{m} = \frac{z - \gamma}{n}$$

should lie in the plane $Ax + By + Cz = D$.

EXERCISES 19 (f)

1. Prove that the common perpendicular to two opposite edges of a regular tetrahedron is inclined at an angle of $45°$ to each of the other four edges.
   (O.C.)

2. AB and CD are two given skew lines and a third line cuts them at X and Y. Find, for different positions of X and Y, the locus of a point Z dividing XY internally in a given ratio. (L.U.)

3. A rectangular swimming bath is 15 m long and 5·4 m wide and the bottom slopes uniformly from a depth of 0·9 m at one end to 1·8 m at the other. Find

   (i) the length of the diagonal joining opposite bottom corners at the deep and shallow ends,
   (ii) the angle which this diagonal makes with the diagonal of the deep end wall which it meets. (O.C.)

4. Three edges AB, AC, AD of a cube are produced to P, Q, R respectively so that $AP = AQ = AR = 3AB/2$. Show that the plane PQR is parallel to the plane BCD and that the section of the cube by the plane PQR is a regular hexagon.

5. Prove that the line joining the mid-points of one pair of adjacent edges of a tetrahedron is equal and parallel to the line joining the mid-points of the opposite pair of edges. Prove also that the lines joining the mid-points of opposite edges are concurrent and bisect each other.

6. If a straight line is parallel to two planes, prove that it is parallel to their line of intersection.

7. Each of three concurrent straight lines OA, OB, OC is perpendicular to a fourth line. Prove that the lines OA, OB, OC are in the same plane. (O.C.)

8. OA, OB, OC are straight lines mutually at right angles, OD is perpendicular to BC and OE to AD. Show that OE is perpendicular to the plane ABC.
   (O.C.)

9. $ABCDEF$ is a regular hexagon of side a and is the base of a hexagonal pyramid, vertex V. If each of the edges VA, VB, ..., VF is of length $2a$, show that the shortest distance between AB and VE is $(2\sqrt{15}a)/5$. (L.U.)

10. AB is the common perpendicular to two skew lines AC, BD making an angle of θ with one another. Show that

$$CD^2 = AB^2 + AC^2 + BD^2 - 2AC \cdot BD \cos \theta.$$

11. $ABCD$ is the floor of a rectangular room and $A'B'C'D'$ is the ceiling, A' being vertically above A, etc. If $AB = 10·2$ m, $BC = 4·8$ m and $AA' = 3·6$ m, find the shortest distance between AB and DB'.

12. *ABCD* is a tetrahedron in which $AD = BC = a, BD = CA = b, CD = AB = c$. Find the length of the shortest distance between *AD* and *BC*.

13. *PN* is a line perpendicular to a plane *NAB*, *A*, *B* being points in this plane such that $AB = 4·13$, the angles $NAB = 63° 30'$, $NBA = 41° 45'$, $PAN = 27° 12'$. Calculate the length of the perpendicular *PN* and the angle *PBN*.
(L.U.)

14. *OA*, *OB*, *OC* are adjacent edges of a cubical block of side 4 m. *P*, *Q*, *R* are the middle points of these edges. The corner *O* is removed by cutting through the plane *PQR*, and the other corners are treated similarly. Show that the inclination of any triangular face to an adjacent square face of the resulting solid body is 125° 16'.

15. The base of a pyramid is a square *ABCD* of side 4 m. The height of the vertex *V* above the plane of the base is also 4 m and all the edges *VA*, *VB*, *VC*, *VD* are equal. Find the angle between the edge *VA* and the base.

16. A rectangular hoarding 3 m high and 9 m long faces due north. Calculate the area of its shadow on the ground when the sun is
> (i) due south at an elevation of 50°,
> (ii) south-west at an elevation of 30°.

17. The hypotenuse *BC* of a right-angled triangle *ABC* lies in a horizontal plane and its sides *AC*, *AB* are inclined to this plane at angles α and β respectively. Show that the inclination of the plane of triangle *ABC* to the horizontal plane is $\sin^{-1}\{\sqrt{(\sin^2 \alpha + \sin^2 \beta)}\}$.

18. Show that the diagonal of a cube is equally inclined to all the edges of the cube and find the angle of inclination.

19. *OA*, *OB*, *OC* are three mutually perpendicular lines and $OA = a$, $OB = b$, $OC = c$. Show that the angle between the planes *OBC* and *ABC* is

$$\tan^{-1}\left\{\frac{a\sqrt{(b^2 + c^2)}}{bc}\right\}.$$
(L.U.)

20. The corners *A*, *B*, *C* of a triangle in an inclined plane are at heights 6, 2 and 11 m respectively above a certain horizontal plane and the sides are $a = 18$ m, $b = 16$ m, $c = 21$ m. Find the lengths of the sides of the projected triangle and thence find, to three figures, the cosine of the angle between the two planes.
(L.U.)

21. Two straight lines *m*, *n* are skew. Show that the centre of the sphere which touches *m* at a given point *P* and *n* at a given point *Q* may be obtained as the intersection of three planes.
(L.U.)

22. Three spheres, each of radius *a*, rest on the horizontal base of a cylindrical tin. Each sphere touches the other two and also the curved surface of the tin. Find the radius of the tin. If a fourth sphere, also of radius *a*, resting on the three spheres just touches the top of the tin, find the height of the tin.
(L.U.)

23. A liqueur glass of height 0·125 m is of diameter 0·075 m at the top and 0·05 m at the bottom. Obtain the radius of the sphere which (*a*) just touches the bottom of the glass and the sloping sides, (*b*) fits into the glass exactly half-way down the slope.

24. n spheres form a ring on a horizontal plane, their centres being at the corners of a regular polygon, and each sphere touches two others. Another sphere touches all the spheres of the ring and rests on the same horizontal plane. If r denotes the radius of a sphere of the ring and R the radius of the other sphere, prove that $r = 4R \sin^2 (\pi/n)$.   (L.U.)

25. Through a fixed point O at distance c from the centre of a sphere of radius a, three planes are drawn at right angles to one another so that all three planes intersect the sphere. Show that the sum of the areas of the three circles of intersection of the planes with the sphere is $\pi(3a^2 - c^2)$.

26. The vertices of a triangle are the points $A(-1, 2, 3)$, $B(2, 1, 3)$ and $C(3, 1, 4)$. Find the value of $\cos C$.

27. A plane makes intercepts OA, OB, OC of lengths a, b, c on the coordinate axes. Show that the area of the triangle ABC is

$$\tfrac{1}{2}(b^2c^2 + c^2a^2 + a^2b^2)^{\frac{1}{2}}.$$

28. The equations of two straight lines are $x = ay + b$, $z = cy + d$ and $x = Ay + B$, $z = Cy + D$. Show that the two lines are perpendicular if $aA + cC + 1 = 0$.

29. Find the equations of the straight line passing through the point (α, β, γ) which is perpendicular to each of the lines

$$\frac{x}{l_1} = \frac{y}{m_1} = \frac{z}{n_1} \quad \text{and} \quad \frac{x}{l_2} = \frac{y}{m_2} = \frac{z}{n_2}.$$

30. A plane is perpendicular to each of the planes $x + y + 3z = 0$, $3x - 2y + 4z = 0$ and passes through the point $(1, 1, 1)$. Find

   (i) the direction-cosines of the normal to the plane,
   (ii) the equation of the plane.   (N.U.)

CHAPTER 20

MENSURATION OF SIMPLE SOLID FIGURES

20.1. Introduction

It has been assumed (Chapter 4) that the student is familiar with the idea of the circular measure of an angle and we assume here that the following formulae in the mensuration of the circle are also known:—

(i) for a sector of a circle of radius r, in which the circular measure of the angle between the bounding radii is θ, the length of arc $= r\theta$ and the area of the sector $= \frac{1}{2}r^2\theta$;

(ii) for a circle of radius r (the particular case of (i) above in which $\theta = 2\pi$), the circumference $= 2\pi r$ and the area $= \pi r^2$.

It is assumed also that the area of a trapezium, in which a, b are the lengths of the parallel sides and h is the height, is known to be $\frac{1}{2}(a + b)h$. Two particular cases of this formula to be especially noted are:—

(i) if $a = b$, the trapezium is a rectangle of sides b, h and area bh,

(ii) if $a = 0$, the trapezium is a triangle of base b, height h and area $\frac{1}{2}bh$.

The contents of this chapter include the mensuration of the simpler solid figures such as the prism, pyramid, cylinder, cone and sphere. The methods of the calculus are used where these seem to be suitable.

20.2. The volume of a right prism

The unit of volume is the cube whose edges are of unit length. A rectangular parallelepiped whose edges are of lengths a, b, c can be subdivided into abc cubes with edges of unit length and the volume of such a parallelepiped is therefore abc.

A *right* prism is one whose end faces are congruent polygons and whose remaining faces are *rectangles*. Since any polygon can be subdivided into a number of triangles, such a prism can be considered to be composed of a number of right triangular prisms, and it is fundamental to obtain a formula for the volume of such a prism.

Fig. 183 shows the end elevation of a right triangular prism, AD being the perpendicular from A to the base BC of the triangular end face ABC. Suppose that the prism is cut by planes parallel to the rectangular face of which BC is one side and let PQ, GH be the lines of intersection of two such planes with the end face ABC. Through

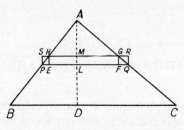

Fig. 183

P, Q draws PS, QR perpendicular to GH to meet it at S and R and through G, H draw GF, HE perpendicular to PQ to meet it at F and E. Let PQ, GH meet AD at L and M, and let $AM = x$, $AL = x + \delta x$.

From the similar triangles AHM, ABD, $AM/AD = AH/AB$, and from the similar triangles AHG, ABC, $AH/AB = HG/BC$. Hence $HG/BC = AM/AD$, giving, since $AM = x$,

$$HG = x \cdot \frac{BC}{AD}.$$

Similarly,

$$PQ = (x + \delta x)\frac{BC}{AD}.$$

The element of volume δV of the prism of which $PQGH$ is the end elevation lies between the volumes of the rectangular parallelepipeds of which $EFGH$ and $PQRS$ are end elevations. We have therefore, if h is the length of the prism,

$$HG \cdot ML \cdot h < \delta V < PQ \cdot ML \cdot h.$$

Substituting for HG, PQ and noting that $ML = \delta x$, these inequalities can be written

$$\frac{BC}{AD} \cdot h \cdot x \, \delta x < \delta V < \frac{BC}{AD} \cdot h \cdot (x + \delta x)\delta x,$$

from which it follows by arguments similar to those used in Chapter 10 that

$$\frac{dV}{dx} = \frac{BC}{AD} \cdot h \cdot x.$$

The total volume V is therefore given by

$$V = \int_0^{AD} \frac{BC}{AD} \cdot h \cdot x \cdot dx$$

$$= \frac{BC}{AD} \cdot h \left[\frac{x^2}{2}\right]_0^{AD} = \tfrac{1}{2}BC \cdot AD \cdot h$$

$$= \text{area of end face} \times \text{length}.$$

By subdividing any right prism into its component triangular prisms, it follows that *the volume of a right prism is equal to the product of its length and the area of its end face*. This result has been used in the derivation of formula (12.1), viz., the volume V of a solid body, the area of whose cross-section by a plane at distance x from the origin is a function $S(x)$ of x, is given by

$$V = \int_a^b S(x)dx, \qquad (20.1)$$

where a, b are the end values of x for the solid under discussion.

20.3. The volume of an oblique prism

Fig. 184 shows an oblique prism whose end faces are of area B and whose altitude (measured perpendicular to the planes of the end faces) is h.

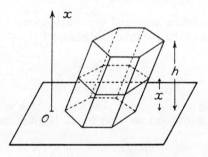

FIG. 184

Taking the x-axis perpendicular to the planes of the end faces, the cross-section of the prism at any altitude x for which $0 < x < h$ is also of area B, and (20.1) gives for the volume V of the prism,

$$V = \int_0^h B\,dx = Bh. \qquad (20.2)$$

Hence *the volume of any prism is equal to the product of the area of the base and the altitude*.

20.4. The volume of a pyramid

In Fig. 185, V is the vertex and $ABCD \ldots$ the base of a pyramid of which $A'B'C'D' \ldots$ is a plane section parallel to the base. VP is the perpendicular from V to the plane $ABCD \ldots$, cutting the plane $A'B'C'D' \ldots$ at P'. By similar triangles we have

$$\frac{A'B'}{AB} = \frac{VA'}{VA} = \frac{VP'}{VP},$$

and then, by theorem 1 (§ 18.6),

$$\frac{\text{area } A'B'C'D' \ldots}{\text{area } ABCD \ldots} = \frac{A'B'^2}{AB^2} = \frac{VP'^2}{VP^2}. \tag{20.3}$$

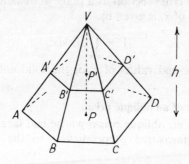

Fig. 185

If the origin is taken at V and the x-axis along VP, the area $S(x)$ of a cross-section of the pyramid at distance $VP' = x$ from V is given by (20.3) as

$$S(x) = \frac{B}{h^2} . x^2, \tag{20.4}$$

where B is the area of the base $ABCD \ldots$ and h is the altitude VP of the prism. Substitution for $S(x)$ in (20.1) gives for the volume V of the pyramid

$$V = \int_0^h \frac{B}{h^2} x^2 \, dx = \frac{B}{h^2} \left[\frac{x^3}{3} \right]_0^h$$

$$= \tfrac{1}{3} Bh. \tag{20.5}$$

Hence *the volume of a pyramid is equal to the product of one third the area of the base and the altitude.* Since a tetrahedron is a pyramid with a triangular base, the same formula applies for its volume.

Example 1 *A pyramid stands on a square base and its top is cut away by a plane parallel to the base and 0·06 m from it. If the area of the top of the remaining frustum is one-quarter of the area of the base and the volume of the frustum is $3·5 \times 10^{-4}$ m^3, find the length of the edge of the base.* (L.U.)

In Fig. 186, $VABCD$ is the pyramid, $A'B'C'D'$ is the top of the frustum and VP is perpendicular to the plane $ABCD$ meeting $A'B'C'D'$ at P'. We have shown in equation (20.4) that the areas of parallel sections of a pyramid are proportional to the squares of their distances from the vertex, so that $VP'^2/VP^2 =$ ratio of areas $A'B'C'D'$ and $ABCD = 1/4$ and hence $VP' = \tfrac{1}{2}VP$. Since $P'P = 0·06$ m it follows that $VP = 0·12$ m. If a is the length of an edge of the square base, its area is a^2 and the volume of the pyramids $VABCD$, $VA'B'C'D'$ are respectively $\tfrac{1}{3}a^2 . VP$ and $\tfrac{1}{3}.\left(\dfrac{a^2}{4}\right).VP'$ or $0·04a^2$ and $0·005a^2$ when we substitute $VP = 0·12$, $VP' = 0·06$.

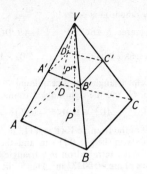

FIG. 186

The difference in the volumes of these pyramids is the volume of the frustum $A'B'C'D'ABCD$, so that

$$0.04a^2 - 0.005a^2 = 3.5 \times 10^{-4}$$

leading to $a = 0.1$ m.

Example 2. *The base of a tetrahedron is an equilateral triangle whose sides are each 0.08 m in length. The remaining edges of the tetrahedron are each 0.12 m in length. Calculate (i) the height, (ii) the volume, (iii) the total surface area.* (L.U.)

Let ABC be the base and V the vertex of the tetrahedron (Fig. 187). VP is perpendicular to the base and, from symmetry, P is the point of intersection of

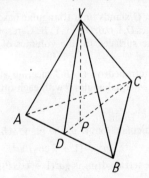

FIG. 187

the medians of the triangle ABC. If D is the mid-point of the edge AB,

$$DC^2 = AC^2 - AD^2 = 0.0064 - 0.0016 = 0.0048,$$

so that $DC = 0.06928$ m and $PC = \frac{2}{3}DC = 0.04619$ m.

From the right-angled triangle VPC,

$$VP^2 = VC^2 - PC^2 = 0.0144 - 0.0021 = 0.0123,$$

so that the height VP of the tetrahedron is 0.111 m.

The volume V of the tetrahedron is given by

$$V = \tfrac{1}{3}.\text{area } \triangle ABC.VP = \tfrac{1}{3}.\tfrac{1}{2}.AB.DC.VP$$

$$= \tfrac{1}{3} \times \tfrac{1}{2} \times 0.08 \times 0.06928 \times 0.111 = 1.02 \times 10^{-4} \text{ m}^3.$$

Since VAB is an isosceles triangle, the line joining V to the mid-point D of AB is perpendicular to AB and the right-angled triangle VAD gives

$$VD^2 = VA^2 - AD^2 = 0.0144 - 0.0016 = 0.0128,$$

so that the altitude VD of the triangle VAB is 0.113 m. The three sloping faces are triangles of base 0.08 m, height 0.113 m and therefore each is of area 0.00452 m^2. The base of the tetrahedron is a triangle of base 0.08 m, height 0.06928 m and therefore of area 0.00277 m^2. Hence the total surface area

$$= 0.00277 + 3 \times 0.00452 = 0.01633 \text{ m}^2.$$

EXERCISES 20 (a)

1. A rectangular swimming-bath is 15 m long and 5·4 m wide and the bottom slopes uniformly from a depth of 0·9 m at one end to 1·8 m at the other. Find the volume of the bath.

2. OA, OB, OC are adjacent edges of a cubical block of side 0·04 m. P, Q, R are the middle points of these edges. The corner O is removed by cutting through the plane PQR and the other corners are treated similarly. Find the volume of the remaining solid.

3. A pyramid with vertex O stands on a triangular base ABC. A plane parallel to the base cuts the edges OA, OB, OC in A', B', C' respectively. If $OA' = OA/3$, find the ratio of (i) the surfaces, (ii) the volumes of the pyramids $OA'B'C'$, $OABC$.   (O.C.)

4. The three edges of a tetrahedron $OABC$ meeting at the vertex O have the same length a and make equal angles θ with each other. Prove the following results:—

   (i) $AB = BC = CA = 2a \sin \tfrac{1}{2}\theta$,
   (ii) if p is the perpendicular from O to the plane ABC, then

   $$3p^2 = (1 + 2 \cos \theta)a^2,$$

   (iii) the volume of the tetrahedron is $\tfrac{1}{6}a^3(1 - \cos \theta)(1 + 2 \cos \theta)^{1/2}$.   (L.U.)

5. A pyramid stands on a square base and every edge has the same length a. Prove that the length of a diagonal of the base is twice the height of the pyramid.
   Calculate, in terms of a, (i) the volume, (ii) the total surface area, including the base, of the pyramid.   (L.U.)

6. A right pyramid stands on a square base of side $2a$, any sloping face being inclined at $60°$ to the base. Through one edge of the base a plane is drawn, at $30°$ to the base, cutting the pyramid. Find the volume of the new pyramid formed.   (L.U.)

7. The base of a tetrahedron is an equilateral triangle of side $4a\sqrt{3}$; the remaining three edges are equal in length. A sphere of radius a touches

each of the four faces internally. Calculate the volume of the tetrahedron.

(L.U.)

8.   The areas of the top and bottom of the frustum of a pyramid are respectively 24 m² and 54 m², and their distance apart is 10 m. Find the volume of the frustum.   (L.U.)

20.5. The volume and curved surface of a cylinder

The surface generated by a straight line which intersects the circumference of a circle and which is always perpendicular to the plane of the circle is known as a *right circular* cylindrical surface. The space bounded by such a surface and two planes perpendicular to the generators is a *right circular* cylinder and its *axis* is the line joining the centres of its circular ends. In what follows the word cylinder will be used to denote briefly such a body.

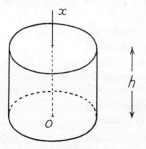

FIG. 188

Taking the origin at the centre of the circular base and the x-axis perpendicular to its plane, the area of any cross-section of the cylinder is πr^2, where r is the radius of the circular base. Hence, if the altitude of the cylinder is h, equation (20.1) gives for its volume V,

$$V = \int_0^h \pi r^2 \, dx = \pi r^2 h, \qquad (20.6)$$

so that *the volume of a cylinder is equal to the product of the area of its base and its altitude.*

It has been shown in § 12.8 that the area of the curved surface of a frustum of a right circular cone is given by

$\frac{1}{2}$(sum of circumferences of the circular ends) × the slant height.

The cylinder is a special case of such a frustum in which the circumference of each circular end is $2\pi r$ and the height is h, r and h being respectively the radius of the base and the altitude of the cylinder. Hence the area S of the curved surface of a cylinder is given by

$$S = 2\pi r h. \qquad (20.7)$$

Example 3. *The height of a cylinder is h and the radius of its base is r. If V is its volume and A the total area of its surface, show that*

$$2\pi r^3 - rA + 2V = 0. \hspace{2cm} \text{(O.C.)}$$

The total area of surface is that of the curved surface and the two plane ends, so that

$$A = 2\pi rh + 2\pi r^2.$$

Also $V = \pi r^2 h$, giving $h = V/\pi r^2$, and substitution in the expression for A gives

$$A = 2\pi r\left(\frac{V}{\pi r^2}\right) + 2\pi r^2,$$

from which the required result follows immediately.

EXERCISES 20 (b)

1. Three solid spheres of radii $2a$, a, a are placed inside a right circular cylinder of radius $2a$ whose length is such that each sphere touches the other two and touches one of the plane circular ends of the cylinder. Find the volume of the cylinder. (L.U.)

2. The height of a cylinder is $0·21$ m and its volume is $1·056 \times 10^{-3}$ m^3. Find the area of its curved surface.

3. The axes of three cylinders are the sides of a plane triangle and each cylindrical surface passes through the opposite vertex of the triangle. Show that the curved surfaces of the three cylinders are equal in area.

4. A cylinder of height $3a$ is inscribed in a sphere of radius $2a$. Find the *total* surface area of the cylinder.

5. A rectangular piece of paper, $0·11$ m by $0·06$ m, is curved so as to form the curved surface of a cylinder. Find the volumes of the two cylinders which can be so formed.

6. The radius of the base and height of a cylinder are respectively r and h. The radius and length are increased by *small* amounts ρ and λ respectively. Show that the volume of the cylinder is increased by

$$\frac{2\rho}{r} + \frac{\lambda}{h},$$

of itself approximately. (O.C.)

20.6. The volume and curved surface of a cone

The solid generated by the revolution of a right-angled triangle about one of the sides containing the right angle is known as a *right circular cone*. The side of the triangle about which the rotation takes place is the *axis* of the cone. In what follows we shall use the word cone to denote briefly such a body.

In Fig. 189, the vertex O of the cone is taken as origin and its axis as the axis of x. The semi-vertical angle of the cone is α and Q is the centre of a circular section at distance $OQ = x$ below O. The radius PQ of such a section is clearly $x \tan \alpha$, so that the area of the section is $\pi x^2 \tan^2 \alpha$. If h is the height of the cone, equation (20.1) gives for

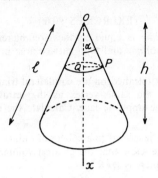

FIG. 189

its volume V,

$$V = \int_0^h \pi x^2 \tan^2 \alpha \, dx = \pi \tan^2 \alpha \left[x^3/3 \right]_0^h$$

$$= \tfrac{1}{3}\pi h^3 \tan^2 \alpha. \qquad (20.8)$$

This formula can be cast into an alternative form by observing that, if r is the radius of the circular base of the cone,

$$r = h \tan \alpha, \qquad (20.9)$$

and, elimination of $\tan \alpha$ between (20.8) and (20.9), gives

$$V = \tfrac{1}{3}\pi r^2 h. \qquad (20.10)$$

The formula given in § 12.8 for the surface of a frustum of a cone can be adapted to give the surface of the complete cone by observing that the circumference of one of the circular ends of the frustum is, in this case, zero. Hence, if l is the *slant* height (see Fig. 189), the curved surface S is given by

$$S = \pi r l. \qquad (20.11)$$

From the diagram $l = h \sec \alpha$, and this, together with (20.9), enables the surface area to be expressed in the alternative form

$$S = \pi h^2 \tan \alpha \sec \alpha. \qquad (20.12)$$

Example 4. *A cone of height h is cut into two portions by a plane parallel to the base. Find the distance of this plane from the vertex, if the product of the volumes of the two portions is to be a maximum.* (L.U.)

Let α be the semi-vertical angle of the cone and x the distance from the vertex of the cutting plane. Then the volumes of the complete cone and the upper portion are respectively $\tfrac{1}{3}\pi h^3 \tan^2 \alpha$ and $\tfrac{1}{3}\pi x^3 \tan^2 \alpha$. The volumes of the two portions into which the cone is cut are therefore $\tfrac{1}{3}\pi x^3 \tan^2 \alpha$ and $\tfrac{1}{3}\pi(h^3 - x^3)$ $\tan^2 \alpha$. The product of these two volumes will be a maximum when the function

$$F(x) \equiv x^3(h^3 - x^3)$$

is a maximum. This occurs when

$$\frac{dF}{dx} \equiv 3h^3 x^2 - 6x^5 = 0,$$

or when, $x = h/\sqrt[3]{2}$.

EXERCISES 20 (c)

1. The altitude of a cone is equal to the circumference of its base. Find expressions for the volume and *total* surface area in terms of the radius r of its base.

2. The faces of a regular tetrahedron are equilateral triangles of side a. A cone is inscribed, having the same vertex as the tetrahedron and base the inscribed circle of the opposite face. Calculate the curved surface of this cone.   (L.U.)

3. A cone of height h is inscribed in a sphere of radius R. Find an expression for its volume. Hence show that the greatest volume of a cone which can be inscribed in the sphere is $\frac{32}{81}\pi R^3$.   (O.C.)

4. A cone is cut into two parts of equal volume by a plane parallel to its base. Find the ratio of the curved surface area of the part which contains the original vertex to the curved surface area of the other part.   (L.U.)

5. The radii of the ends of a frustum of a cone are $a-b$ and $a+b$ and its slant height is $2h$. Show that:—

   (i) its volume $= \dfrac{2\pi}{3}(3a^2 + b^2)\sqrt{(h^2 - b^2)}$,

   (ii) its *total* surface area $= 2\pi(a^2 + 2ah + b^2)$.

6. A bell tent consists of a conical part above and a cylindrical part near the ground. Show that, for a given volume and given radius of the circular base, the area of canvas used is a minimum when the semi-vertical angle of the cone is $\cos^{-1}(2/3)$.

7. If R, r are the radii of the larger and smaller faces respectively of the frustrum of a cone of height h, show that its volume is $\pi h(R^2 + Rr + r^2)/3$ and its curved surface is

$$\pi(R + r)\sqrt{\{h^2 + (R - r)^2\}}.$$

8. A cone is inscribed in a sphere of radius a. If the vertical angle of the cone is 2θ, prove that its volume is

$$\tfrac{8}{3}\pi a^3 \sin^2 \theta \cos^4 \theta.$$

By putting $x = \cos^2 \theta$ find the value of x, and hence the value of θ, for which the volume of the cone is a maximum.   (L.U.)

20.7. The mensuration of the sphere

In Fig. 190, O is the centre of a sphere of radius r, OA is a vertical radius which is taken as the x-axis, and PQR is a horizontal circular section whose centre N is at depth x below O. From the right-angled triangle ONP,

$$NP^2 = OP^2 - ON^2 = r^2 - x^2.$$

Hence the area of the circular section PQR is $\pi(r^2 - x^2)$, and, by (20.1), the volume V of a frustum of the sphere bounded by parallel planes at depths h, k $(h < k)$ below O is given by

$$V = \int_h^k \pi(r^2 - x^2)dx$$

$$= \pi r^2(k - h) - \tfrac{1}{3}\pi(k^3 - h^3). \qquad (20.13)$$

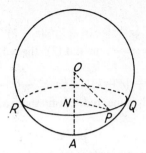

FIG. 190

By writing $h = 0$, $k = r$ in this formula, the volume of a hemisphere is given by

$$\tfrac{2}{3}\pi r^3,$$

and hence the volume of a complete sphere is

$$\tfrac{4}{3}\pi r^3. \tag{20.14}$$

Formula (20.13) applies, of course, to the volume of a frustum whose limiting planes are on the *same* side of the centre of the sphere. If the planes be at distances h, k from the centre but on *opposite* sides of this point, the frustum can be divided into two by a plane through the centre. The volume of the first frustum is then

$$\pi r^2(k - 0) - \tfrac{1}{3}\pi(k^3 - 0^3) = \pi r^2 k - \tfrac{1}{3}\pi k^3,$$

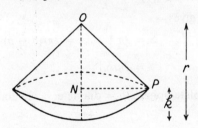

FIG. 191

and the volume of the second is similarly

$$\pi r^2 h - \tfrac{1}{3}\pi h^3.$$

Hence the volume of the whole frustum is

$$\pi r^2(k + h) - \tfrac{1}{3}\pi(k^3 + h^3). \tag{20.15}$$

The volume of a sector of a sphere can be found by dividing it into a cone and a spherical cap. Thus, Fig. 191, if r is the radius of the sphere and k is the height of the spherical cap, the height of the cone is clearly $r - k$. Since $OP = r$, the right-angled triangle ONP gives $NP^2 = OP^2 - ON^2 = r^2 - (r - k)^2$, so that the volume of the conical portion is

$$\tfrac{1}{3}\pi\{r^2 - (r - k)^2\}(r - k).$$

By writing $k = r$, $h = r - k$ in (20.13), the volume of the spherical cap is

$$\pi r^2\{r - (r - k)\} - \tfrac{1}{3}\pi\{r^3 - (r - k)^3\},$$

and, by addition and some reduction, the volume of the sector is found to be

$$\tfrac{2}{3}\pi r^2 k. \tag{20.16}$$

The area of the curved surface of a spherical frustum or zone of radius r in which the perpendicular distance between the parallel plane ends is h is the area of the surface of revolution obtained by rotating the arc of the circle $x^2 + y^2 = r^2$ included between the points for which $x = a$, $x = b$, where $b - a = h$, about the x-axis. By equation (12.27) of § 12.8, the surface area S is given by

$$S = 2\pi \int_a^b y\sqrt{\{1 + (dy/dx)^2\}}dx$$

$$= 2\pi \int_a^b \sqrt{\{y^2 + (y\,dy/dx)^2\}}dx. \tag{20.17}$$

Since $x^2 + y^2 = r^2$, $2x + 2y(dy/dx) = 0$ and

$$y^2 + \left(y\frac{dy}{dx}\right)^2 . = y^2 + (-x)^2 . = y^2 + x^2 = r^2.$$

Hence (20.17) gives

$$S = 2\pi \int_a^b r\,dx = 2\pi r(b - a)$$

$$= 2\pi rh, \tag{20.18}$$

since $b - a = h$.

For a complete sphere, $h = \text{diameter} = 2r$ and the surface area is

$$4\pi r^2. \tag{20.19}$$

Example 5. *A plane cuts a sphere of radius r into two segments whose curved surfaces are in the ratio* 3:1. *Find the distance of the plane from the centre of the sphere. Prove that the volume of the larger segment is* $9\pi r^3/8$.   (L.U.)

Let x = distance of cutting plane from the centre of the sphere. Then the width of the two zones are respectively $r + x$ and $r - x$. Hence, by (20.18),

$$\frac{2\pi r(r + x)}{2\pi r(r - x)} = \frac{3}{1}$$

leading to $x = \tfrac{1}{2}r$.

The larger segment consists of a hemisphere and a frustum whose bounding planes are at distance 0 and $r/2$ from the centre. By (20.13), the volume is

$$\tfrac{2}{3}\pi r^3 + \pi r^2(\tfrac{1}{2}r - 0) - \tfrac{1}{3}\pi(\tfrac{1}{8}r^3 - 0^3),$$

or $9\pi r^3/8$.

Example 6. *A cylindrical hole is bored through a solid sphere of radius r, the axis of the cylinder coinciding with a diameter of the sphere. Prove that, if l is the length of the resulting hole, the volume of the remainder of the sphere is $\pi l^3/6$.* (L.U.)

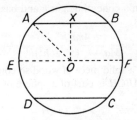

FIG. 192

Fig. 192 shows a section through the axis EF of the cylindrical hole. O is the centre of the sphere, $ABCD$ the section of the hole and X the mid-point of AB. The right-angled triangle AOX gives

$$OX^2 = AO^2 - AX^2 = r^2 - \frac{l^2}{4},$$

so that the volume of the cylinder of which $ABCD$ is a section is

$$\pi l\left(r^2 - \frac{l^2}{4}\right).$$

The volume of each of the spherical caps AED, CFB is given by writing $k = r$, $h = l/2$ in equation (20.13). The volume of each is therefore

$$\pi r^2\left(r - \frac{l}{2}\right) - \frac{\pi}{3}\left(r^3 - \frac{l^3}{8}\right).$$

Hence, the required volume remaining

$$= \frac{4}{3}\pi r^3 - \pi l\left(r^2 - \frac{l^2}{4}\right) - 2\pi r^2\left(r - \frac{l}{2}\right) + \frac{2\pi}{3}\left(r^3 - \frac{l^3}{8}\right).$$

$$= \pi l^3/6.$$

EXERCISES 20 (d)

1. Prove the equivalence of the two formulae

$$\frac{\pi h^2(3R - h)}{3} \quad \text{and} \quad \frac{\pi h(h^2 + 3r^2)}{6},$$

for the volume of a cap of height h cut from a sphere of radius R, r being the radius of the plane base of the cap. [Equation (20.13) may be assumed.]
(L.U.)

2. A solid sphere of radius 0·1 m is divided by a plane into two parts, the volume of one part being half that of the other. Find the distance of the plane from the centre of the sphere, correct to the nearest millimetre.
(Q.E.)

3. A spherical iron shell of outside diameter 0·356 m weighs 68 kg. Calculate its thickness, assuming it to be uniform, if the density of the iron is 7690 kg/m³.   (Q.E.)

4. A sphere is inscribed in a cone of height h and base-radius a. Show that the volume of the sphere is

$$\frac{4\pi a^3 h^3}{3\{a + \sqrt{(a^2 + h^2)}\}^3}.$$ (L.U.)

5. A top is in the form of a solid piece of wood so constructed that the portion $OABC$ is a cone and $ABCH$ is part of a sphere of radius a.

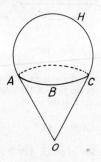

FIG. 193

The cone is such that its generators OA, OB, OC, etc., are tangents to the sphere at A, B, C, etc. Prove that the area of the surface of the top is

$$\pi a^2 (1 + \sin \theta)^2 \operatorname{cosec} \theta,$$

where θ is the semi-vertical angle of the cone.   (L.U.)

6. A hollow cone of height $3a$ and vertical angle 60° has its axis vertical and vertex downwards. A sphere of radius $2a$ rests on the cone. Prove that the surface area of the part of the sphere within the cone is one-quarter of the total surface area of the sphere.   (L.U.)

7. A cone of height h and a hemisphere are on the same side of their common circular base of radius $r(h > r)$. Prove that the area of that part of the surface of the hemisphere which is outside the cone is $4\pi r^3 h/(h^2 + r^2)$.   (L.U.)

8. A sphere rests in a horizontal circular hole of radius $2a$ and the lowest point of the sphere is a below the plane of the hole. Calculate the area of the part of the surface of the sphere below the hole and the volume of this part of the sphere.   (L.U.)

20.8. Summary of some mensuration formulae

Some of the more important formulae are here collected for easy reference.

Circle.

Circumference $= 2\pi r$. r = radius.
Area $= \pi r^2$.

Length of arc $= r\theta$. θ = angle between extreme radii.
Area of sector $= \frac{1}{2}r^2\theta$.

Trapezium.
Area $= \frac{1}{2}(a + b)h$. a, b the parallel sides,
    h = altitude.

Prism.
Volume $= Bh$. B = area of base,
    h = altitude.

Pyramid.
Volume $= \frac{1}{3}Bh$. B = area of base,
    h = altitude.

Cylinder.
Volume $= \pi r^2 h$. r = radius of base,
Area of curved surface $= 2\pi rh$. h = altitude.

Cone.
Volume $= \frac{1}{3}\pi r^2 h = \frac{1}{3}\pi h^3 \tan^2 \alpha$. r = radius of base,
Area of curved surface $= \pi rl$ h = altitude,
    $\quad = \pi h^2 \tan \alpha \sec \alpha$. l = slant height,
    α = semi-vertical angle.

Sphere.
Volume $= \frac{4}{3}\pi r^3$. r = radius.
Area of surface $= 4\pi r^2$.
Surface of zone $= 2\pi rh$. h = height of zone.

EXERCISES 20 (e)

1. A circle is divided by a chord into two portions whose areas are in the ratio $2:1$. Find an equation for the angle θ subtended by the chord at the centre. (L.U.)

2. Two wheels of diameters $2\cdot4$ m and $1\cdot8$ m are in the same plane and their centres are $2\cdot7$ m apart. Calculate the least length of the belt that will go round the wheels. (Q.E.)

3. AP and AQ are equal chords of a circle. If the area of the circle enclosed between AP and AQ is twice the area of the triangle APQ, prove that
$$\sin 2\theta + \sin \theta = \theta,$$
where θ is the radian measure of the angle PAQ.
    Determine an approximate value for θ by drawing graphs of the two sides of this equation for values of θ between $\pi/3$ and $\pi/2$. (L.U.)

4. Show that the volume of a regular tetrahedron whose edges are all of length a is $(\sqrt{2}a^3)/12$.

5. A wooden block in the form of a cube of edge $2b$ has each corner cut off by saw cuts through the middle points of the three edges in that corner.

Prove that the total surface area of the block of wood that remains is $b^2(12 + 4\sqrt{3})$. (L.U.)

6. OA, OB, OC are adjacent edges of a cubical block of side a. L, M are the middle points of the edges OA, OB respectively. The tetrahedron $OCLM$ is removed by cutting along the plane LMC. Find, in terms of a, the volume of the solid remaining. (L.U.)

7. Show that the volume of a frustrum of a pyramid is

$$\tfrac{1}{3}h(B_1{}^2 + B_1B_2 + B_2{}^2),$$

where h is the height of the frustrum and $B_1{}^2$, $B_2{}^2$ are the areas of its parallel ends.

8. A closed cylinder of height $3a$ is inscribed in a sphere of radius $2a$. Find the area of the whole surface of the cylinder including its plane ends.

9. Out of a wooden cone of radius a and height h is carved a solid composed of a hemisphere of radius $a/2$ and a cylinder of the same radius, their circular bases coinciding and the axis of the solid coinciding with the axis of the cone. If the hemisphere touches the base of the cone and the rim of the opposite end of the cylinder lies on the curved surface of the cone, find the volume of the solid and express it as a fraction of the volume of the cone. (L.U.)

10. Show that the altitude of the cone of greatest volume which can be inscribed in a sphere of radius a is $4a/3$. If in this cone there is now described the cylinder of greatest volume, show that its volume is 32/243 of the volume of the sphere. (L.U.)

11. The radii of the circular ends of a frustrum of a cone are a and b. If a sphere can be inscribed in the frustrum to touch the plane ends and also to touch the curved surface, show that the area of the curved surface exceeds the area of the surface of the sphere by $\pi(a - b)^2$. (L.U.)

12. A solid sphere of radius r is divided by a plane at a distance x from its centre. From the *larger* portion a conical hole is drilled out, the vertex of the cone being at the centre of the sphere and its plane base being coincident with the circle of division of the sphere. Find the volume of the remaining solid and show that, if $x = 3r/5$, its total surface area is equal to that of the original sphere. (L.U.)

13. A lead sphere of diameter 0·1 m is melted and cast into a solid cone. Find the height of the cone if the area of its curved surface is as small as possible. (L.U.)

14. A cube stands on a plane and is enclosed by a hollow cone which also stands on the plane. Prove that the volume of the cone is a minimum when the angle at its vertex is $2\tan^{-1}\{1/(2\sqrt{2})\}$. (L.U.)

15. Two floating buoys are made, one spherical and the other in the form of a cone. Both are 1·5 m high and the diameter of the base of the cone is also 1·5 m. Calculate the ratio of the surface areas. The conical buoy is arranged to float with vertex upwards and base horizontal, and half the volume is submerged. Find the height of the vertex above the surface. (L.U.)

16. The semi-vertical angle of a cone is $\tan^{-1}(3/4)$ and the radius of its base is r. If its curved area is $8/15$ of that of a hemisphere of radius R, find the ratio $R:r$. (L.U.)

17. Three spheres each of radius a rest on a horizontal plane with their centres at the vertices of an equilateral triangle of side $2a$. A fourth equal sphere rests symmetrically on top of the other three. Find the height of the highest point of the fourth sphere above the horizontal plane, and prove that the ratio of the volume of the tetrahedron whose vertices are the centres of the spheres to the volume of a sphere is $1:\pi\sqrt{2}$. (L.U.)

18. The figure represents a lens which is bounded by two equal spherical surfaces of radius 1.44 m, and whose maximum thickness is 0.01 m. Calculate,

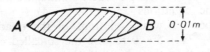

A B $0.01\,m$

FIG. 194

(i) the diameter of the lens, i.e., the length AB, (ii) its total surface area, (iii) its volume. (L.U.)

19. Two spheres of radii a and b cut orthogonally. Find the area of the first sphere which lies outside the second.

20. The distance between the planes of two circles is λ times the sum of their radii. Show that the surface areas of the spherical and conical zones bounded by them are in the ratio $\sqrt{(1+\lambda^2)}:1$.

21. AB is a diameter of a sphere and N is a point on AB. Prove that the volume of the space included between the spheres on AB, AN, NB as diameters respectively is equal to half the volume of a cylinder, the radius of whose base is NU and whose altitude is AB, where NU is drawn perpendicular to AB to meet the outer sphere in U. (L.U.)

22. A regular tetrahedron is inscribed in a sphere of radius R. Prove that

   (i) the height of the tetrahedron is $4R/3$,
   (ii) the plane of the base of the tetrahedron cuts the sphere into segments having volumes in the ratio $7:20$. (L.U.)

23. A wine glass has the shape of a cone with semi-vertical angle $30°$ and vertical depth a. The glass is completely filled with liquid and a spherical ball is then gently lowered into the liquid until it rests in contact with the inner surface of the cone. Prove that the greatest overflow occurs when the radius of the ball is $a/2$.

   (Assume that the ball is *not* completely immersed.) (L.U.)

24. A sector is cut from a piece of paper of radius r and formed into a cone of semi-vertical angle α. For what value of α will the volume of the cone be a maximum?

25. Assuming that the earth is a sphere of radius 6400 kilometres, find the area of the portion of it visible to an observer at an altitude of 3 kilometres.

CHAPTER 21

COMPLEX NUMBERS

21.1. Introduction

Equations such as $x - 2 = 0$, $2x - 6 = 0$ are soluble in terms of positive integers while the equations $x + 2 = 0$, $2x + 6 = 0$ are soluble only if the number system includes negative integers. Similarly equations like $5x = 17$ have solutions only if the number system contains rational numbers of the form p/q where p and q are integers. Solutions of the quadratic equations $x^2 = 5$, $x^2 - 4x - 1 = 0$ require the irrational number $\sqrt{5}$ while solutions of the quadratic equations $x^2 + 5 = 0$, $x^2 - 2x + 5 = 0$ require the so-called *complex numbers* and such numbers form the subject of the present chapter.

Complex numbers are introduced here by first constructing an algebra of *ordered pairs* of real numbers and, if a and b are the real numbers involved, the complex number is (at present) denoted by the symbol $[a, b]$. An example of an ordered pair of real numbers has, in fact, already been afforded by the rectangular Cartesian coordinates which fix the position of a point in a plane. Thus the point $(2, 5)$ is one with abscissa 2 and ordinate 5 while the point $(5, 2)$ has abscissa 5 and ordinate 2. In the same way, although the real numbers involved need no geometrical significance, the complex number $[a, b]$ differs from the complex number $[b, a]$ unless $a = b$.

The generalisation of the number system to include complex numbers is analogous to the introduction of the rational numbers p/q, for these are also ordered pairs of numbers. The algebra of rational numbers is subject to the rules that if p/q and r/s are two such numbers then, $p/q = r/s$ if and only if $ps = qr$, $(p/q) + (r/q) = (p + r)/q$ and $(p/q) \times (r/s) = (pr/qs)$. Similarly the logical introduction of complex numbers requires the definitions of certain⁰ elementary operations such as equality, addition, multiplication, etc.

In contrast to the rule of equality given above for rational numbers, two complex numbers are said to be equal if, and only if, they are identical; that is

$$[a, b] = [c, d] \quad \text{if and only if } a = c \text{ and } b = d. \qquad (21.1)$$

The meanings given to addition and subtraction are that

$$[a, b] \pm [c, d] = [a \pm c, b \pm d], \qquad (21.2)$$

and, for multiplication by a real number k,

$$k[a, b] = [a, b]k = [ka, kb]. \qquad (21.3)$$

It is a consequence of these definitions that any complex number

$[a, b]$ can be expressed in the form

$$[a, b] = a[1, 0] + b[0, 1] \tag{21.4}$$

and, just as the product of two real numbers is found from a multiplication table, equation (21.4) enables the product of two complex numbers to be found from a multiplication table of the special complex numbers $[1, 0]$ and $[0, 1]$.

The cummutative, associative and distributive laws of multiplication are postulated to apply in the algebra of complex numbers. To link it with the algebra of real numbers, the special complex number $[1, 0]$ is made to behave like the real number unity so that multiplication by $[1, 0]$ leaves any number unchanged. Thus

$$[1, 0][0, 1] = [0, 1][1, 0] = [0, 1] \tag{21.5}$$

and

$$[1, 0]^2 = [1, 0] = 1. \tag{21.6}$$

It follows from (21.3) that the complex number $[a, 0]$ behaves exactly like the real number a. If the other special complex number $[0, 1]$ is denoted by the symbol i, any complex number $[a, b]$ can be written, from equation (21.4), in the form $a + bi$.

The complex number $[1, 0]$ is, from (21.6), such that its square is unity; if the square of the other special complex number $[0, 1]$ is taken as -1, we have

$$i^2 = [0, 1]^2 = [-1, 0] = -1, \tag{21.7}$$

and a genuine extension of the number system is obtained. Using (21.7), the product of two complex numbers is then given by

$$\begin{aligned}
[a, b] \times [c, d] &= (a + bi)(c + di) \\
&= ac + bdi^2 + (ad + bc)i \\
&= ac - bd + (ad + bc)i \\
&= [ac - bd, ad + bc].
\end{aligned} \tag{21.8}$$

If x is a real number, $x^2 - 2x + 5 = (x - 1)^2 + 2^2 > 0$ and the equation $x^2 - 2x + 5 = 0$ has no solution in terms of real numbers. If, however, x is the complex number $a + bi$ we have, using the relation $i^2 = -1$,

$$(a + bi)^2 - 2(a + bi) + 5 = a^2 - 2a + 5 - b^2 + (2ab - 2b)i$$

and the quadratic equation is satisfied if

$$b^2 = a^2 - 2a + 5 \quad and \quad 2ab - 2b = 0.$$

These simultaneous equations are satisfied by $a = 1, b = \pm 2$ so that, in an algebra admitting complex numbers, the quadratic equation $x^2 - 2x + 5 = 0$ has solutions $1 \pm 2i$. The same result is obtained from the usual formula (1.2) for the roots of a quadratic if we also use the relation $i^2 = -1$.

The foregoing definitions of equality, addition and multiplication for the ordered pairs $[a, b]$, $[c, d]$ have been made so that the operations of an algebra using complex numbers are performed in the same way as in the algebra of real numbers. It is possible at this point to drop the symbol $[a, b]$ and use the alternative $a + bi$, treating i as an ordinary number but replacing i^2 by -1 whenever it occurs.

21.2. The geometrical representation of complex numbers

It has already been pointed out that an ordered pair of real numbers can be used to specify the position of a point in a plane and this fact can be used to give a geometrical representation of a complex number. The complex number $z = x + yi$ is represented by a point Z with coordinates (x, y) relative to rectangular axes Ox, Oy with origin O. The usual sign conventions used in graphical work are employed so that, if x and y are both positive the point Z lies in the first of the four quadrants in which the plane is divided by the axes, if $x < 0$ and $y > 0$ the point Z lies in the second quadrant and so on. Examples are given in Fig. 195 in which the points Z_1, Z_2, Z_3 represent respectively the

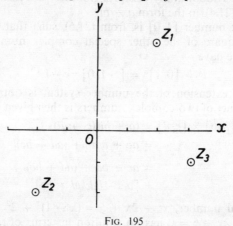

FIG. 195

complex numbers $2 + 3i$, $-2 - 2i$ and $3 - i$. This representation of complex numbers was originally due to the mathematician J. R. Argand and a diagram such as Fig. 195 is often referred to as the *Argand diagram*.

With the notation $z = x + yi$, it is convenient (but not altogether desirable) to refer to x and y respectively as the *real* and *imaginary* parts of the complex number z and it is often useful to express this by the symbols $x = R(z)$, $y = I(z)$. Complex numbers for which $y = 0$ are often said to be "purely real" and those for which $x = 0$ to be "purely imaginary". The representative points of such numbers will lie respectively on the axes Ox, Oy of the Argand diagram and these axes are often referred to as the "real axis" and the "imaginary axis". This terminology is convenient but it is not completely satisfactory for there is nothing imaginary about y (or the axis Oy)—it is just as real as x (or the axis Ox).

If, in Fig. 196, the point Z represents the complex number $z = x + yi$ and if the length OZ is denoted by r and the angle xOZ by θ, we have

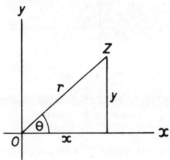

FIG. 196

$$x = r \cos \theta, \quad y = r \sin \theta, \left.\begin{array}{c} \\ \end{array}\right\}$$
$$r = +\sqrt{(x^2 + y^2)}, $$
$$\cos \theta = \frac{x}{r}, \quad \sin \theta = \frac{y}{r}. \qquad (21.9)$$

The length r is called the *modulus* of the complex number $x + yi$ and is denoted by the symbol $|x + yi|$ or $|z|$. The angle θ is called the *argument* or *amplitude* of the complex number and is denoted by $\arg(x + yi)$ or $\text{amp}(x + yi)$. The complex number $z = x + yi$ is expressed in terms of its modulus r and argument θ by the relation

$$z = x + yi = r(\cos \theta + i \sin \theta) \qquad (21.10)$$

and it should be noted that r is always taken as being positive. The last of equations (21.9) show that there is one value of θ for which $-\pi < \theta \leqslant \pi$ but that any value θ differing from this by an even integral multiple of π would give the same representative point Z in Fig. 196. It is usual to take the value of θ in the range $-\pi < \theta \leqslant \pi$ as the *principal value* of the argument of the number z. When determining the principal value of the argument, it is necessary for θ to satisfy both the equations

$$\cos \theta = \frac{x}{r}, \quad \sin \theta = \frac{y}{r},$$

since the single equation $\tan \theta = y/x$ into which these equations can be combined leads to two possible values of θ in the range.

Example 1. *Find the modulus and argument of the complex number* $-2 + 3i$.

In the Argand diagram (Fig. 197), the representative point Z has coordinates

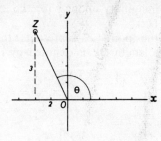

FIG. 197

$(-2, 3)$. From the figure, if r and θ are the required modulus and argument,

$$r \cos \theta = -2, \quad r \sin \theta = 3.$$

Hence

$$r = +\sqrt{\{(-2)^2 + (3)^2\}} = \sqrt{(13)} = 3\cdot61$$

and

$$\cos \theta = \frac{-2}{\sqrt{(13)}}, \quad \sin \theta = \frac{3}{\sqrt{(13)}}.$$

From the latter two equations, $\theta = 123° \, 42' = 2\cdot159$ radians.

21.3. Conjugate complex numbers

The two complex numbers $x + yi$, $x - yi$, which differ only in the sign of their imaginary parts are said to be *conjugate*. If z denotes the complex number $x + yi$, it is convenient to denote the conjugate number $x - yi$ by $\bar{z}$ and immediate consequences are that

$$z + \bar{z} = x + yi + x - yi = 2x,$$
$$z\bar{z} = (x + yi)(x - yi) = x^2 - y^2i^2 = x^2 + y^2.$$

Hence both *the sum and the product of two conjugate complex numbers are real quantities.*

The second of the above facts is useful in the reduction of a fraction of the form $(a + bi)/(c + di)$. If the numerator and denominator are multiplied by the conjugate $c - di$ of the denominator, the real quantity $c^2 + d^2$ is obtained in the resulting denominator; the details are shown in the following example.

Example 2. *Express* $\dfrac{2 + 3i}{3 + 4i}$ *in the form* $a + bi$.   (N.U.)

$$\frac{2 + 3i}{3 + 4i} = \frac{(2 + 3i)(3 - 4i)}{(3 + 4i)(3 - 4i)} = \frac{6 - 8i + 9i - 12i^2}{9 - 16i^2}$$

$$= \frac{18}{25} + \frac{1}{25}i, \quad \text{since } i^2 = -1.$$

21.4. The manipulation of complex numbers

If i is treated as an ordinary number and use is made of the relation $i^2 = -1$, complex numbers obey all the ordinary rules of algebra. Thus, we can write

$$i^3 = i \times i^2 = -i, \quad i^4 = i^2 \times i^2 = -1 \times -1 = 1,$$
$$i^5 = i^4 \times i = i, \quad \text{etc.,}$$

and these relations are often useful in the manipulation of complex numbers. The equality of two complex numbers $a + bi$, $c + di$ implies [from equation (21.1)] that $a = c$ and $b = d$; in other words, if two complex numbers are equal, *we can equate their real parts and we can equate their imaginary parts.*

Some applications of these remarks are given in the two examples which follow.

Example 3. *Simplify* $(2 + i)^4 - (2 - i)^4$.

The quantity in question

$$= 2^4 + 4.2^3 i + 6.2^2 i^2 + 4.2 i^3 + i^4 - 2^4 + 4.2^3 i - 6.2^2 i^2 + 4.2 i^3 - i^4$$
$$= 64i + 16i^3 = 64i - 16i = 48i.$$

Example 4. *Express* $\sqrt{(3 + 4i)}$ *in the form* $a + bi$.

If $\sqrt{(3 + 4i)} = a + bi$, then

$$3 + 4i = (a + bi)^2 = a^2 - b^2 + 2abi.$$

Equating the real and imaginary parts

$$a^2 - b^2 = 3, \quad ab = 2.$$

The solution of these simultaneous algebraical equations is

$$a = \pm 2, \quad b = \pm 1, \quad \text{or} \quad a = \pm i, \quad b = \pm 2i$$

and both solutions lead to the result $\pm(2 + i)$ for $\sqrt{(3 + 4i)}$.

EXERCISES 21 (a)

1. Find the moduli and arguments of the four quantities:—

    (a) -1, (b) i, (c) $3 + 4i$, (d) $-i - \sqrt{3}$

    and show in the Argand diagram the positions of the points representing the quantities.

2. Find the real and imaginary parts of $(2 + 3i)/(4 + 5i)$.

3. Express in the form $a + bi$:—

    (i) $(2 + i)^2$, (ii) $(1 + i)^4$, (iii) $(1 - i)^2/(1 + i)$.

4. Simplify:—

    (i) $(3 + i)^2 + (3 - i)^2$, (ii) $\dfrac{3 + 4i}{3 - 4i} - \dfrac{3 - 4i}{4 + 4i}$

5. Find the modulus and argument of the complex quantity $(1 + 2i)/(3 + 4i)$.

6. Find the moduli and arguments of $z_1 = (2 - i)/(3i - 1)$, $z_2 = (i - 3)/(2 + i)$ and of $z_1 + z_2$. (O.C.)

7. Given that $f(z) = (7 - z)/(1 - z^2)$ where $z = 1 + 2i$, show that $|z| = 2|f(z)|$.

8. Find the real and imaginary parts of the complex number z when

$$\text{(i)} \ \frac{z}{z + 1} = 1 + 2i, \quad \text{(ii)} \ \frac{z + i}{z + 1} = \frac{z + i}{z - 3}.$$

9. If $z = x + yi$ and $\bar{z}$ is the conjugate of z, find the values of x and y such that

$$\frac{1}{z} + \frac{2}{\bar{z}} = 1 + i.$$

10. If x, y, a and b are real numbers and if

$$x + yi = \frac{a}{b + \cos\theta + i\sin\theta}$$

show that $(b^2 - 1)(x^2 + y^2) + a^2 = 2abx$.

21.5. Addition of complex numbers in the Argand diagram

In the Argand diagram (Fig. 198), let Z_1 and Z_2 represent the complex numbers $z_1 = x_1 + y_1 i$ and $z_2 = x_2 + y_2 i$. If P is the remain-

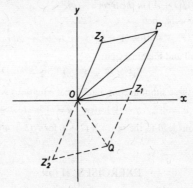

Fig. 198

ing vertex of the parallelogram of which OZ_1 and OZ_2 are adjacent sides, it is easy to show that the coordinates of P are $(x_1 + x_2, y_1 + y_2)$ and hence that P represents the complex number

$$z_1 + z_2 = x_1 + x_2 + (y_1 + y_2)i.$$

If Z_2O is produced to a point Z_2' such that $Z_2'O = OZ_2$, the coordinates of Z_2' will be $(-x_2, -y_2)$ and Z_2' will represent the complex number $-z_2$. If the parallelogram $OZ_2'QZ_1$ is completed, the point Q will represent the complex number $z_1 + (-z_2)$, that is, $z_1 - z_2$.

These constructions give a useful method of fixing the positions in the Argand diagram of points representing the sum or difference of two complex numbers and the following important results can be deduced. Since the moduli of the complex numbers z_1, z_2 and $z_1 + z_2$ are respectively represented by OZ_1, OZ_2, OP and since $OZ_2 = Z_1P$, $OZ_1 + Z_1P \geqslant OP$, it follows that

$$|z_1| + |z_2| \geqslant |z_1 + z_2|. \tag{21.11}$$

Thus *the sum of the moduli of two complex numbers is not less than modulus of the sum of the numbers.* The sign of equality in (21.11) occurs when Z_2 lies on OZ_1 and this happens when the arguments of the complex numbers z_1, z_2 are equal. Returning to Fig. 198, we have $OZ_1 \leqslant OP + PZ_1$, so that $|z_1| \leqslant |z_1 + z_2| + |z_2|$ and hence

$$|z_1| - |z_2| \leqslant |z_1 + z_2|, \tag{21.12}$$

and other results can be derived in similar ways.

Again using Fig. 198, Z_1Q is equal and parallel to OZ_2' and, since $OZ_2 = OZ_2'$, Z_1Q is equal and parallel to OZ_2. Hence OQZ_1Z_2 is a parallelogram and $Z_2Z_1 = OQ$. But OQ measures the modulus of the complex number $z_1 - z_2$ so that, if Z_1 and Z_2 respectively represent the complex numbers z_1 and z_2, then $|z_1 - z_2|$ is equal to the length Z_2Z_1. Again, since Z_2Z_1 is parallel to OQ, the direction Z_2Z_1 can be used to measure the argument of $z_1 - z_2$. The vector $\overrightarrow{Z_2Z_1}$ can therefore be taken as completely representing the complex number $z_1 - z_2$ and this vectorial representation is often useful in working exercises. For example, if A and Z are the points in the Argand diagram representing respectively a fixed complex number a and a variable complex number z, then if $|z - a| = \text{constant} = c$, the locus of the point Z is a circle centre A and radius c.

Example 5. *z is a variable complex number satisfying the relation* $|z - 3| = k|z + 3|$ *where k is a positive real constant. Show that, if* $k \neq 1$, *the locus of the point representing the complex number z in the Argand diagram is a circle. What is the locus when* $k = 1$?

In Fig. 199, Z represents the complex number z and the points A, B represent

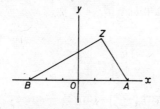

FIG. 199

respectively the numbers 3 and -3. Then $|z - 3|$ is equal to the length AZ and $|z + 3|$ is equal to the length BZ. Hence the point Z moves so that $AZ = kBZ$

and this shows that the locus is a circle except when $k = 1$. For this special value of k, the locus is such that $AZ = BZ$ and is therefore the axis Oy of the Argand diagram.

21.6. Products and quotients of complex numbers

Let z_1, z_2 be complex numbers with moduli r_1, r_2 and arguments θ_1, θ_2. Then

$$z_1 z_2 = r_1(\cos \theta_1 + i \sin \theta_1)r_2(\cos \theta_2 + i \sin \theta_2)$$
$$= r_1 r_2 \{\cos \theta_1 \cos \theta_2 - \sin \theta_1 \sin \theta_2$$
$$+ i(\sin \theta_1 \cos \theta_2 + \cos \theta_1 \sin \theta_2)\},$$

on performing the multiplication and using the relation $i^2 = -1$. Using the addition formulae for the sine and cosine, this can be written

$$z_1 z_2 = r_1 r_2 \{\cos (\theta_1 + \theta_2) + i \sin (\theta_1 + \theta_2)\} \qquad (21.13)$$

and this result can be expressed in the form

$$\left. \begin{array}{c} |z_1 z_2| = |z_1| . |z_2|, \\ \arg(z_1 z_2) = \arg z_1 + \arg z_2 \end{array} \right\} \qquad (21.14)$$

Thus *the modulus of the product of two complex numbers is equal to the product of their moduli* and *the argument of the product is equal to the sum of the arguments*. The reader should note, however, that the second of these statements is not necessarily true of the principal values of the arguments (for example, $\theta_1 + \theta_2$ may exceed π).

The quotient of two complex numbers can be handled in a similar way. Thus

$$\frac{z_1}{z_2} = \frac{r_1(\cos \theta_1 + i \sin \theta_1)}{r_2(\cos \theta_2 + i \sin \theta_2)}$$
$$= \frac{r_1(\cos \theta_1 + i \sin \theta_1)(\cos \theta_2 - i \sin \theta_2)}{r_2(\cos \theta_2 + i \sin \theta_2)(\cos \theta_2 - i \sin \theta_2)}$$
$$= \frac{r_1}{r_2} \left\{ \frac{\cos \theta_1 \cos \theta_2 + \sin \theta_1 \sin \theta_2 + i(\sin \theta_1 \cos \theta_2 - \cos \theta_1 \sin \theta_2)}{\cos^2 \theta_2 + \sin^2 \theta_2} \right\}$$
$$= \frac{r_1}{r_2} \{\cos (\theta_1 - \theta_2) + i \sin (\theta_1 - \theta_2)\}.$$

Hence

$$\left. \begin{array}{c} \left| \dfrac{z_1}{z_2} \right| = \dfrac{|z_1|}{|z_2|}, \\ \arg\left(\dfrac{z_1}{z_2} \right) = \arg z_1 - \arg z_2, \end{array} \right\} \qquad (21.15)$$

and *the modulus of the quotient of two complex numbers is therefore equal to the quotient of their moduli while the argument of the quotient is equal to the difference of the arguments*.

Example 6. *If* $arg\left(\dfrac{z-1}{z+1}\right) = \dfrac{\pi}{2}$, *find the locus of the point which represents z in the Argand diagram.*

(O.C.)

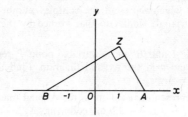

FIG. 200

In Fig. 200, Z represents the complex number z and A, B represent the numbers 1 and -1. The vectors $\overrightarrow{AZ}$, $\overrightarrow{BZ}$ represent completely the complex numbers $z - 1$, $z + 1$ so that $\arg(z - 1)$ is given by the angle xAZ and $\arg(z + 1)$ by the angle xBZ. Since

$$\frac{\pi}{2} = \arg\left(\frac{z-1}{z+1}\right) = \arg(z - 1) - \arg(z + 1),$$

we therefore have angle $xAZ -$ angle $xBZ = 90°$ and, since angle $xAZ =$ angle $xBZ +$ angle BZA, it follows that the angle BZA is a right angle and that the point Z lies on a circle with AB as a diameter.

Example 7. *Use the results of § 21.6 to find the modulus and argument of* $1/(1 - i)^2$.

The modulus r and argument θ of $1 - i$ are given by $r^2 = 1^2 + (-1)^2 = 2$ and $r\cos\theta = 1$, $r\sin\theta = -1$. Hence $r = \sqrt{2}$ and $\theta = -\frac{1}{4}\pi$. Using (21.14), the modulus of $(1 - i)^2$ is $\sqrt{2} \times \sqrt{2} = 2$ and its argument is $-\frac{1}{4}\pi - \frac{1}{4}\pi = -\frac{1}{2}\pi$. Since the modulus and argument of 1 are respectively 1 and 0, use of (21.15) then gives the modulus of $1/(1 - i)^2$ as $1/2$ and its argument as $0 - (-\frac{1}{2}\pi) = \frac{1}{2}\pi$.

EXERCISES 21 (b)

1. O is the origin in the Argand diagram and P, Q, R represent respectively the complex numbers $3 + 4i$, $4 + 6i$, $1 + 2i$. Show that $OPQR$ is a parallelogram and find its area.

2. Find the modulus of the complex number

$$\frac{(2 - 3i)(3 + 4i)}{(6 + 4i)(16 - 8i)}.$$

3. P and Q represent the complex numbers z_1, z_2 in the Argand diagram and O is the origin. If $|z_1 - z_2| = |z_1 + z_2|$, show that OP is perpendicular to OQ.

4. If $z_1 = 2 + i$, $z_2 = -2 + 4i$ and

$$\frac{1}{z_3} = \frac{1}{z_1} + \frac{1}{z_2},$$

find z_3. If z_1, z_2, z_3 are represented on the Argand diagram by P, Q, R respectively and O is the origin, show that OR is perpendicular to PQ.

5. If a is a real number and b is a complex number, show that the points in the Argand diagram which represent the complex numbers $az_1 + b$, $az_2 + b$, $az_3 + b$ form a triangle similar to that formed by the points representing z_1, z_2, z_3.

6. Two fixed points A and B, and a variable point P represent the complex numbers z_1, z_2 and z in the Argand diagram. Find the locus of P given that $\arg(z - z_1) = \arg z_2$.

7. Show that the representative points in the Argand diagram of the complex numbers $1 + 6i$, $3 + 10i$, $4 + 12i$ are collinear.

8. If a, b are the complex numbers represented by the points A, B in the Argand diagram, what geometrical quantities correspond to the modulus and argument of b/a?

21.7. De Moivre's theorem

When n is a positive integer, De Moivre's theorem states that $(\cos\theta + i\sin\theta)^n = \cos n\theta + i\sin n\theta$. If we assume the truth of this, we have

$$(\cos\theta + i\sin\theta)^{n+1} = (\cos\theta + i\sin\theta)^n(\cos\theta + i\sin\theta)$$
$$= (\cos n\theta + i\sin n\theta)(\cos\theta + i\sin\theta)$$
$$= \cos n\theta\cos\theta - \sin n\theta\sin\theta$$
$$+ i(\sin n\theta\cos\theta + \cos n\theta\sin\theta)$$
$$= \cos(n+1)\theta + i\sin(n+1)\theta.$$

If then the theorem is true when the index is n, it is also true when the index is $(n + 1)$. But

$$(\cos\theta + i\sin\theta)^2 = \cos^2\theta - \sin^2\theta + 2i\sin\theta\cos\theta$$
$$= \cos 2\theta + i\sin 2\theta,$$

so that the theorem is true when $n = 2$. It follows that it is true when $n = 3, 4, \ldots$ and we have established by the method of induction that

$$(\cos\theta + i\sin\theta)^n = \cos n\theta + i\sin n\theta \qquad (21.16)$$

whenever n is a positive integer.

When n is a negative integer, put $n = -m$ so that m is a positive integer. Then

$$(\cos\theta + i\sin\theta)^n = (\cos\theta + i\sin\theta)^{-m} = \frac{1}{(\cos\theta + i\sin\theta)^m}$$
$$= \frac{1}{\cos m\theta + i\sin m\theta}, \quad \text{by (21.16).} \qquad (21.17)$$

But,

$$\frac{1}{\cos m\theta + i\sin m\theta} = \frac{\cos m\theta - i\sin m\theta}{(\cos m\theta + i\sin m\theta)(\cos m\theta - i\sin m\theta)}$$

$$= \frac{\cos m\theta - i \sin m\theta}{\cos^2 m\theta + \sin^2 m\theta}$$

$$= \cos(-m\theta) + i \sin(-m\theta)$$

$$= \cos n\theta + i \sin n\theta$$

and substitution in (21.17) shows that De Moivre's theorem remains valid when n is a negative integer.

Next suppose that n is a fraction; put $n = p/q$ where p and q are integers, no loss of generality occurring if q is taken as positive. By (21.16), since q is a positive integer,

$$\left(\cos \frac{p}{q}\theta + i \sin \frac{p}{q}\theta\right)^q = \cos p\theta + i \sin p\theta,$$

and, since p is a positive or negative integer,

$$\cos p\theta + i \sin p\theta = (\cos \theta + i \sin \theta)^p.$$

Hence

$$\left(\cos \frac{p}{q}\theta + i \sin \frac{p}{q}\theta\right)^q = (\cos \theta + i \sin \theta)^p$$

and it follows that $\cos(p\theta/q) + i \sin(p\theta/q)$ is a qth root of $(\cos \theta + i \sin \theta)^p$.

It has been shown in the last paragraph that, when n is a fraction p/q, $\cos(p\theta/q) + i \sin(p\theta/q)$ is one of the values of $(\cos \theta + i \sin \theta)^{p/q}$ and it is useful to find the other values of this quantity. To do this, suppose that $\rho(\cos \alpha + i \sin \alpha)$ represents any value of $(\cos \theta + i \sin \theta)^{p/q}$ so that

$$\rho^q(\cos \alpha + i \sin \alpha)^q = (\cos \theta + i \sin \theta)^p.$$

Since p and q are integers, this can be written

$$\rho^q(\cos q\alpha + i \sin q\alpha) = \cos p\theta + i \sin p\theta$$

and, by equating the real and imaginary parts,

$$\rho^q \cos q\alpha = \cos p\theta, \quad \rho^q \sin q\alpha = \sin p\theta.$$

By squaring and adding we have $\rho = 1$ and the above relations reduce to $\cos q\alpha = \cos p\theta$, $\sin q\alpha = \sin p\theta$ so that $q\alpha = 2k\pi + p\theta$ where k is zero or any integer. Taking $k = 0, 1, 2, \ldots, (q - 1)$ in succession

$$\cos \left(\frac{p\theta}{q} + \frac{2k\pi}{q}\right) + i \sin \left(\frac{p\theta}{q} + \frac{2k\pi}{q}\right)$$

are all values of $(\cos \theta + i \sin \theta)^{p/q}$. These values are all distinct and there are no further values given by other values of k since any other integral value of k will differ from any one of $0, 1, 2, \ldots, (q - 1)$ by a multiple of q. Using the convenient notation

$$\operatorname{cis} \theta = \cos \theta + i \sin \theta \qquad (21.18)$$

we can therefore write

$$\operatorname{cis}^{p/q} \theta = \operatorname{cis}\left(\frac{p\theta}{q} + \frac{2k\pi}{q}\right), \quad k = 0, 1, 2, \ldots, (q-1), \quad (21.19)$$

and this is De Moivre's theorem for fractional indices.

Example 8. *If n is an integer and $z = \cos\theta + \sin\theta$, show that*

$$2\cos n\theta = z^n + \frac{1}{z^n}, \quad 2i\sin n\theta = z^n - \frac{1}{z^n}. \quad (21.20)$$

Use these results to establish the formula

$$8\cos^4\theta = \cos 4\theta + 4\cos 2\theta + 3.$$

Using De Moivre's theorem we have

$$\cos n\theta + i\sin n\theta = (\cos\theta + i\sin\theta)^n = z^n.$$

Also,

$$\cos n\theta - i\sin n\theta = \cos(-n\theta) + i\sin(-n\theta)$$
$$= (\cos\theta + i\sin\theta)^{-n} = z^{-n},$$

and the results (21.20) follow by addition and subtraction.
Taking $n = 1$ in (21.20),

$$(2\cos\theta)^4 = \left(z + \frac{1}{z}\right)^4 = z^4 + 4z^2 + 6 + \frac{4}{z^2} + \frac{1}{z^4}$$

$$= \left(z^4 + \frac{1}{z^4}\right) + 4\left(z^2 + \frac{1}{z^2}\right) + 6$$

$$= 2\cos 4\theta + 8\cos 2\theta + 6,$$

where we have used (21.20) with $n = 4$ and $n = 2$ in the last step. The required result then follows on division by 2.

21.8. The cube root of unity

We can use De Moivre's theorem in the form (21.19) to find the three cube roots of unity. Thus, taking $p = 1$, $q = 3$, $\theta = 0$ we have, since $\cos 0 = 1$, $\sin 0 = 0$, $\operatorname{cis} 0 = 1$,

$$(1)^{\frac{1}{3}} = \operatorname{cis}\left(\frac{2k\pi}{3}\right) \text{ where } k = 0, 1, 2. \quad (21.21)$$

With $k = 0$ the right-hand side of (21.21) is $\operatorname{cis} 0 = 1$, while with $k = 1$ and 2, it gives respectively $\operatorname{cis}(2\pi/3)$ and $\operatorname{cis}(4\pi/3)$. These are

$$\cos\frac{2\pi}{3} + i\sin\frac{2\pi}{3} = -\frac{1}{2} + i\frac{\sqrt{3}}{2}$$

and

$$\cos\frac{4\pi}{3} + i\sin\frac{4\pi}{3} = -\frac{1}{2} - i\frac{\sqrt{3}}{2},$$

so that the required cube roots are $1, \frac{1}{2}(-1 \pm i\sqrt{3})$.

If ω is used to denote the complex root $\frac{1}{2}(-1 + i\sqrt{3})$, then
$$\omega^2 = \tfrac{1}{4}(-1 + i\sqrt{3})^2 = \tfrac{1}{4}(1 - 2i\sqrt{3} - 3) = \tfrac{1}{2}(-1 - i\sqrt{3})$$
and this is the second complex cube root of unity. Hence we can write the three cube roots of unity in the form
$$1, \omega, \omega^2 \quad \text{where } \omega = \tfrac{1}{2}(-1 + i\sqrt{3}). \tag{21.22}$$
It should be noted that (by definition) $\omega^3 = 1$, that
$$\omega^4 = \omega \times \omega^3 = \omega, \quad \omega^5 = \omega^2 \times \omega^3 = \omega^2, \quad \text{etc.,}$$
and that
$$1 + \omega + \omega^2 = 1 + \tfrac{1}{2}(-1 + i\sqrt{3}) + \tfrac{1}{2}(-1 - i\sqrt{3}) = 0. \tag{21.23}$$
These relations are often useful in working examples.

Example 9. *If ω is a complex cube root of unity, form the quadratic equation whose roots are ω and $1/\omega$.*

The product of the roots is $\omega \times (1/\omega) = 1$ and their sum is
$$\omega + \frac{1}{\omega} = \frac{\omega^2 + 1}{\omega} = -\frac{\omega}{\omega} = -1,$$
since $\omega^2 + 1 = -\omega$ by equation (21.23). Hence the required quadratic equation is $x^2 + x + 1 = 0$.

Example 10. *If ω is a complex cube root of unity and if $x = a + b$, $y = a\omega + b\omega^2$, $z = a\omega^2 + b\omega^4$, show that $x^2 + y^2 + z^2 = 6ab$.*

We have $x^2 = a^2 + 2ab + b^2$, $y^2 = a^2\omega^2 + 2ab\omega^3 + b^2\omega^4 = a^2\omega^2 + 2ab + b^2\omega$, $z^2 = a^2\omega^4 + 2ab\omega^6 + b^2\omega^8 = a^2\omega + 2ab + b^2\omega^2$. Hence
$$x^2 + y^2 + z^2 = a^2(1 + \omega^2 + \omega) + 6ab + b^2(1 + \omega + \omega^2) = 6ab,$$
when use is made of equation (21.23).

EXERCISES 21 (c)

1. Use De Moivre's theorem to show that
$$\frac{(\cos 3\theta + i\sin 3\theta)^5 (\cos\theta - i\sin\theta)^3}{(\cos 5\theta + i\sin 5\theta)^7 (\cos 2\theta - i\sin 2\theta)^5} = \cos 13\theta - i\sin 13\theta.$$

2. Use De Moivre's theorem to show that
$$\cos 4\theta = \cos^4\theta - 6\cos^2\theta \sin^2\theta + \sin^4\theta,$$
$$\sin 4\theta = 4\cos^3\theta \sin\theta - 4\cos\theta \sin^3\theta.$$

3. Use the method of § 21.8 to find the four fourth roots of unity.

4. Show that
$$\left(\frac{1 + \sin\theta + i\cos\theta}{1 + \sin\theta - i\cos\theta}\right)^n = \cos n\left(\frac{\pi}{2} - \theta\right) + i\sin n\left(\frac{\pi}{2} - \theta\right).$$

5. Show that
$$\frac{(\cos\theta + i\sin\theta)^2}{(\sin\phi + i\cos\phi)^5} = \sin(2\theta + 5\phi) - i\cos(2\theta + 5\phi).$$

6. Use De Moivre's theorem to find the value of
$$\left\{ \frac{\sqrt{(-3)} - 1}{\sqrt{(-3)} + 1} \right\}^6.$$

7. If ω is a complex cube root of unity, show that
$$(1 - \omega + \omega^2)(1 + \omega - \omega^2) = 4.$$

8. Find the two square roots of i and the four values of $(-16)^{\frac{1}{4}}$.

9. Find the three roots of the equation $(1 - z)^3 = z^3$.

10. If ω is a complex cube root of unity, show that
    (i) $(1 + \omega - \omega^2)^3 - (1 - \omega + \omega^2)^3 = 0$,
    (ii) $(a + b\omega + c\omega^2)(a + b\omega^2 + c\omega) = a^2 + b^2 + c^2 - bc - ca - ab$.

EXERCISES 21 (d)

1. Express $(5 + 4i)(3 + 2i)$ in the form $a + bi$ where a and b are real. Deduce a pair of factors of $7 - 22i$ and hence express $7^2 + 22^2$ as the product of two positive integers.   (N.U.)

2. If $z = x + yi$ and $z^5 = 1$, show that $4x(y^4 - x^4) = 1$.

3. Express $(1 + 2i)^2$ and $(1 + 2i)^3$ in the form $a + bi$ where a and b are real. Hence find real numbers r and s for which $1 + 2i$ is a root of the equation $z^3 + rz^2 - 7z + s = 0$.   (N.U.)

4. Find the square root of $5 + 12i$.

5. Find two real numbers x and y such that
$$(1 + i)x + 2(1 - 2i)y = 3.$$

6. Express in the form $a + bi$,
$$\frac{(4 + 3i)\sqrt{(3 + 4i)}}{3 + i}.$$

7. Find the modulus and argument of $(2 - i)^2(3i - 1)/(i + 3)$.   (O.C.)

8. If $z_1 = (1 + 7i)/(1 - i)$ and $z_2 = (17 - 7i)/(2 + 2i)$ find the moduli of $z_1, z_2, z_1 + z_2$ and $z_1 z_2$.   (O.C.)

9. Prove that if $(z - 6i)/(z + 8)$ is real, the locus of the point representing the complex number z in the Argand diagram is a straight line.   (O.C.)

10. Prove that if $(z - 2i)/(2z - 1)$ is purely imaginary, the locus of the point representing z in the Argand diagram is a circle and find its centre and radius.   (O.C.)

11. If z is a complex number and
$$\left| \frac{z - 1}{z + 1} \right| = 2,$$
find the equation of the curve in the Argand diagram on which the point representing z lies.   (O.C.)

12. If z_1 and z_2 are two complex numbers such that $|z_1 - z_2| = |z_1 + z_2|$, show that the difference of their arguments is $\pi/2$ or $3\pi/2$.

13. If z is represented by a point on the circle of radius a which touches the y-axis of the Argand diagram at the origin O and lies in the first and fourth quadrants, prove that $z - 2a = iz \tan (\arg z)$. (O.C.)

14. Find the modulus and argument of $(a + ib)^2$ where a and b are real. Deduce that

$$\tan^{-1}\left(\frac{2ab}{a^2 - b^2}\right) = 2 \tan^{-1}\left(\frac{b}{a}\right).$$

15. If $z^3 = 2(i - 1)$, find all the possible values of the real part of $2z$.

16. The complex numbers $z - 2$ and $z - 2i$ have arguments which are (i) equal and (ii) differ by $\frac{1}{2}\pi$ and each argument lies between $-\pi$ and π. In each case find the locus of the point which represents z in the Argand diagram and illustrate by a sketch. (N.U.)

17. If $z = \cos\theta + i\sin\theta$ and $0 < \theta < \pi$, find the values of the modulus and argument of $2/(1 - z^2)$.

18. Complex numbers z_1 and z_2 are given by the formulae

$$z_1 = R_1 + i\omega L, \quad z_2 = R_2 - \frac{i}{\omega C}$$

where R_1, R_2, ω, L and C are all real. If z is given by $z^{-1} = z_1^{-1} + z_2^{-1}$, find the value of ω for which z is real. (O.C.)

19. If $z = 1 + i$, mark on the Argand diagram the points A, B, C, D, representing z, z^2, z^3, z^4. Find the moduli and arguments of $z^3 - z^4$ and $z^2 - z^4$. Show that the angle BDC is $\arg \{(z^3 - z^4)/(z^2 - z^4)\}$ and that the angles BDC and ACB are equal. (L.U.)

20. Prove that if $Z = X + iY = (z - 1)(z + 2)$ and z is a complex number with modulus 1 and argument θ,

(i) $Y/X = -3 \cot\frac{1}{2}\theta$, (ii) $(X + 1)^2 + Y^2 = 1$. (O.C.)

21. O is the origin and A represents the number $z = 1$ in the Argand diagram. If P represents a variable complex number z, prove that PO is perpendicular to PA if the real part of $(z - 1)/z$ is zero. Deduce that, if $z = (1 + iw)^{-1}$, where w is a variable real number, then the point representing z describes a circle of unit diameter. (L.U.)

22. Use De Moivre's theorem to find the four fourth roots of $8(-1 + i\sqrt{3})$ in the form $a + ib$, giving a and b correct to 2 decimal places. (N.U.)

23. Use De Moivre's theorem to show that

$$\frac{\cos 5x}{\cos x} = 1 - 12 \sin^2 x + 16 \sin^4 x.$$

24. If ω is a complex cube root of unity, show that
$$(2 + 5\omega + 2\omega^2)^6 = (2 + 2\omega + 5\omega^2)^6 = 729.$$

25. ω is a complex cube root of unity and a, b, c are real quantities. Show that $(a + \omega b + \omega^2 c)^3$ is only real if a, b and c are not all different.

ANSWERS TO THE EXERCISES

Exercises 1 (a). p. 16.
1. (i) $3/4$, $-1/2$. (ii) $2\cdot643$, $0\cdot757$.
3. $x = 1$, $2\cdot5 \leqslant x \leqslant 4$.
4. $2q^2 = 9pr$.
5. $c^3x^2 + b(b^2 - 3ac)x + a^3 = 0$.

Exercises 1 (b). p. 18.
1. $6\cdot194$, $1\cdot357$.
2. -2.
3. 1, -3, $-1 \pm \sqrt{5}$.
4. 2, -1; $-1/3$, $-17/3$.
5. ±2, ±1; ±1, ±2.
6. $5/2$, $-1/2$; 4, 1.

Exercises 1 (c). p. 20.
1. $\pm(\sqrt{3} + \sqrt{2})$.
2. $\pm(\sqrt{12} - \sqrt{6})$.
3. $\pm\{\sqrt{(a + \tfrac{1}{2}b)} + \sqrt{(\tfrac{1}{2}b)}\}$.
4. $69/578$, $37/578$.

Exercises 1 (d). p. 20.
2. $x > 6$ and $-6 < x < -2$.
3. 3, $-1/2$.
4. $2x^2 - 14x + 7 = 0$.
7. $q^2x^2 - (p^2 - 2q)(q^2 + 1)x + (q^2 + 1)^2 = 0$.
8. -11, $33/25$.
10. $1 \pm \sqrt{3}/2$.
11. $x^2 + a(p + q)x + b(p^2 + q^2) + (a^2 - 2b)pq = 0$; 5, $1/5$.
12. $x^2 + 9x + 64 = 0$.
14. -2, -1, $1/3$, $4/3$.
15. 9.
16. 4, 7.
17. 4, $-2/3$; -10, $-23/24$.
18. -1, 7; 1, 5; 5, 1; 7, -1.
19. ±1, ±2; $\pm\sqrt{6}$, $\pm\sqrt{(3/2)}$.
20. a, a.
21. $5/12$, $1/12$; $7/12$, $-5/12$.
22. 8, $-8/3$; 2, 4; 4, 2; $-1 \pm \sqrt{(11/3)}$, $-1 \mp \sqrt{(11/3)}$.
24. $(7/2) + 2\sqrt{3}$.
25. $1/2$, 1, 1, 2.

Exercises 2 (a). p. 25.
1. (i) $1/(4b^3x^3)$. (ii) y.
2. (i) $1/512$. (ii) $3/2$.
3. x^2y^3/z^9.
5. $\sqrt{x/y}$.
6. 4.

Exercises 2 (b). p. 29.
3. $3\cdot2$.
4. (i) $0\cdot3557$. (ii) $0\cdot0305$.
5. (a) 2, -4. (b) $-0\cdot603$.
6. $1\cdot768$.
7. -4.
8. $2\cdot71$, $1\cdot71$.

Exercises 2 (c). p. 32.
1. $p = -5$, $q = 8$; $x - 4$.
2. $3(a - b)(b - c)(c - a)$.
3. $\lambda = 20$, $\mu = -39$.
4. $a = 3 - 6n$, $b = 3n^2 - 3n + 1$.
5. $A = 0$, $B = -2/5$, $C = -3/5$.

Exercises 2 (d). p. 35.

1. $\dfrac{3}{5 - x} - \dfrac{2}{5 + x}$.

2. $x - 1 + \dfrac{8}{3(x + 2)} + \dfrac{1}{3(x - 1)}$.

3. $\dfrac{1}{x + 1} - \dfrac{2}{(x + 1)^2} + \dfrac{1}{(x + 1)^3}$.

4. $1 - \dfrac{1}{x + 1} + \dfrac{1}{(x + 1)^2}$.

5. $\dfrac{2x - 3}{x^2 + 1} - \dfrac{2}{x - 4}$.

6. $\dfrac{13x}{9(x^2 + 9)} + \dfrac{7}{9(x - 3)} + \dfrac{7}{9(x + 3)}$.

7. $\dfrac{3x}{x^2 + 2x - 5} - \dfrac{1}{x - 3}$.

8. $\dfrac{5}{9(y - 2)} - \dfrac{5}{9(y + 1)} + \dfrac{1}{3(y + 1)^2}$.

Exercises 2 (e). p. 35.

1. 0.

3. 3^n.

4. (i) 3/2. (ii) 3.

6. (i) 0·06424. (ii) −0·3107.

8. 1, 2.

9. 0, 1.

10. $a = 5/2, b = 9/2$.

11. 512.

13. $a = 31, b = -12$.

15. $-(b - c)(c - a)(a - b)(a^2 + b^2 + c^2 + bc + ca + ab)$.

16. $a = 1, b = -6, c = 7, d = -1$.

17. $\alpha = 2, \beta = -3; \alpha = -5, \beta = 4$.

18. $a = -5, b = 6; (x - 1)^2(x - 2)(x - 3)$.

19. 2.

20. $x^2 + 2x + 2$.

21. $\dfrac{2}{2x + 1} - \dfrac{1}{x + 1} + \dfrac{2}{(x + 1)^2}$.

22. $\dfrac{1}{12(x - 2)} + \dfrac{11x + 8}{12(x^2 + 2x + 4)}$.

23. $\dfrac{3}{x - 1} + \dfrac{2}{x + 1} + \dfrac{1}{x^2 + 1}$.

24. $\dfrac{2}{3(1 + x)^2} + \dfrac{1}{3(1 - x + x^2)}$.

25. $\dfrac{1}{x + 1} - \dfrac{2}{x - 1} + \dfrac{3x + 2}{x^2 + 4}$.

Exercises 3 (a). p. 40.

1. (i) $-1, 3, 7; 27$. (ii) $1, 3, 9; 2187$. (iii) $-1, +1, -1, +1$.

3. −220.

4. 3/2.

5. 5120.

6. 3, 75.

8. 1/3, 2.

Exercises 3 (b). p. 44.

1. (a) 20. (b) 14·2.

2. 78·7.

3. $\dfrac{3 + x^2}{x^2 - 2x + 3}$.

4. 3.

5. 6, −1/2.

6. Converges when $a < -2$ or $a > 0$, limit of sum $= 1 + a$.
   Also converges when $a = 0$, limit of sum $= 0$.

8. $a = -3, b = 2, c = 5$; sum $= (n/2)(11 - n - 2n^2)$.

Exercises 3 (c). p. 49.

1. 1365.

2. (i) 2520. (ii) 5040.

3. 6.

4. 1024.

5. 256. Sum $= 711040$.

6. 182.

7. (i) 60. (ii) 60.

8. 37 to 5.

9. 1/4.

Exercises 3 (d). p. 54.

2. $-5^5/(2^7 . 3^{3·5})$.

3. 567/16, 6th.

4. 8.

6. 7.

7. $\dfrac{1}{1 + x^2} + \dfrac{1}{(1 - x)^2} - \dfrac{2}{1 - x}$; $2x^3 + 4x^4 + 4x^5 + 4x^6 + 6x^7$.

8. 0·7930, 0·7929.

Exercises 3 (e). p. 55.

1. $(m - n)\{a + \tfrac{1}{2}(m + n - 1)(b - a)\}$; 77/75.

2. $(3a - c)/(a + c)$.

5. 25.

6. £1071.   7. £147, £1193.
9. 11.   12. $26^3 \times 10^3$.
13. 186.   14. (i) 360. (ii) 144.
16. 15 to 7.   17. 16/21.
19. 1001, 2002, 3003.   20. $a = -1$. $-80, 432$.
23. $1 + \frac{11}{2}x + \frac{119}{8}x^2 + \frac{6}{1}\frac{7}{6}\frac{5}{}x^3$.

24. $\dfrac{3}{(1-x)^2} - \dfrac{1}{1-x} - \dfrac{1}{1+x}$.

Exercises 4 (a). p. 63.

1. $\cos\theta = \sqrt{15}/4$, $\tan\theta = 1/\sqrt{15}$, $\operatorname{cosec}\theta = 4$, $\sec\theta = 4/\sqrt{15}$, $\cot\theta = \sqrt{15}$.
2. $\sin\theta = \pm 3/5$, $\cos\theta = \pm 4/5$.
3. (i) $\sin\theta = 4/5$, $\tan\theta = -4/3$. (ii) $\sin\theta = -4/5$, $\tan\theta = 4/3$.

Exercises 4 (b). p. 68.

1. (i) $-0{\cdot}5299$. (ii) $-0{\cdot}3420$. (iii) $2{\cdot}1445$. (iv) $-2{\cdot}5593$.
2. $\sin 2A = -0{\cdot}96$, $\cos 2A = 0{\cdot}28$.   3. 0°. 30°; 60°, 90°; 0°, 150°.
4. $69^\circ\, 39'$.   5. $1{\cdot}166$.
6. $21^\circ\, 28'$, 90°, $158^\circ\, 32'$.

Exercises 4 (c). p. 72.

1. $n \times 180^\circ + (-1)^n \times 18^\circ$, $n \times 180^\circ - (-1)^n \times 54^\circ$.
2. $14^\circ\, 2'$, $123^\circ\, 41'$, $194^\circ\, 2'$, $303^\circ\, 41'$.
3. $37^\circ\, 55'$, $25^\circ\, 37'$; $154^\circ\, 23'$, $142^\circ\, 5'$.   4. $\left(\dfrac{2n}{3} + \dfrac{1}{4}\right)\pi$, $\left(\dfrac{1}{4} - 2n\right)\pi$.
5. $n \times 180^\circ + 35^\circ$, $n \times 180^\circ + 45^\circ$.   6. $n \times 360^\circ + 210^\circ$.

Exercises 4 (d). p. 72.

1. $\cos\theta = \dfrac{\pm 2ab}{a^2 + b^2}$, $\tan\theta = \pm\dfrac{a^2 - b^2}{2ab}$.
2. 0.
3. $\sin\theta = t/\sqrt{(1 + t^2)}$, $\cos\theta = 1/\sqrt{(1 + t^2)}$, $\operatorname{cosec}\theta = \sqrt{(1 + t^2)}/t$, $\sec\theta = \sqrt{(1 + t^2)}$, $\cot\theta = 1/t$.
4. $-1/2$, $-2/\sqrt{3}$.   6. $\pm 1/\sqrt{2}$.
13. -4°, 50°.   14. 0, $\pi/4$, $\pi/2$.
15. $57^\circ\, 54'$, $122^\circ\, 6'$.   16. 0°, 63°, 135°, 281°, 360°.
17. $-1{\cdot}27$, $0{\cdot}11$ radians.   18. 60°, $70^\circ\, 32'$, $289^\circ\, 28'$, 300°.
19. $(2n + 1)\pi/10$.
20. $n \times 360^\circ \pm 149^\circ\, 21'$, $n \times 360^\circ \pm 78^\circ\, 50'$.
21. $r = 4{\cdot}717$, $\theta = 148^\circ$.   22. 30°, 45°, 150°, 210°, 225°, 330°.
23. $n \times 180^\circ + 63^\circ\, 26'$.
24. $(2n + 1)\pi/10$, provided n is not of the form $5r + 2$ where r is zero or an integer.
25. $\pi/6$, $5\pi/6$, $7\pi/6$, $11\pi/6$.

Exercises 5 (b). p. 81.

2. $\dfrac{(1 + t)(3 + t)}{1 + t^2}$.

Exercises 5 (c). p. 85.

3. $\cos(A - B) = 1 - \frac{1}{2}(p^2 + q^2)$, $\sin(A + B) = \dfrac{-2pq}{p^2 + q^2}$.
4. 0°, 90°, 120°, 135°, 240°, 315°, 360°.
5. $2n\pi$, $(2n - 1)\pi/2$, $(2n + 1)\pi/5$.   6. $(4n \pm 1)\pi/(p + q)$, $(4n \pm 1)\pi/(p - q)$.
7. $36^\circ\, 52'$, $126^\circ\, 52'$.   8. $306^\circ\, 52'$.

Exercises 5 (d). p. 88.

1. $\pi/4, 5\pi/6, \pi/3, \pi/6$.   7. $1/2$.

Exercises 5 (e). p. 92.

1. $28° 41'$.   2. $2·91$ m

5. $0·00775$ m.

Exercises 5 (f). p. 92.

2. $45°$.

4. $\dfrac{\tan A + \tan B + \tan C - \tan A \tan B \tan C}{1 - \tan B \tan C - \tan C \tan A - \tan A \tan B}$.

14. $0°, 10°, 45°, 50°, 90°, 130°, 135°, 170°, 180°$.

15. $63° 26', 161° 34', 243° 26', 341° 34'$.

16. $139° 48', 287° 35'$.   17. $n \times 36° + 9°, 45° - n \times 180°$.

18. $131° 48', 11° 48'; 168° 12', 48° 12'$.

19. $60°, 63° 26', 243° 26', 300°$.

20. $38° 23', 111° 37', 218° 23', 291° 37'$.

21. $22° 38', 36° 50'; n \times 360° + 22° 38', n \times 360° + 36° 50'$.

24. $1/6$.

Exercises 6 (a). p. 97.

2. $(4ab)/(a + b)^2$.

Exercises 6 (d). p. 110.

1. $A = 83° 31', a = 17·5$ m, $b = 10$ m.

2. No solution.

3. $A = 38° 26', C = 90°, c = 27·35$ m.

4. $A = 33° 16', C = 44° 19', a = 60·21$ m.

5. $A = 100°, B = 45° 40', a = 21·5$ m; $A = 11° 20', B = 134° 20', a = 4·3$ m.

6. $A = 103° 2', B = 29° 58', c = 5·25$ m.

7. $A = 108°, B = 37° 59', C = 34° 1'$.

8. $c = 7·93, R = 4·39$.

Exercises 6 (e). p. 113.

1. $14° 11', 98° 25'$.   2. $35° 48', 46° 3', 98° 9'$.

3. $A = 56° 24', B = 22° 8', C = 101° 28', c = 5·686$ m.

4. $B = 85° 37', C = 43° 23', BD = 3·365$.

6. 62190 m^2.

Exercises 6 (f). p. 115.

2. $\tan^{-1}\left(\dfrac{\tan \alpha \sin \phi}{\sin \theta}\right)$ 3. $3·54$ m, $57° 41'$.

4. $21·9°$ (approx.).   6. $90·8$ m.

7. $15·32$ km, S. $34° 25'$ E.   8. 1011 m.

Exercises 6 (g). p. 116.

1. $9·79$ m.   2. $\sin \theta = 0·8126, a = 95·7$.

3. $180° - 2A, 180° - 2B, 180° - 2C$.

6. $26, 30$ m.   12. $b = 7·22, c = 5·55$.

13. $A = 116° 9', B = 11° 51', a = 17·09, b = 3·91$.

14. $108° 42', 48° 46', 22° 32'$, area $= 2·595$ m^2

16. 624 m^2.   17. $A = 75° 43', c = 17·76$ m.

18. $74° 7', 49° 29'$.

19. $C = 28°\ 30'$, $A = B = 75°\ 45'$, sides 32, 65, 65 m.
20. Angle $ABC = 75°\ 31'$, area $= 4\cdot13\ m^2$.
21. $6\cdot61$ km, S. $68°$ E.   22. $114\cdot7$ m.
23. 931 m.   24. $467\cdot9$, $784\cdot7$ m.

Exercises 7 (a). p. 124.

1. $9, -1, 0$.   2. $0, -3; 1/2, -2$.
3. $1/2, -1/2; n\pi + (-1)^n(\pi/6)$.
4. (i) $y = \pm\sqrt{(x - x^2)/2}$.   (ii) $y = (-1 \pm \sqrt{5})(x/2)$.
5. $3, 2\cdot5, 2\cdot25, 2\cdot125$ m/s.
6. $2 - 4x - 2\,\delta x, 2 - 4x$.   7. $3x^2$.

Exercises 7 (b). p. 128.

1. $15x^2$.   2. $4x^3 - 2x$.
3. $-1/x^3$.   4. $2\cos 2x$.
5. $-3\sin 3x$.   6. $1 + \cos x$.

Exercises 7 (c). p. 129.

1. $3t^2 - 2t, 0, 2/3$ seconds.   2. $32\ m/s^2$.
3. $1\cdot3 \times 10^{-5}\ m^2/s$.   4. $40\ mm^2$.
5. $0\cdot2$ per cent., $100\cdot2$.   6. $\pi\ m^2$.

Exercises 7 (d). p. 130.

1. $0, 8\cdot402, 29\cdot61$.   2. $y = (2x^3)/(4x^2 + 3), 0\cdot842$.
3. $ax^2 + (2a + b)x + a + b + c$.   4. $f(x) = -\frac{5}{4}x^3 + \frac{23}{12}x^2 + \frac{37}{6}x + 9$.
6. $1/(2\sqrt{x})$.   7. $2ax + b, x = -b/2a, y = (4ac - b^2)/4a$.
8. $-3, 2, -6$.   9. 8.
10. 5.   11. 1.
12. $-1, 1/2$.   13. (i) $8x + 1$.   (ii) $-1/x^2$.
14. (i) $a\cos ax$.   (ii) $-\sin 2x$.   15. $-1/(x + 2)^2$.
16. $6x + \cos x, 1$.   17. $0\cdot4012$ cubic metres per minute.
18. (i) 300.   (ii) 432.   20. 1 second, $1\ m/s^2$.
21. $75°\ 58'$.   23. $-3, \pm6$.
25. $8\cdot3$ m, 83 m/s.

Exercises 8 (a). p. 135.

1. $12x^2 - \cos x$.   2. $10\cos 2x$.
3. $(1 - 3x^2)\cos x + x^3\sin x$.   4. $1 - 2x$.
5. $-2x(1 + 4x^2)$.   6. $x^2(3\cos x - x\sin x)$.
7. $81x^2 + 108x + 36$.
8. $x\{2\sin x\cos x + x(\cos^2 x - \sin^2 x)\}$.   9. $-2\sin x\cos x$.
10. $40x^3 - 2\sin x\cos x + \cos x - x\sin x$.

Exercises 8 (b). p.137.

1. $\dfrac{1 - x^2}{(1 + x^2)^2}$.   2. $\dfrac{-5}{(1 + 2x)^2}$.

3. $\dfrac{-5x}{(1 + 2x^2)^2}$.   4. $\dfrac{8x}{(3 - 2x^2)^3}$.

5. $\dfrac{2\cos x}{(1 - \sin x)^2}$.   6. $\dfrac{2}{(\sin x + \cos x)^2}$.

7. $\cot x - x\,\operatorname{cosec}^2 x$.   8. $2\sec^2 x\tan x$.

9. $\dfrac{\cos^3 x - \sin^3 x}{(\sin x + \cos x)^2}$.   10. $-2\cot x\,\operatorname{cosec}^2 x$.

Exercises 8 (c). p. 141.

1. $12(4x - 5)^2$.

2. $5x^4(x + 3)^4(2x + 3)$.

3. $2\cos 2(x - \alpha)$.

4. $2\sec^2 2x$.

5. $3\sec 3x \tan 3x$.

6. $6\tan(3x + 1)\sec^2(3x + 1)$.

7. $-3\sin^2(2 - x)\cos(2 - x)$.

8. $2x(1 - x)(1 - 2x)$.

9. $3\sin^2 x \sin 4x$.

10. $4\sec^2 x \tan x$.

11. $\sin^{m-1}\theta \cos^{n-1}\theta(m\cos^2\theta - n\sin^2\theta)$.

12. $-(a/t^2)\sin(2a/t)$.

13. (i) $m\sin^{m-1} x \cos x$. (ii) $mx^{m-1}\cos(x^m)$. (iii) $-\sin x \cos(\cos x)$.

Exercises 8 (d). p. 143.

1. $6\left(x^4 - \dfrac{1}{x^2}\right)^2\left(2x^3 + \dfrac{1}{x^3}\right)$.

2. $30x^2(2 - 5x^3)^{-3}$.

3. $\dfrac{1}{2\sqrt{(1 + x)}}$.

4. $\dfrac{-1}{2(1 + x)^{3/2}}$.

5. $\dfrac{x - 1 - 2x^2}{\sqrt{(1 + x^2)}}$.

6. $\dfrac{\cos\sqrt{x}}{2\sqrt{x}}$.

Exercises 8 (e). p. 145.

5. $\sin^{-1} x + x/\sqrt{(1 - x^2)}$.

Exercises 8 (f). p. 148.

1. $\dfrac{1 - 3x^2 y^2}{2x^3 y}$.

2. $\dfrac{\cos 2x}{y}$.

3. $-\dfrac{2x + y}{x + 2y}$.

4. $\dfrac{r}{2}\tan\theta$.

7. $(8 + 4x)\cos x + (1 - 4x - x^2)\sin x$.

Exercises 8 (g). p. 149.

1. (i) $2x^2 - 3x$. (ii) $(9t^{1/2})/2 - t^{-1/2} + 6$. (iii) $-2\theta^{-3} + 2\theta^{-3/2}$.

2. (i) $x\cos x$. (ii) $6x - 10$.

3. (i) $\cos t \sin 3t + 3\sin t \cos 3t$. (ii) $2t\sin^{-1} t + t^2/\sqrt{(1 - t^2)}$.

4. (i) $x^2\sin x$. (ii) $\dfrac{1}{x^2}\tan x + \left(1 - \dfrac{1}{x}\right)\sec^2 x$.

5. (i) $x(1 - 3x)^2(2 - 15x)$. (ii) $2\theta\cos 2\theta(\cos 2\theta - 2\theta\sin 2\theta)$.

6. (i) $\dfrac{2x^3 + 3x^2 + 1}{(x + 1)^2}$. (ii) $\dfrac{2x + 3\sqrt{x}}{2(x + 2\sqrt{x} + 1)}$. (iii) $\dfrac{-1}{\sqrt{x}(1 + \sqrt{x})^2}$.

7. (i) $\dfrac{2x - x^2 - 4}{(x^2 - 4)^2}$. (ii) $\dfrac{x^2 - 6x + 10}{(x - 3)^2}$. (iii) $-\dfrac{(1 + x)\sin x + \cos x}{(1 + x)^2}$.

8. (i) $\dfrac{\sin x - x\cos x}{\sin^2 x}$. (ii) $\dfrac{x - \sin x \cos x}{x^2\cos^2 x}$. (iii) $\dfrac{2\cos x + 3}{(2 + 3\cos x)^2}$.

9. (i) $nx^{n-1}(\tan nx + x\sec^2 nx)$. (ii) $\dfrac{-2x\cos 2}{\sin^2(x^2 - 1)}$.

10. (i) $y^2 + 2xy\dfrac{dy}{dx}$. (ii) $6(3y + 2)\dfrac{dy}{dx}$.

(iii) $\dfrac{y - x(dy/dx)}{y^2}$. (iv) $\dfrac{x(dy/dx) - y}{x^2}$.

11. (i) $6(x^2 - x)^5 (2x - 1)$. (ii) $3 \sin (2 - 3x)$. (iii) $2x (\sin 2x + x \cos 2x)$.

12. (i) $\dfrac{1}{(1 + t^2)^{3/2}}$. (ii) $-(1 + \theta)^{-1} (2\theta + \theta^2)^{-1/2}$.

(iii) $2x(x^2 - 2)(x^2 - 1)^{-3/2}$.

14. (i) $\dfrac{\sec^2 (x/2)}{2\sqrt{\{1 - \tan^2 (x/2)\}}}$. (ii) $\dfrac{\cos (x/2)}{2\{1 + \sin^2 (x/2)\}}$.

15. (i) $\dfrac{2}{\sqrt{(1 - t^2)}}$. (ii) $\dfrac{3t}{\sqrt{\{t(1 - t^3)\}}}$.

16. (i) $\dfrac{1}{x^2} \sin \left(\dfrac{1}{x}\right)$. (ii) $2x \sec^2 (x^2)$. (iii) $\dfrac{-2}{1 + x^2}$.

21. $2x/(x + y)$.

23. (i) $\dfrac{2 - x}{y}$. (ii) $\dfrac{6x - y}{8y + x}$.

24. (i) $\dfrac{2x \sin y + y \sin x}{\cos x - x^2 \cos y}$. (ii) $\dfrac{\cos y - y^2 \cos x}{x \sin y + 2y \sin x}$.

25. 1.

26. (i) $x(6 - x^2) \sin x + 6x^2 \cos x$. (ii) $\dfrac{2}{(1 + x^2)^2}$. (iii) $\dfrac{2}{(1 + x)^3}$.

28. $\dfrac{x \cos x - 2 \sin x}{x^3}, \dfrac{(6 - x^2) \sin x - 4x \cos x}{x^4}$.

Exercises 9 (a). p. 155.

1. $0 \cdot 00764$ m/s.
2. $5 \cdot 28$ m^2/s, $55 \cdot 41$ m^3/s.
3. $0 \cdot 5$ mm/s.

Exercises 9 (b). p. 159.

1. 3 (min.), 7/3 (max.).
2. -4 (min.), 0 (max.).
3. $-3, -1/3$.
4. $-3 + 2\sqrt{2}, -2\sqrt{2} - 3$.
5. $-3, 1$ (min.), 3/2 (max.).
6. 25.

Exercises 9 (c). p. 162.

2. Each $0 \cdot 06$ m.
3. $4 \cdot 32 \times 10^{-4}$ m^3.
5. $26 \cdot 3$ m.
7. $0 \cdot 1188$ m^2.
8. $1/20(\pi + 4)$ m.

Exercises 9 (d). p. 167.

1. 0, 2/3.
2. $(2n + 1)\pi$, n an integer.
3. $x = -2$ gives a minimum, $x = 2$ gives a point of inflexion.
6. (i) ± 1, infinite. (ii) $\pm 1/\sqrt{2}$.

Exercises 9 (e). p. 167.

1. (i) $25\pi/16$. (ii) $9\pi/4$. $1 \cdot 44$ times.
2. 2 per cent./second.
3. $1 \cdot 6$ mm/s.
4. After 2 and 7/2 seconds. 0 m/s, -6 m/s^2;
   9/2 m/s, 12 m/s^2 Min. vel. $-3/2$ m/s.
5. 8 m/s^2.
6. $x = -c/2b, y = -c^2/4ab$. A minimum.

7. 1 (min.), -1 (max.).

10. $\left(\dfrac{2}{n+1}\right)^{\frac{n}{n-1}}$

11. Length $= 64/x^2$, area $= (256/x) + 2x^2$.
12. Sum of volumes $= (l^3 - 3l^2x + 3x^2l)/1728$.
    Minimum volume $= l^3/6912$.
13. 0·04 m.
14. 0·224 m^3, 1·25 m.
15. 4·8 km/h, 4·2 km.
16. £1·79.
18. Square of side $r\sqrt{2}$.
19. 0·3 m.
20. $\sqrt[3]{2} = 1·26$.
21. -17 (min.), 1·519 (max.), $x = 4/3$.
22. $-b/3a$.
25. $-1/2$ (min.), 1/2 (max.).

Exercises 10 (a). p. 174.

1. $(3x^{7/3}/7) + C$.
2. $(-3/x) + C$.
3. $x + x^2 + (x^3/3) + C$.
4. $(x^3/3) + 2x - (1/x) + C$.
5. $x - (1/x) + C$.
6. $\sin x - \cos x + C$.
7. $2\tan x + \tan^{-1} x + C$.
8. $(x^6/6) + x^4 - (2x^3/3) + (x^2/2) - 3x + C$.
9. $t^3 - (t^2/2) + 7t + C$.
10. $C - 2\theta - (1/\theta) - (1/2\theta^2)$.
11. $(ax^2/2) + (bx^3/3) + (cx^4/4) + C$.
12. $\theta^2 + \sin \theta + C$.
13. $\tan \theta - \theta + C$.
14. $(x + \sin x)/2 + C$.
15. $\tan^{-1} x + C$.

Exercises 10 (b). p. 176.

1. $y = x^2 - x + 1$.
2. $y = 1 + x - (2x^3/3)$.
3. 8/3.
4. 13 m from the origin.
5. $r = \dfrac{dx}{dt}, f = \dfrac{d^2x}{dt^2}, x = 8t^3 - 12t^2 + 6t - 1$.
6. At distance 39·33 from the origin, 27·33.

Exercises 10 (c). p. 183.

1. $343\frac{1}{3}$.
2. 4/3.
3. 81/10.
4. 16.
5. $\pi/2$.
6. 1.
7. $1 - (\pi/4)$.
8. $(\pi/3) + (1/2)$.
9. $41\frac{1}{3}$.
10. 1/4.
11. 1/6.
12. $y = 3x^2 - x^3, 6\frac{3}{4}$.
13. $\sqrt{2} - 1$.

Exercises 10 (d). p. 186.

1. $729\pi/35$.
2. $\pi/14$.
3. $625\pi/12$.
4. $8\pi/5$.
5. $593\pi/40$.
6. $\pi^2/2$.

Exercises 10 (e). p. 186.

1. (i) $(3x^4/2) + 5x^3 - 6x^2 + C$.   (ii) $(2x^{3/2}/3) - (4x^{5/2}/5) + (2x^{7/2}/7) + C$.
2. (i) $3x^3 - 12x^2 + 16x + C$.   (ii) $(x^3/3) + x^2 - 3x + C$.
5. $\frac{1}{2}(\theta - \sin \theta) + C$.
6. $6y = 6 + 9x + 3x^2 - x^3, 9$.
7. $(5/2) - \cos t$.
8. $3y = x^3 - 6x^2 + 9x + 3, 10/3$.
9. 5 m/s, 5/3 m.
10. $\dfrac{2M - W}{2EI}, \dfrac{3M - W}{6EI}$.
11. (i) 29/6.   (ii) 30057/1120.

12. (i) $1/6$. (ii) 0.

13. (i) $\pi/6$. (ii) $\frac{1}{2}(b^2 - a^2) + \sin b - \sin a$.

14. 256.

15. $2, -3$.

16. $y = 9 + 6x - 3x^2$, 32.

17. $27a^2/64$.

18. 55·5.

19. 128/5, $c = 3\cdot032$.

21. $1\cdot26 \times 10^{-3}$ m³.

22. Height $= r^2/4a$, vol. $= \pi r^4/8a$.

24. $9\pi/4$.

Exercises 11 (*a*). p. 192.

1. $-\frac{1}{4}(2 - x)^4 + C$.

2. $-\frac{1}{11}(1 - x)^{11} + C$.

3. $-\dfrac{1}{4(2x - 1)^2} + C$.

4. $\frac{2}{5}\sqrt{(5x - 7)} + C$.

5. $1/4$.

6. $1/2$.

7. (i) $\frac{1}{2}(\sin 2x + \cos 2x) + C$. (ii) $\frac{1}{2}\tan(2x - 1) + C$.

8. $\pi/24$.

9. $\sin^{-1}(x/2) + C$.

10. $\dfrac{2}{\sqrt{3}}\tan^{-1}\left(\dfrac{2x + 1}{\sqrt{3}}\right) + C$.

11. $\sin^{-1}\left(\dfrac{x - 2}{3}\right) + C$.

12. $\dfrac{1}{\sqrt{17}}\tan^{-1}\dfrac{3x - 2}{\sqrt{17}} + C$.

Exercises 11 (*b*). p. 194.

1. $\pi/2$.

2. $\frac{1}{2}\alpha + \frac{1}{4}\sin 2\alpha$.

3. $5\pi/4$.

4. $1 + (\pi/2)$.

5. $9/8$.

6. $1/4$.

7. $1/2$.

8. $\sqrt{2}/4$.

9. $5/6$.

10. 0.

Exercises 11 (*c*). p. 200.

1. $\frac{1}{2}\sin(x^2 - 1) + C$.

2. $-\dfrac{1}{9(x^3 + 8)^3} + C$.

3. $2\sin\sqrt{x} + C$.

4. $-\frac{1}{5}\cos^5 x + C$.

5. $\frac{1}{6}\cos^3 2x - \frac{1}{2}\cos 2x + C$.

6. $-\frac{1}{9}\sqrt{(1 - 9x^2)} - \frac{1}{3}\sin^{-1} 3x + C$.

7. $\frac{1}{4}\tan^{-1} x^4 + C$.

8. $\frac{1}{2}x\sqrt{(4 - x^2)} + 2\sin^{-1}\left(\dfrac{x}{2}\right) + C$.

9. 1·06.

10. 0·2165.

11. 1·172.

12. $a^3/3$.

13. $\pi/4$.

14. $\pi a^2/4$.

15. $2/3$.

16. 2.

Exercises 11 (*d*). p. 204.

1. $-x\cos x + \sin x + C$.

2. $\frac{1}{2}x + \frac{1}{4}\sin 2x + C$.

3. $\frac{1}{2}\sin^2 x + C$.

4. $x\sin^{-1} x + \sqrt{(1 - x^2)} + C$.

5. $-\sqrt{(1 - x^2)}\sin^{-1} x + x + C$.

6. -2π.

8. (i) $(\pi^2 + 8\pi - 32)/(2\sqrt{2})$. (ii) 1.

Exercises 11 (*e*). p. 209.

1. 0·1054, 0·1054.

2. 0·6931.

3. 1·1.

4. 74·84.

5. 0·5235.

Exercises 11 (*f*). p. 210.

1. $\frac{1}{3}(x^2 - 3)^3 + C$.

2. $\frac{1}{2}\tan^2 x + C$.

3. $-\sqrt{(4 - x^2)} + C$.

4. $\frac{1}{2}\tan^{-1}\left(\dfrac{x + 1}{2}\right) + C$.

5. $-\operatorname{cosec}\theta - \sin\theta + C.$

6. $\frac{1}{3}\tan^{-1}\left(\dfrac{2x+1}{3}\right) + C.$

7. $-\dfrac{\sqrt{(4+x^2)}}{4x} + C.$

8. $\frac{1}{3}\tan^3\theta + C.$

9. $-\frac{1}{4}(\cos 2x + \sqrt{3}x) + C.$

10. $\frac{2}{15}(x+2)^{3/2}(3x-4) + C.$

11. $1/6.$

12. $435/237160.$

13. $4/3.$

14. $0.$

15. $8/105.$

16. $(4\sqrt{2} + 2 - \pi\sqrt{2})/8.$

17. $1/12.$

18. $\pi/4.$

19. $(\pi^2 + 4)/16.$

20. $2a^7/35.$

22. $-9/8.$

23. (i) $x = \tan\theta.$ (ii) $x = \sin\theta.$

24. $16\cdot65.$

25. $577.$

Exercises 12 (a). p. 216.

1. $64/3.$

2. $128/81.$

3. $7a^2/9.$

4. $1/6.$

5. $2\pi.$

6. $\pi(5a + \frac{2}{3}a^3).$

7. $\pi\left(ax^2 - \dfrac{x^3}{3}\right).$

Exercises 12 (b). p. 219.

1. $1/2.$

2. $71\cdot5.$

3. $1\cdot978.$

4. $1.$

Exercises 12 (c). p. 224.

1. $3l/4.$

2. $\bar{x} = 10\cdot8, \bar{y} = 4\cdot5.$

3. $\bar{x} = 635/217, \bar{y} = 1275/496.$

4. $85/31.$

5. $50\pi a^3.$ C. of G. is on x-axis at distance $10a/3$ from origin.

6. On y-axis at distance $3\cdot6$ from origin.

Exercises 12 (d). p. 229.

1. $3Ml^2/5,$ where $M = $ mass, $l = $ length of rod.

2. $\rho a^4/20.$

3. $Mb^2/9,$ where $M = $ mass.

4. $Mh^2/2,$ where $M = $ mass.

5. $a^2/3.$

6. $d^2 + (b^2/4).$

Exercises 12 (e). p. 232.

1. $12\cdot1.$

2. $\pi r/2.$

3. $12.$

4. $2\pi r^2.$

5. $241\cdot2.$

6. $3\pi.$

Exercises 12 (f). p. 233.

1. $79/6.$ Results differ because part of curve lies below x-axis.

2. $8/3.$

3. $(96\sqrt{5})/5.$

4. $8a^2/3, 32\pi a^3/3.$

5. $(3\pi^2/2) + 4\pi.$

6. $2\cdot25 \times 10^{-3}$ m$^3.$

7. (i) $0.$ (ii) $2/\pi.$

11. $\bar{x} = 4a/3\pi, \bar{y} = 4b/3\pi.$

12. $\bar{x} = a\pi/2, \bar{y} = a\pi/8.$

13. $16/15, \bar{x} = 4/7, \bar{y} = 0.$

14. $\bar{x} = 3a/8, \bar{y} = 0.$

15. $\bar{x} = 3/2, \bar{y} = 0.$

16. $\rho a^4/3.$

17. $2/9.$

20. $3Ma^2/5.$

21. $a(\alpha - \beta).$

22. $108.$

23. $\frac{1}{2}\{(2b^{2/3} - a^{2/3})^{3/2} - a\}.$

24. $3\pi.$

25. $\pi(10\sqrt{10} - 1)/27.$

Exercises 13 (a). p. 240.

1. (i) $x(1 + 2\log_e x).$ (ii) $-1/x.$

2. $\log_e x$, $x(\log_e x - 1) + C$. 3. 2 (approx.).

4. (i) $2x/(x^4 - 1)$. (ii) $-\operatorname{cosec} x$.

5. $1/\sqrt{(x^2 + 1)}$, $\log_e \{x + \sqrt{(x^2 + 1)}\} + C$.

Exercises 13 (b). p. 244.

1. (i) $e^{2x}(2 \sin 3x + 3 \cos 3x)$. (ii) $2xe^{x^2}$.

5. (i) $(1 - e^{-3})/3$. (ii) $\frac{1}{2}(e^2 - e^{-2}) - 2$. (iii) 2.

7. (i) $(x^2 - 2x + 2)e^x + C$. (ii) $\frac{1}{2}e^x(\sin x + \cos x) + C$.

8. e^{-1}

Exercises 13 (c). p. 250.

1. $\frac{1}{2}\log_e (x^2 + 1) + C$. 2. $\frac{1}{2}\log_e (4x^3 + x^2 - 2x + 3) + C$.

3. $3 \log_e 2$. 4. $\frac{1}{2}\log_e (x^2 + a^2) + \tan^{-1}\left(\dfrac{x}{a}\right) + C$.

5. $\log_e (e^x + e^{-x}) + C$. 6. $\frac{1}{4}\log_e (3 + 4 \tan x) + C$.

7. $\log_e (\log_e t) + C$.

8. $2 \log_e (2t^2 + 2t + 1) - 7 \tan^{-1}(2t + 1) + C$.

9. $\dfrac{x}{2} + \dfrac{3}{4}\log_e (2x - 3) + C$. 10. $2x - \log_e (x + 2) + C$.

11. $-\dfrac{x^3}{6} - \dfrac{x^2}{2} - 2x - 4 \log_e (4 - 2x) + C$.

12. $\log_e\left(\dfrac{x - 2}{x + 3}\right) + C$. 13. $\log_e \{(x + 1)^2(x - 2)^4\} + C$.

14. $\dfrac{x^2}{2} + \dfrac{3}{2}\log_e (x - 1) - \frac{1}{2}\log_e (x + 1) + C$.

15. $4 \log_e (x - 3) - \dfrac{15}{x - 3} + C$.

16. $2 \log_e (x - 3) - \log_e (x^2 + 4) + \dfrac{1}{2}\tan^{-1}\left(\dfrac{x}{2}\right) + C$.

17. $\log_e (4/3)$. 18. $(1/2) + \log_e (3/4)$.

19. $\dfrac{x^3}{3}\left(\log_e x - \dfrac{1}{3}\right) + C$.

20. $\dfrac{x^3}{3}\tan^{-1} x - \dfrac{x^2}{6} + \dfrac{1}{6}\log_e (x^2 + 1) + C$.

21. $x \tan x + \log_e \cos x + C$.

22. $\left(\dfrac{x^2}{2} - 8\right)\log_e (x + 4) - \dfrac{x^2}{4} + 2x + C$.

23. $\left(\dfrac{x}{x + 1}\right)\log_e x - \log_e (x + 1) + C$.

24. $\tan x (\log_e \tan x - 1) + C$.

Exercises 13 (d). p. 255.

1. $-4(3 - 22x + 36x^2)$. 2. $\dfrac{3(1 - 5x)}{2(1 - x)^4(1 - 3x)^{1/2}}$.

3. $10^x \log_e 10$.

4. $(\log_e x)^x \left\{\log_e (\log_e x) + \dfrac{1}{\log_e x}\right\}$.

9. $A = 2, B = 8, C = -9$. 10. $-2/3 < x \leqslant 2/3$.

Exercises 13 (e). p. 256.

1. (i) $1/(\sin x \cos x)$. (ii) $\sec x$. 2. e^{-1}.

3. (i) $\dfrac{1}{x(1 + x)}$. (ii) $\dfrac{1}{2\sqrt{(x^2 - 1)}}$.

6. (i) $\frac{1}{3}e^{3x} + 2\log_e(x - 1) + C.$ (ii) $\frac{1}{2}(1 - e^{-\pi/2}) + \log_e 2.$

8. $\alpha = -A/2, \beta = \sqrt{(B - A^2/4)}.$ 10. $2\pi(e^2 - e^{-2} + 4).$

11. (i) $\log_e(e^x - e^{-x}) + C.$ (ii) $e^{-1}\log_e(x^e + e^x) + C.$

13. (i) $x - \log_e(x + 2) + C.$ (ii) $\dfrac{x^2}{2} - 4x + 5\log_e(x + 1) + C.$

14. (i) $\frac{3}{2}\log_e 2.$ (ii) $\frac{3}{4}\log_e(t - 1) - \frac{5}{12}\log_e(t + 3) - \frac{1}{3}\log_e t + C.$

15. (i) $\frac{1}{3}\log_e\{x(x^2 - 3)^4\} + C.$ (ii) $\frac{1}{4}\log_e\left(\dfrac{x + 1}{x - 1}\right) - \dfrac{x}{2(x^2 - 1)} + C.$

16. (i) $x + \frac{4}{3}\log_e(x + 2) - \frac{25}{3}\log_e(x + 5) + C.$

   (ii) $3\log_e(x - 2) - \frac{1}{2}\log_e(2x + 1) + \dfrac{1}{2x + 1} + C.$

17. (i) $\frac{1}{4}\log_e(\frac{5}{3}).$ (ii) $\frac{1}{2}\log_e(\frac{5}{2}).$

18. $\log_e(11/6).$

21. (i) $\dfrac{1 + 3x^2 - 2x^4}{(1 - x^2)^{3/2}}.$ (ii) $(5^{1 + \sin^2 x})\sin 2x . \log_e 5.$

23. $147/40.$

25. $\dfrac{2^n - 1}{n . 6^n}.$

Exercises 14 (a). p. 263.

2. (i) (a) $(2, 2\sqrt{3}).$ (b) $\{(5\sqrt{2})/2, -(5\sqrt{2})/2\}.$
   (ii) (a) $(\sqrt{2}, \pi/4).$ (b) $(13, 112° 37').$

4. (i) $\sqrt{(a^2 + b^2)}.$ (ii) $\sqrt{(a^2 + b^2)}.$ (iii) $\sqrt{(2p^2 + q^2 + r^2 - 2pr - 2pq)}.$

5. $4\sqrt{5}, 5, \sqrt{17}.$ 8. (5, 5).

Exercises 14 (b). p. 267.

1. $(2, 5/3), (-20, 31).$ 2. $\sqrt{41}, (-1/3, 5/3).$

5. (i) 1.   (ii) $a^2.$ 7. 21.

8. $(8 - 3\sqrt{3})/4.$

Exercises 14 (c). p. 271.

1. $x^2 + y^2 - 6x - 8y = 0.$ 2. $x^2 + y^2 - 8x - 8y + 7 = 0.$

3. $2x - y - 2 = 0.$ 4. $y = \pm 3x.$

5. $(-1/11, -19/11).$ 6. $(0, 0), (1, 2); \sqrt{5}.$

7. $(2, 0), (2, 10); (2, 5).$

Exercises 14 (d). p. 271.

1. $6 + 2\sqrt{3}.$ 2. $\sqrt{68}, \sqrt{50}, \sqrt{106}.$

3. $(13/3, 8/3).$ 5. $(1/6, 1/2).$

6. $\dfrac{\sqrt{\{r_1{}^2 + r_2{}^2 - 2r_1r_2\cos(\theta_1 - \theta_2)\}}}{2\sin(\theta_1 - \theta_2)}.$

7. $(1, 1), (-5/2, 9/2), (9/2, -5/2).$

8. $(8, 0).$ 10. $51\cdot 8, 165\cdot 3, 209\cdot 5, 2500.$

11. 10.   12. 4.

13. $(4, 2), (0, 4).$ 14. $8x + 6y = 63.$

15. $15x^2 - y^2 + 2ax = a^2.$

17. $2x - y - 16 = 0, 2x - y + 8 = 0.$

18. $x^2 + y^2 - 12x - 18y + 109 = 0.$

19. $3x - y - 1 = 0.$

20. $2x^2 + 2y^2 \pm x \pm 13y \mp 13 = 0.$

21. $(11\sqrt{2})/5, (11\sqrt{13})/10, (11\sqrt{17})/10.$

22. (5, 6).

23. $a = 6, b = 3, AB = 1/12.$

Exercises 15 (a). p. 280.

1. (i) $\sqrt{3}x + y + 3 = 0.$ (ii) $3x - 5y + 15 = 0.$ (iii) $3x + 2y = 0.$
2. $4/3.$
3. (i) $13, -65/12.$ (ii) $5/12.$ (iii) $5.$
4. $(1, 1), 2x + 3y = 5.$ 5. $3/\sqrt{5}.$
6. $5x + y = 9.$ 7. $x + y = 8.$
8. $2x - y = 4.$
9. $x \cos \alpha + y \sin \alpha = \pm c \sqrt{(\sin \alpha \cos \alpha)}.$
10. $a^2 + b^2 = a^2 b^2.$ 11. $x - 2y + 2 = 0, 4x - y = 6.$
12. $y + mx = 0, (a + b/m)y - bx + ab = 0; \left(\dfrac{ab}{am + 2b}, \dfrac{-mab}{am + 2b} \right).$

Exercises 15 (b). p. 285.

1. $135°.$
2. $AB, 3x + 4y = 19; BC, x - 3y = 2; CA, 5x - 2y + 3 = 0,$
   $\tan B = 13/9, \tan C = 13/11.$
3. $y + 2 = 0, y - \sqrt{3}x + 3\sqrt{3} + 2 = 0.$
4. $29y - 2x = 0.$
5. (i) $x = 2.$ (ii) $4x + 3y + 1 = 0.$ (iii) $3x - 4y = 0.$
6. $4x + 3y = 11.$
7. $x - y + 1 = 0; (2, 3); x + y = 5.$
8. (i) $7x - 2y = 7.$ (ii) $3x + y = 2.$ (iii) $(11/13, -7/13).$
9. $(7, 7).$

Exercises 15 (c). p. 288.

1. (i) $1.$ (ii) 0 (point lies on line).
2. $h \cos \alpha + k \sin \alpha - p.$
3. Point is centre of an escribed circle.
4. $6x - 4y = 3.$
5. $2x - 159y = 11, 34x + 27y = 57.$
6. $x - 2y + 1 = 0, 2x + y = 3.$
7. $4x - 7y + 3 = 0, 49x + 28y = 12; 2x + 3y = 5, 15x - 10y + 8 = 0,$
   $x - 3y + 2 = 0, 3x + y + 1 = 0.$
8. $B(x - h) - A(y - k) = \pm(Ax + By + C).$

Exercises 15 (d). p. 292.

1. $x + 9y = 11, 2x - 4y + 11 = 0.$ 2. $x = 1, y = 1.$
3. (i) $x + y = 1.$ (ii) $x + 3y = 1.$ 4. $23x + 23y = 11.$
5. $18x + 29y + 4 = 0.$ 6. $45°.$
8. $2x - y = 0, 2x - 11y = 0.$

Exercises 15 (e). p. 297.

1. $m = 0.031, c = 2.5.$ 2. $m = 2.3, c = -10.$
3. $a = 3, u = 100; 100.$ 4. $\alpha = 2, \beta = \frac{1}{2}.$
5. $\alpha = \frac{1}{3}, c = 4.$ 7. $C = 10, n = -\frac{3}{2}.$
8. $C = 0.000136, n = 0.546.$ 9. $k = 5 \times 10^{-7}, n = 4.$
10. $A = \frac{3}{2}, n = \frac{1}{2}; m = \frac{1}{2}, c = 1; x = 1$ and $4.$

Exercises 15 (f). p. 298.

1. $x + y = 6.$
2. $AB = 10$, equation $4x - 3y + 5 = 0$; equation to $BC, 3x + 4y + 10 = 0$; C is point
   $(-10, 5)$ or $(6, -7).$
3. $4x + 3y + 1 = 0, 4x + 3y + 11 = 0, 4x + 3y = 9.$
4. $4x + 3y = 9, 4x + 3y + 11 = 0; 27/8, 121/24.$

5. $85°\ 14'$.   6. $ax + by = a^2$.

7. $(0, 1/m)$.

8. $3y - 4x + 6 = 0$; $(6·6, 6·8)$, $(3, 2)$; $8·4$.

11. $x + 7y + 20 = 0$, $7x - y = 10$.

12. $(139/74, 353/74)$.

13. $3x + y = 6$, $4x + y = 8$; $x = 1$.

14. $x - y = 0$, $x + 2y = 0$; $(0, 0)$, $(2, 2)$, $(-2, 1)$, $(-1/3, 13/3)$; area $= 53/6$.

15. $x + 2y = 8$, $2x - y + 9 = 0$; $(-5, -1)$.

17. $2x - 5y = 18$.   19. $(8, 3)$.

21. $4x - 3y + 2 = 0$.   22. $13x - 11y = 9$, $5x - y = 0$.

24. $x^2 - 4xy - y^2 = 0$.   25. $ab(x^2 - y^2) + (b^2 - a^2)xy = 0$.

26. $A = 271$, $C = 82$.   27. $A = 60$, $B = 0·017$.

28. $A = 6·4$, $n = 3$.   29. $a = 5·99$, $b = 1·20$.

30. $k = 210$, $n = 0·3$.

Exercises 16 (a). p. 304.

1. $(5, -6)$, $\sqrt{61}$.

2. (i) $x^2 + y^2 + 10x + 12y - 39 = 0$.

   (ii) $x^2 + y^2 - 2ax + 2by + 2b^2 = 0$.

3. $x^2 + y^2 - 2x - 2y - 3 = 0$.

4. $x^2 + y^2 - 10y = 0$.   5. $x^2 + y^2 - 11x - 7y + 30 = 0$.

6. $x + 5y + 2 = 0$.   7. $x^2 + y^2 - 5x - y + 4 = 0$.

8. $x^2 + y^2 + 5x - 5y = 0$.

Exercises 16 (b). p. 308.

1. $3x + y = 19$, $y = 2$.

2. (i) $gx + fy = 0$.   (ii) $x + 2y = 7$.

3. $4x - 3y = 25$, $4x - 3y + 25 = 0$.

4. $3x + 2y = 0$.

5. $2x^2 + 2y^2 - 8x + 4y + 5 = 0$.   6. $12/5$.

7. $y = 0$, $3y = 4x$; $(3, 0)$, $(9/5, 12/5)$.

8. $2\sqrt{\left(R^2 - \dfrac{a^2b^2}{a^2 + b^2}\right)}$.

Exercises 16 (c). p. 312.

1. 3.   2. 12.

3. (i) $(10, 2)$.   (ii) 2.   5. $x + 3y = 7$.

6. $3x - y = 1$, $x - 3y + 13 = 0$; $(2, 5)$; $x^2 + y^2 - 4x - 10y + 19 = 0$.

Exercises 16 (d). p. 314.

1. $x - y = 3$.

2. $5x + y + 2 = 0$, $2x - 10y + 21 = 0$.

5. $x^2 + y^2 + 4x - 7y + 5 = 0$.

6. $x^2 + y^2 - 20x = 0$, $x^2 + y^2 - 4x - 16y + 64 = 0$.

Exercises 16 (e). p. 314.

1. (i) 2, $(1, 3)$.

   (ii) $(6, 3)$, $(-2, -1)$; $x^2 + y^2 - 23x + 36y - 15 = 0$.

2. $(13/3, 16/3)$, $(10\sqrt{2}/3)$.

3. $x^2 + y^2 - 6x - 4y - 28 = 0$.   4. $x^2 + y^2 - 4x + 2y - 20 = 0$.

5. $3x^2 + 3y^2 - 26x - 16y + 61 = 0$.

6. $x^2 + y^2 + 15y = 0$, $3(x^2 + y^2) - 20y = 0$.

7. $(1, 4)$, 4.   8. $-40/9$.

10. $3x - 4y = 25$, $x = \pm 5$.   11. $3x^2 + 3y^2 \pm 8ay + 4a^2 = 0$.

12. $x^2 + y^2 - 10x - 5y + 25 = 0, (3, 4)$.
13. $40y = 9x$.
14. $(-1, 3), \sqrt{10}/2; x - 3y + 15 = 0; \sqrt{2}$.
15. $(x - 4)^2 + (y \pm 3)^2 = 9, (x + 8)^2 + (y \pm 15)^2 = 225$.
16. $(6, 0), (0, 3), (0, 4); 4x + 7y = 73; \sqrt{11}$.
17. $(1, 0); \sqrt{3}/2$.   18. 7.
19. $3x^2 + 3y^2 - 6x - 10y + 3 = 0$.
20. $5x^2 + 5y^2 + 24x - 36y = 0; 2x = 3y$.
21. 24/5.
22. (i) $(-3\cdot4, 2\cdot2)$.   (ii) $4x + 3y + 7 = 0$.   (iii) 289/24.
23. $x^2 + y^2 + 8x + 10y - 59 = 0, x^2 + y^2 - 8x - 6y + 21 = 0$.
24. $x^2 + y^2 - 11x + 11y + 4 = 0$.
25. $(1/2, 1), \sqrt{5}/2$.

Exercises 17 (a). p. 321.

1. $6x - 4y = 51$.
2. $x - 2y + 144 = 0, 2x + y + 18 = 0; (-36, 54)$.
3. $(a/16, -a/2), (a/4, a); 16x + 4y + a = 0, 4x - 2y + a = 0, (-a/8, a/4)$.
4. $x + y = 3$.
5. $4y = x^2 - 4x + 16; y = x, y + 3x = 0$.
6. $90°, \tan^{-1}\left\{\dfrac{3}{2}\left(\dfrac{a^{1/3}b^{1/3}}{a^{2/3} + b^{2/3}}\right)\right\}$.

Exercises 17 (b). p. 323.

1. $-1/t_1$.
3. $xy = y(h - 2a) + 2ak$.
4. $h = 0, k = 2t; x = 0$.

Exercises 17 (c). p. 330.

1. $2; 1/2; 3$.
2. $x^2 + 2y^2 = 100$.
3. $\dfrac{x^2}{25} + \dfrac{y^2}{16} = 1$.
5. $2x - y = \pm6$.
6. $8x - 3y = 36, 3x - 2y = 18$.
10. $\dfrac{x}{a}\cos\phi + \dfrac{y}{b}\sin\phi = 1$.

Exercises 17 (d). p. 336.

1. $(\pm\sqrt{13}, 0); \sqrt{13}/3; 8/3$.
2. $3x^2 - y^2 = 3a^2$.
3. $x - 2y + 1 = 0, 2x + y + 7 = 0$.
4. $30x - 24y \pm \sqrt{161} = 0$.
7. $(t^2 + 1)(x/a) - (t^2 - 1)(y/b) = 2t$.
8. $\dfrac{x}{a}\sec\phi - \dfrac{y}{b}\tan\phi = 1$.

Exercises 17 (e). p. 341.

3. $\sqrt{2}$.
4. $x + t^2y = 2ct$.
5. $3\sqrt{5}$.
8. A rectangular hyperbola with the edges of the corner as asymptotes.

Exercises 17 (f). p. 341.

1. $(1, 0); x + 1 = 0$.
2. $a^2t^3/2$.
3. $(at_1t_2, a\{t_1 + t_2\})$.
7. $-t - (2/t)$.
10. $9x - 4y + 4 = 0, x - 4y + 36 = 0$.
11. $\sqrt{(7/3)}x + \sqrt{(20/3)}y = 9, 4\sqrt{(7/3)}x + \sqrt{(20/3)}y = 16; 36° 29'$.
12. $(\{a + b\}\cos\theta, \{a - b\}\sin\theta)$.

14. $(5, 15/2)$.
18. $x^2 - y^2 = 32$.

17. $1/\sqrt{2}$.

Exercises 19 (c). p. 381.

1. $32°\ 54'$.
3. (i) $54°\ 44'$.　(ii) $65°\ 54'$.
6. (i) $65°\ 23'$.　(ii) $100°$.　(iii) $49°\ 14'$.
8. $AB = 17,\ BC = 20,\ CA = 29$ m.　$(5\sqrt{3})/12$.

2. $60°$.
4. (i) $77°\ 59'$.　(ii) $81°\ 26'$.

Exercises 19 (d). p. 384.

3. $r\sqrt{(d^2 - r^2)}/d,\ (d^2 - r^2)/d$.

5. $0\cdot0125$ m.

Exercises 19 (e). p. 391.

1. $7;\ x^2 + y^2 + z^2 + 2x - 4y - 16z + 20 = 0$.
2. $x^2 - y^2 + z^2 = 0$.
6. $\frac{1}{3}, \frac{2}{3}, -\frac{2}{3}; 9, \frac{9}{2}, -\frac{9}{2}$.
8. $x + y + 2z = 3, 60°$.
10. $Al + Bm + Cn = 0$ and $A\alpha + B\beta + C\gamma = D$.

4. $72°\ 14'$.
7. $2x + 3y - z = 14$.
9. 13.

Exercises 19 (f). p. 392.

2. Plane parallel to AB, CD.
3. (i) $15\cdot97$ m.　(ii) $70°\ 12'$.
12. $\sqrt{\{(b^2 + c^2 - a^2)/2\}}$.
15. $54°\ 44'$.
18. $54°\ 44'$.
20. $20\cdot62, 15\cdot2, 15\cdot59$ m, $0\cdot843$.
22. $a(3 + 2\sqrt{3})/3, 2a(3 + \sqrt{6})/3$.
23. (a) $0\cdot022$ m.　(b) $0\cdot0314$ m.
29. $\dfrac{x - \alpha}{m_1 n_2 - m_2 n_1} = \dfrac{y - \beta}{n_1 l_2 - n_2 l_1} = \dfrac{z - \gamma}{l_1 m_2 - l_2 m_1}$.
30. (i) $-2/\sqrt{6}, -1/\sqrt{6}, 1/\sqrt{6}$;
　　(ii) $2x + y - z = 2$.

11. $2\cdot88$ m.
13. $1\cdot465, 20°\ 56'$.
16. (i) $22\cdot7$.　(ii) $33\cdot1$ m^2.

26. $5/6$.

Exercises 20 (a). p. 400.

1. $109\cdot35$ m^3.
3. (i) $1:9$.　(ii) $1:27$.
6. $(\sqrt{3a^3})/2$.
8. 380 m^3.

2. $5\cdot33 \times 10^{-5}$ m^3.
5. (i) $(\sqrt{2a^3})/6$.　(ii) $(1 + \sqrt{3})a^2$.
7. $18\cdot5a^3$.

Exercises 20 (b). p. 402.

1. $4(3 + 2\sqrt{2})\pi a^3$.
4. $(7/2 + 3\sqrt{7})\pi a^2$.

2. $5\cdot28 \times 10^{-2}$ m^2.
5. $5\cdot78 \times 10^{-5}, 3\cdot15 \times 10^{-5}$ m^3.

Exercises 20 (c). p. 404.

1. $(2\pi^2 r^3)/3;\ \{1 + \sqrt{(4\pi^2 + 1)}\}\pi r^2$.
2. $\pi a^2/4$.
4. $1\cdot7$.

3. $(2R - h)(\pi h^2/3)$.
8. $x = 2/3, \theta = 35°\ 15'$.

Exercises 20 (d). p. 407.

2. $0\cdot023$ m.
8. $5\pi a^2, 13\pi a^3/6$.

3. $0\cdot026$ m.

Exercises 20 (e). p. 409.

1. $\sin\theta = \theta - (2\pi/3)$.
3. $1\cdot37$ radians.

2. $12\cdot03$ m.
6. $(23a^3)/24$.

8. $\frac{1}{2}\pi(7 + 6\sqrt{7})a^2$.

9. $\pi a^2(3h - a)/24$; $(3h - a)/(8h)$.

12. $2\pi r^2(r + x)/3$.

13. 0.1 m.

15. $4:(1 + \sqrt{5})$; 1.19 m.

16. $5:4$.

17. $a(6 + 2\sqrt{6})/3$.

18. (i) 0.24 m.   (ii) 0.0905 m^2.   (iii) 0.000266 m^3.

19. $2\pi a^2\left\{1 + \dfrac{a}{\sqrt{(a^2 + b^2)}}\right\}$.

24. $54° 44'$.

25. 1.21×10^5 km^3 (approx.).

Exercises 21 (a). p. 417.

1. (a) $1, \pi$.   (b) $1, \frac{1}{2}\pi$.   (c) $5, 0.927$.   (d) $2, -5\pi/6$.

2. $23/41, 2/41$.

3. (i) $3 + 4i$.   (ii) -4.   (iii) $-1 - i$.

4. (i) 16.   (ii) $48i/25$.

5. $1/\sqrt{5}, 0.1798$.

6. (i) $1/\sqrt{2}, 5\pi/4$.   (ii) $\sqrt{2}, 3\pi/4$.   (iii) $\frac{1}{2}\sqrt{10}, 2.82$.

8. (i) $-1, \frac{1}{2}$.   (ii) $\frac{1}{5}, -\frac{2}{3}$.

9. $\frac{3}{10}, \frac{9}{10}$.

Exercises 21 (b). p. 421.

1. 2

2. $\frac{5}{34}$.

4. $1.2 + 1.6i$.

6. A line through A parallel to OB, O being the origin.

8. OB/OA, angle BOA.

Exercises 21 (c). p. 425.

3. $\pm 1, \pm i$.

6. 1.

8. $\pm(1 + i)/\sqrt{2}, \pm\sqrt{2} \pm i\sqrt{2}$.

9. $\frac{1}{2}, \frac{1}{2}(1 \pm i\sqrt{3})$.

Exercises 21 (d). p. 426.

1. $7 + 22i, (5 - 4i)(3 - 2i), 13 \times 41$.

3. $-3 + 4i, -11 - 2i$; $r = 4, s = 30$.

4. $\pm(3 + 2i)$.

5. $x = 2, y = \frac{1}{2}$.

6. $\pm 5(1 + i)/2$.

7. $5, 0.6435$.

8. $5, 6.5, 2.061, 32.5$.

10. Centre $\frac{1}{4} + i$, radius $\frac{1}{4}\sqrt{17}$.

11. $(x + \frac{5}{3})^2 + y^2 = \frac{16}{9}$.

14. $a^2 + b^2, \tan^{-1}\left(\dfrac{2ab}{a^2 - b^2}\right)$.

15. $2, -1 \pm \sqrt{3}$.

16. (i) $x + y = 2$.   (ii) $(x - 1)^2 + (y - 1)^2 = 2$.

17. $\operatorname{cosec} \theta, \frac{1}{2}\pi - \theta$.

18. $\left\{\dfrac{L - CR_1{}^2}{CL(L - CR_2{}^2)}\right\}^{1/2}$.

19. $2\sqrt{2}, \frac{1}{4}\pi$; $2\sqrt{5}, 0.4637$.

22. $\pm(1.73 + i), \pm(1 - 1.73i)$.

INDEX

Acceleration, 154.
Addition of complex numbers, 418.
Addition theorems, for sine and cosine, 75; for tangent, 77; general proof of, 262.
Ambiguous case, in solution of triangles, 106.
Amplitude, of complex number, 415; of periodic function, 67.
Angle, between line and plane, 378; between two lines, 281, 387; of depression, 113; of elevation, 113.
Angles, multiple, 78; negative, 59; of triangle in terms of sides, 102; small, 89; submultiple, 80.
Answers to exercises, 428.
Apollonius, circle of, 353.
Applications, of differential calculus, 152; of integral calculus, 212.
Approximate methods of integration, 204; Simpson's rule, 208; trapezoidal rule, 206.
Approximations, 128; successive, 252.
Arc, length of, 229.
Area, calculation of, 181; further examples, 212; of quadrilateral, 266; of sector of circle, 395; of surface of revolution, 231; of trapezium, 395; of triangle, 97; of triangle with given vertices, 265.
Argand diagram, 414.
Argument, of complex number, 415.
Arithmetic mean, 37.
Arithmetical progression, 37.
Asymptotes, of hyperbola, 337.
Auxiliary circle, of ellipse, 329.

Binomial theorem, for fractional and negative indices, 53; for positive integral index, 50.
Bisectors, of angles between given lines, 287.

Cartesian coordinates, 260; relation with polar, 260.
Centre, of ellipse, 325; of gravity, 219; of hyperbola, 332; of mass, 219; of similitude, 351.
Centroid of triangle, 356.
Ceva's theorem, 360; converse of, 360.
Change of variable, integration by, 195, 199.
Circle, Apollonius, 353; auxiliary, 329; director, 328; escribed, 100; general equation to, 302; geometrical theorems on, 344; inscribed, 98; intersections with straight line, 306; tangent to, 305; through intersections of given circles, 313; through three given points, 304;

whose diameter is join of given points, 303; with given centre and radius, 302.
Circles, orthogonal, 311; radical axis to, 312.
Circumcentre, of triangle, 355.
Combinations, 45.
Common logarithms, 27.
Complex numbers, 412; addition of, 418; argument of, 415; conjugate, 416; geometrical representation of, 414; imaginary part of, 414; manipulation of, 417; modulus of, 415; principal value of argument of, 415; products and quotients of, 420; real part of, 414.
Compound interest, 41.
Cone, surface and volume of, 402.
Conic section, definition of, 317.
Conical surface, definition of, 368.
Conjugate complex numbers, 416.
Convergence of geometric series, 42.
Coordinates, of point dividing join of two points, 263; in three dimensions, 385; polar, 260; rectangular, 260; systems of, 259.
Coplanar lines, definition of, 369.
Cosine, addition theorem for, 76; differential coefficient of, 126; formula for triangle, 95.
Cube roots of unity, 424.
Cuboid, definition of, 367.
Curve sketching, 165.
Cylinder, surface and volume of, 401.
Cylindrical surface, definition of, 368.

De Moivre's theorem, 422.
Definite integral, 181; evaluated by change of variable, 199.
Dependent variable, 120.
Depression, angle of, 113.
Derivative, 125; higher, 148.
Derived function, 125.
Desargues' theorem, 366.
Diameter, of ellipse, 325.
Differential calculus, some applications of, 152.
Differential coefficient, as rate measurer, 128, 152; of cos x, 126; of cosec x, 137; of cot x, 137; of e^x, 243; of function of function, 137; of implicit functions, 146; of inverse functions, 143; of product, 133; of quotient, 135; of sec x, 137; of sin x, 126; of $\sin^{-1} x$, 144; of sum, 132; of tan x, 136; of $\tan^{-1} x$, 144; of x^n, 125, 141.
Differentiation, from first principles, 124; logarithmic, 251.
Dihedral angle, definition of, 378.
Direction cosines, 386.